Intelligent Communication Networks

With the advent of Big Data, conventional communication networks are often limited in their inability to handle complex and voluminous data and information as far as effective processing, transmission, and reception are concerned. This book discusses the evolution of computational intelligence techniques in handling intelligent communication networks.

- Provides a detailed theoretical foundation of machine learning and computational intelligence algorithms.
- Highlights the state of art machine learning-based solutions for communication networks.
- Presents video demonstrations and code snippets on each chapter for easy understanding of the concepts.
- Discusses applications including resource allocation, spectrum management, channel estimation, and physical layer of wireless networks.
- Demonstrates applications of machine learning techniques for optical networks.

The text is primarily intended for senior undergraduate and graduate students and academic researchers in fields of electrical engineering, electronics and communication engineering, and computer engineering.

Intelligent Communication Networks

Research and Applications

Edited by

Rajarshi Mahapatra
Siddhartha Bhattacharyya
Avishek Nag

CRC Press is an imprint of the
Taylor & Francis Group, an informa business

Designed cover image: Shutterstock

First edition published 2024
by CRC Press
2385 NW Executive Center Drive, Suite 320, Boca Raton FL 33431

and by CRC Press
4 Park Square, Milton Park, Abingdon, Oxon, OX14 4RN

CRC Press is an imprint of Taylor & Francis Group, LLC.

© 2024 Taylor & Francis Group, LLC

© 2024 selection and editorial matter, Rajarshi Mahapatra, Siddhartha Bhattacharyya, and Avishek Nag; individual chapters, the contributors

Reasonable efforts have been made to publish reliable data and information, but the author and publisher cannot assume responsibility for the validity of all materials or the consequences of their use. The authors and publishers have attempted to trace the copyright holders of all material reproduced in this publication and apologize to copyright holders if permission to publish in this form has not been obtained. If any copyright material has not been acknowledged, please write and let us know so we may rectify in any future reprint.

Except as permitted under U.S. Copyright Law, no part of this book may be reprinted, reproduced, transmitted, or utilized in any form by any electronic, mechanical, or other means, now known or hereafter invented, including photocopying, microfilming, and recording, or in any information storage or retrieval system, without written permission from the publishers.

For permission to photocopy or use material electronically from this work, access www.copyright.com or contact the Copyright Clearance Center, Inc. (CCC), 222 Rosewood Drive, Danvers, MA 01923, 978-750-8400. For works that are not available on CCC, please contact mpkbookspermissions@tandf.co.uk.

Trademark notice: Product or corporate names may be trademarks or registered trademarks and are used only for identification and explanation without intent to infringe.

ISBN: 978-1-032-30021-4 (hbk)
ISBN: 978-1-032-30026-9 (pbk)
ISBN: 978-1-003-30311-4 (ebk)

DOI: 10.1201/9781003303114

Typeset in Times
by SPi Technologies India Pvt Ltd (Straive)

Dedication

Rajarshi Mahapatra dedicates this volume to his wife, Sarmila.

Avishek Nag dedicates this volume to his late parents, his wife, and his son.

Siddhartha Bhattacharyya dedicates this volume to his loving wife Rashni.

Contents

Preface

In the present scenario, machine learning algorithms are used in almost every application, and communication network is one of them. Providing affordable, high-speed communication is challenging and becoming more complex every day. Increased complexity of networks in terms of the coexistence of different technologies, high volume of users, and different management and operations stakeholders can be handled better using machine learning methodologies. There is a dearth of literature to provide a comprehensive overview of the theory, applications, and research of machine learning for communication networks, including wireless, optical, and computer networks.

This book will provide a comprehensive and highly accessible treatment to the theory, applications, and current research developments to the technological aspects related to machine learning for wireless communications, optical networks, and computer networks. With the advent of Big Data, conventional communication networks are often limited in their inability to handle complex and voluminous data and information regarding effective processing, transmission, and reception. The huge explosion of data calls for real-time processing and delivery.

Existing communication systems are burdened with bandwidth restrictions. This book explores the different facets of computational intelligence in evolving intelligent communication networks which would become more efficient, robust, and fail-safe. Thus, the book intends to overcome existing limitations of conventional communication networks by introducing intelligent algorithms in the current set-up, thereby enabling intelligent decision-making in conventional communication network frameworks and architectures. Optimization of the relevant network architectures is also a concern, which this book will address.

This book comprises ten well-versed contributory chapters covering the applications of machine learning and computational intelligence to different facets of communication networks.

Deep reinforcement learning (DRL) has emerged as a powerful technique for solving complex decision-making problems in various domains. One such domain is resource allocation, which involves the efficient distribution of limited resources to maximize system performance. Traditional resource allocation approaches often rely on handcrafted heuristics or optimization algorithms, which may struggle to adapt to dynamic and uncertain environments. Chapter 1 focuses on the application of deep reinforcement learning to resource allocation problems. By combining deep neural networks with reinforcement learning algorithms, DRL provides a flexible framework for learning resource allocation policies directly from raw input data. The agent learns to make sequential decisions by interacting with the environment, receiving feedback in the form of rewards and updating its policy to maximize cumulative rewards over time. The chapter further discusses the key components involved in deep reinforcement learning for resource allocation, including the representation of the state and action spaces, reward design, exploration strategies, and network architectures. It also highlights recent advancements in the field, such as multi-agent reinforcement learning, hierarchical reinforcement learning, and the integration of domain knowledge into DRL frameworks.

The requirement to serve many users, increasing data rate, and small coverage cells make fifth-generation (5G) wireless communication networks complex and heterogeneous. To fulfill all these requirements and quality of service (QoS) goals, people use many network cells differently in capacity and coverage. With the increasing number of such smaller cells in 5G, resource allocation with interference coordination (RAIC) has become too complex. Moreover, unplanned heterogeneous network (HetNet) deployments require self-organizing and intelligent RAIC techniques. The non-convexity and complexity make it challenging to obtain an optimal RAIC strategy. Deep Reinforcement Learning (DRL), which uses a deep neural network (DNN) as a predictor, is a relatively new technique that has arisen in recent years to accomplish various difficult tasks efficiently.

Chapter 2 discusses the use of a DRL-based RAIC technique to optimize power and spectrum allocation with interference coordination in downlink 5G HetNets while guaranteeing the individual user QoS goals. Additionally, as 5G networks are naturally dispersed systems, centralized DRL techniques would necessitate a lot of data interchange, whereas fully decentralized approaches would cause poor performance and slower convergence. The chapter also gives the technical details for implementing Federated Deep Reinforcement Learning-Based Resource allocation in Heterogeneous Networks. It provides some simulation results showing that the federated DRL and Q-Learning-based RAIC method performs better, takes less execution time under different coverage and data rate requirements, and converges fast.

The introduction and wide usage of wearable technologies and the integration of wearable sensors, wireless communications, data analytics, and Artificial Intelligence (AI) have brought forward the nano-networks paradigm. This can give consumers more control over their lives to enhance their well-being through personalized real-time data analysis. This data can be possibly used together with virtual and augmented reality environments in various nano-applications such as preventive treatment, diagnostics, and rehabilitation. Internet of Nano-Things (IoNT) introduces significant challenges and opportunities for big data analysis research in various fields, such as next-generation heterogeneous networks, environmental monitoring, and healthcare. Such a huge volume of diverse real-time data may not be handled efficiently by traditional mechanisms, necessitating improved data analysis methods.

Moreover, when the ultra-dense Heterogeneous Networks (HetNets) involved in the emerging cloud-based IoNT paradigm are considered, it is expected to encounter challenges in terms of spectrum sharing and management due to the scarcity of spectral resources. Therefore, Chapter 3 focuses on the challenges introduced and the approaches to take into account when resolving these problems. The chapter also provides an overview of the deployment aspects of IoNT, along with a review of their main application areas in IoT environments.

The metaverse promises many opportunities in marketing, gaming, education, blockchains, and many other industries as it transitions from independent, virtually simulated worlds to a network of 3D-integrated worlds. India is expected to contribute significantly to developing and expanding metaverse technologies. Apart from its engineering prowess, India has a big user base for video games, social media networks, and e-commerce. However, user privacy and confidentiality remain a challenge as the metaverse broadens the scope of virtual reality technologies. Chances of cyber-crimes, digital surveillance, and identity theft will heighten over time. Individuals, as well as governments across the world, are bound to face the challenges of setting appropriate regulations.

In Chapter 4, the authors focus on the limitations and challenges of the metaverse in India and build on past research on privacy and data safety. Here, the definition of metaverse is borrowed from Wikipedia, suggesting it as "a collective virtual shared space, formed by the fusion of virtually enhanced physical reality and physically persistent virtual space, which includes all virtual worlds combined, augmented reality, and the Internet." The study includes the aspects of socialization and Digital Twins intertwined in the metaverse. In conclusion, the chapter also suggests ways in which metaverse technology can be used with appropriate digital safety.

The rapid development of communication technologies and emerging network services provokes data traffic in communication networks, where optical networks act as the backbone for communication networks. Due to the network data traffic increase and network complexity, it is essential to construct an Intelligent Optical Network (ION). Artificial Intelligence (AI) and the evolution of several Machine Learning (ML) schemes have enhanced optical networks to meet their demand and make the network more intelligent. These techniques can be applied in various optical network applications to mitigate nonlinearities present in the fiber, wireless optical channel prediction, performance monitoring, and Quality of Transmission (QoT) estimation.

Chapter 5 deals with various intelligent optical networks and the challenges faced by the optical networks and gives insight and knowledge about techniques and schemes that make optical networks autonomous. Several applications of ION under various network scenarios and glimpses of optical network software tools are discussed. Future directions of ION are presented to advance and bring about technical advancements.

To achieve the diverse requirements for massive connectivity, low latency, high throughput, high reliability, and better fairness beyond fifth-generation (B5G), wireless networks must employ nonorthogonal multiple access (NOMA), a key enabling technology. The primary principle of NOMA accommodates multiple users within a shared resource block. Many recently proposed B5G multiple access systems can be considered specific implementations of the NOMA principle, which serve as a universal framework for such systems.

Chapter 6 summarizes the latest innovations, research findings, and applications in NOMA. It delves into the papers included in this special issue and examines their contributions in the context of existing research. Additionally, it highlights upcoming research challenges associated with NOMA in the context of B5G and future generations.

The complexity of establishing Intelligent Telecommunication Networks (ITN) equipped with artificial intelligence technologies should be considered when implementing automated control systems. Providing a quality signal and keeping strength levels for the transmission of commands to mobile robots is crucial to prevent malfunction during the operation of the mobile robot, especially when sending command signals over lengthy distances and in environments that create substantial radio frequency noise. An ITN provides stability on output signal control tasks and compensates for any strength loss due to noise. Therefore, using fuzzy logic to measure signal loss and distortion, an adaptive neuro-fuzzy inference system (ANFIS) was implemented to compensate for the incoming signal strength to minimize data or information loss. The proposed inference system in Chapter 7 can be used to control the telecommunication system; for example, it can be used to adjust the signal gain level to improve signal quality while transmitting commands to mobile robots.

Sensor networks are becoming increasingly common with the advent of IoT. Such networks include industrial sensor networks, body sensor networks, smart home sensor networks, smart city sensor networks, etc. As wireless sensors are easily deployed and configured, wireless sensor networks (WSNs) are more flexible and thus quite popular, especially in harsh and inaccessible environments. Such WSNs can be enhanced by means of intelligent clustering, routing, and localization. Machine learning can greatly contribute to optimizing the performance of such WSN systems. Machine learning techniques have been used to improve the security, routing, coverage and connectivity, localization, congestion control, data aggregation, quality of service, and energy harvesting to improve the lifetime of WSNs. Machine learning is also being adopted in WSN applications. Chapter 8 presents a review of the research related to adopting machine learning techniques for optimizing WSNs. The chapter also discusses the challenges related to the application of machine learning algorithms in the area of WSNs.

Sixth-generation (6G) wireless networks are envisioned to be ubiquitous. To ensure ubiquity and seamless connectivity, networks with unmanned aerial vehicles (UAVs) are envisioned as a potential solution. However, many challenges must be overcome to make such networks feasible. Recently, soft frequency reuse (SFR) has been exploited to reduce the interference of UAV-assisted networks. However, the existing deep learning-based solution in allocating resource plans among the UAVs in a multi-UAV network is only minimally accurate.

In Chapter 9, the authors propose a machine-learning model to address the low-accuracy issue of UAV-assisted networks with SFR and use nine classifiers to allocate the resource plans. It is found that the random forest provides the highest accuracy. It outperforms the deep learning-based solutions by up to 7.3% more accuracy. They also want the answer to this question: “Should we consider the positions of all UAVs to allocate a resource plan to a specific UAV?” Through simulations, it is also found that the positions of all UAVs are not required to be considered in allocating resource plans in a network with 12 UAVs. However, the positions of all UAVs should be considered for the network with a smaller number of UAVs.

In Computational Intelligence (CI), data processing and machine learning have become driving forces widely used for intelligence application development and hardware products with AI capabilities. In Chapter 10, the authors describe the architecture of CI and its core technologies, including neurocomputing, granular computing, fuzzy sets, evolutionary algorithms, and the technology’s design methodology. Communication is currently being improved by using Artificial Intelligence (AI) techniques to improve communication efficiency. For the development of successful communication network models, CI systems possess characteristics such as adaptability, fault tolerance, high computational speed, and error resilience in the face of noisy input. The chapter discusses neural networks, fuzzy systems, evolutionary computation, artificial immune systems, swarm intelligence, and soft computing. Each field of research is synthesized and compared to provide a clear understanding of existing challenges and identify promising new directions. The authors conclude that communication network systems can be designed and analyzed using CI. The results of this study provide a better understanding of AI techniques for enhancing communication networks and shed light on future research.

Since this book presents an understanding of the fundamental concepts of computational intelligence applied to evolving efficient and fail-safe communication networks, it will benefit graduate students, researchers, and faculty members in academia. The editors would feel rewarded if this book can supplement the knowledge base in this field to further unearth indigenous intelligent models and frameworks for the future.

Rajarshi Mahapatra
Naya Raipur, India

Avishek Nag
Dubin, Ireland

Siddhartha Bhattacharyya
Ostrava, Czech Republic and Zagreb, Croatia

Editors

Rajarshi Mahapatra earned a PhD in electronics and electrical communication engineering from the Indian Institute of Technology Kharagpur, Kharagpur, India, and a postdoctoral degree from the CEA-LETI, Grenoble, France. In his postdoctoral research, he was engaged in FP7 Call4 BeFEMTO and Greentouch projects. He is an Associate Professor with the Department of Electronics and Communication Engineering, Dr. SPM IIIT Naya Raipur. He also served as Dean (Academics) of IIITNR. He previously worked for Collins Aerospace, Hyderabad, on software-defined radio and electronic warfare.

Dr. Mahapatra has worked extensively in the domain of physical layer design and analysis of a wireless communication system. He worked in the fields of cognitive radio, 5G & 6G communication, heterogeneous wireless communication, molecular communication, and energy-efficient communication. His team designed and developed software-defined radio and direction-finding systems for EW applications in Collins Aerospace. He has approximately 18 years of teaching, research, and industry experience. He has guided PhD scholars in the area of wireless communication and published several research papers in various refereed journals and IEEE journals. He is a regular reviewer of premier IEEE Transactions and other peer-reviewed journals and IEEE conferences. He has organized many workshops on 5G and also developed several high-value research labs, including the 5G test bed. He has successfully completed and undertaken high-value sponsored projects in the field of communication systems.

He was awarded a National scholarship, an MHRD scholarship for research, and an EU-FP7 fellowship for a European project. He is a senior member of IEEE and a member of the Communication Society. His research interests include 5G and beyond communication, machine learning for communication, molecular communication, intelligent reflecting surfaces, and optical access networks.

Dr. Avishek Nag is an Assistant Professor in the School of Computer Science at University College Dublin in Ireland. Dr. Nag earned the BE (Honours) degree from Jadavpur University, Kolkata, India, in 2005; the MTech degree from the Indian Institute of Technology, Kharagpur, India, in 2007; and a PhD from the University of California, Davis, in 2012. He worked as a research associate at the CONNECT center for future networks and communication at Trinity College Dublin before joining University College Dublin. Dr. Nag received the Best Paper Award at the 2nd IEEE Advanced Networks and Telecommunication Symposium in 2008 and has published over 35 publications, including journals, conference proceedings, and book chapters with over 950 citations. His research interests include but are not limited to Cross-layer optimization in Wired and Wireless Networks, Network Reliability, Mathematics of Networks (Optimization, Graph Theory), Network Virtualization, Software-Defined Networks, Machine Learning, Data Analytics, Blockchain, and the Internet of Things. Dr. Nag is a senior member of the Institute of Electronics and electrical engineers (IEEE) and also the outreach lead for Ireland for the IEEE UK and Ireland Blockchain Group.

Dr. Siddhartha Bhattacharyya earned a Bachelors in Physics, Bachelors in Optics and Optoelectronics, and Masters in Optics and Optoelectronics from the University of Calcutta, India, in 1995, 1998, and 2000, respectively. He completed his PhD in Computer Science and Engineering from Jadavpur University, India, in 2008. He is the recipient of the University Gold Medal from the University of Calcutta for his Masters. He has received several coveted awards, including the Distinguished HoD Award and Distinguished Professor Award conferred by the Computer Society of India, Mumbai Chapter, India, in 2017, the Honorary Doctorate Award (D. Litt.) from the University of South America, and the South East Asian Regional Computing Confederation (SEARCC) International Digital Award ICT Educator of the Year in 2017. He was appointed as the ACM Distinguished Speaker for 2018–2020. He was inducted into the People of ACM Hall of Fame by ACM, USA, in 2020. He was appointed as the IEEE Computer Society Distinguished Visitor for 2021–2023. He was elected as a full foreign member of the Russian Academy of Natural Sciences (RANS) and the Russian Academy of Engineering (REA). He was also elected a full fellow of the Royal Society for Arts, Manufactures and Commerce (RSA), London, UK.

He currently serves as a Senior Researcher in the Faculty of Electrical Engineering and Computer Science of VSB Technical University of Ostrava, Czech Republic. He also serves as the Scientific Advisor of Algebra University College, Zagreb, Croatia. Prior to this, he served as the Principal of Rajnagar Mahavidyalaya, Rajnagar, Birbhum. He served as a Professor in the Department of Computer Science and Engineering of Christ University, Bangalore. He was the Principal of RCC Institute of Information Technology, Kolkata, India, from 2017 to 2019. He has also served as a Senior Research Scientist in the Faculty of Electrical Engineering and Computer Science of VSB Technical University of Ostrava, Czech Republic (2018–2019). Prior to this, he was the Professor of Information Technology at RCC Institute of Information Technology, Kolkata, India. He served as the Head of the Department from March 2014 to December 2016. Prior to this, he was an Associate Professor of Information Technology at RCC Institute of Information Technology, Kolkata, India, from 2011 to 2014. Before that, he served as an Assistant Professor in Computer Science and Information Technology at the University Institute of Technology, The University of Burdwan, India, from 2005 to 2011. He was a Lecturer in Information Technology at Kalyani Government Engineering College, India, 2001–2005. He is a co-author of six books and the co-editor of 100 books and has more than 400 research publications in international journals and conference proceedings to his credit. He has three PCTs and 20 patents to his credit. He has been a member of the organizing and technical program committees of several national and international conferences. He is the founding Chair of ICCICN 2014, ICRCICN (2015, 2016, 2017, 2018), and ISSIP (2017, 2018) (Kolkata, India). He was the General Chair of several international conferences such as WCNSSP 2016 (Chiang Mai, Thailand), ICACCP (2017, 2019) (Sikkim, India), ICICC 2018 (New Delhi, India), and ICICC 2019 (Ostrava, Czech Republic).

His research interests include hybrid intelligence, pattern recognition, multimedia data processing, social networks, and quantum computing.

Contributors

Aasita Bali
Christ University
Bangalore, India

Abhinav Singh Parihar
IIT Indore
Indore, India

Arishmita Aditya
Christ University
Bangalore, India

C. Nolivos
Southwest State University
Russia

Debabrata Samanta
Rochester Institute of Technology
Kosovo

Doorgesh Sookarah
University of Technology Mauritius
Mauritius

Hadi Zahmatkesh
OsloMet - Oslo Metropolitan
University
Oslo, Norway

Kiran Muloor
LTIMindtree Limited and CHRIST
(Deemed to be University)
Bangalore, India

M. Bobyr
Southwest State University
Russia

S. Manochandar
CARE College of Engineering
Tamil Nadu, India

Md. Imran Ahmed
American International University
Bangladesh

Md. Sakir Hossain
Bangabandhu Sheikh Mujibur Rahman
Aviation and Aerospace University
Lalmonirhat, Bangladesh

Md. Shakhawat Hossain
Independent University
Bangladesh

Mirza Hasibul Hasan
American International University
Bangladesh

Ondrej Krejcar
University of Hradec Kralove
Czech Republic

M. Pradeep Doss
National Institute of Technology
Tiruchirappalli
Tamil Nadu, India

R. K. Jeyachitra
National Institute of Technology
Tiruchirappalli
Tamil Nadu, India

Rajarshi Mahapatra
Dr. SPM IIIT-Naya Raipur
Raipur, Chhattisgarh, India

Rayhan Khan Ridoy
American International University
Bangladesh

Sandhya Armoogum
University of Technology Mauritius
Mauritius

Satish Kumar
Dr. SPM IIIT-Naya Raipur
Raipur, Chhattisgarh, India

Shubham Bisen
IIT Indore
Indore, India

Somesh Kumar Sahu
LTIMindtree Limited
Bangalore, India

Tapan Kumar Behera
Forrester Research
Cambridge, USA

Vimal Bhatia
IIT Indore
Indore, India

and

University of Hradec Kralove
Czech Republic

1 Various Deep Learning-based Resource Allocation Techniques in Wireless Communication System

Satish Kumar and Rajarshi Mahapatra
SPM IIIT-Naya Raipur, Raipur, Chhattisgarh, India

1.1 THE NEXT-GENERATION WIRELESS COMMUNICATION SYSTEMS

Recent wireless technologies, fifth generation (5G) and beyond systems are expected to support a wide range of emerging applications such as augmented reality (AR), virtual reality (VR), autonomous vehicles, and industry 4.0 [1]. The need for "network intelligence" to support very high data rates, extremely low latency, and a variety of other quality-of-service (QoS) needs has been sparked as a result of this. As a consequence, the issues that 5G wireless operators face are the complexity of networks, the diversification of services, and the customization of user experiences. Table 1.1 represents the fourth generation (4G), 5G, and sixth generation (6G) communication system requirements [1]. In addition, it is anticipated that the 6G system would provide 1000 times more simultaneous wireless transmission than the 5G system. Compared to 5G's enhanced mobile broadband (eMBB), 6G is anticipated to provide ubiquitous mobile ultra-broadband (uMUB) services. Ultra-reliable low-latency communications, a primary 5G characteristic, will be a crucial driver in 6G communication offering ultra-high-speed with low-latency communication (uHSLLC) by adding features such as end-to-end delay of less than 1 millisecond [2], greater than 99.9999% reliability [3], and 1 Tbps peak data throughput.

In the 6G communication system, up to 10 million linked devices per square kilometer will be available [3]. 6G is projected to give Gbps coverage everywhere, including novel locations like the sky (10,000 km) and the ocean (20 nautical miles) [3]. In 6G, the volume spectral efficiency, as opposed to the commonly utilized area spectral efficiency, will be significantly higher [2]. The 6G system will have sophisticated battery technology and exceptionally extended battery life for energy gathering. Mobile devices will not require separate charging in 6G systems. Multiple technologies,

DOI: 10.1201/9781003303114-1

TABLE 1.1
4G, 5G, and 6G Communication System [1]

Particulars	4G	5G	6G
Peak Data Rate	1 Gbps	10 Gbps	1 Tbps
End-to-End Latency	100 ms	10 ms	1 ms
Maximum Spectral Efficiency	15bps/Hz	30 bps/Hz	100 bps/Hz
Maximum Mobility	Up to 350 km/hr	Up to 500 km/hr	Up to 1000 km/hr
Satellite Integration	No	NO	Yes
AI	No	Partial	Fully
Autonomus Vehicle	No	Partial	Fully
Architecture	MIMO	Massive-MIMO	Intelligent Surfaces
THz Communication	No	Very Limited	Fully
Service Level	Video	VR, AR	Tactile
Maximum Frequency of Operation	6 GHz	90 GHz	10 THz
Haptic Communication	No	Partial	Fully
XR	No	Partial	Fully

including satellite-based networks, connected intelligence, seamless energy transfer through wire-less transmission, ubiquitous super 3D connectivity, massive multiple-input multiple-output (M-MIMO), 5G New Radio, millimeter wave (mmWave), tera-hertz (THz) communication, etc., are envisioned to support the requirements in new super-dense heterogeneous networks (HetNets)-based next-generation communication environment [1]. Artificial intelligence (AI), Modulation identification, signal decoding, multiple antennas with beamforming and intelligent reflecting surfaces, and resource provisioning in these super-dense HetNets are the main challenges in next-generation wireless networks to provide the best performance while supporting guaranteed QoS [4, 5]. Figure 1.1 represents a general overview of the next-generation (6G) wireless communication system.

These technologies traditionally require time-consuming and computationally expensive optimization algorithms to find the optimal solution. For example, in 6G, synchronization between different base stations (BS) and devices will be crucial to support high data rates and low latency. Higher frequency bands in 6G will require tighter synchronization and beamforming techniques to mitigate the effects of signal propagation and interference. The handover will also be critical in 6G networks, particularly in ultra-dense networks where multiple base stations may be available [6]. New handover algorithms will be needed to support seamless connectivity and minimize latency while considering signal strength, interference, and network load.

Routing is another essential function in 6G networks, particularly as the number of connected devices and services increases. New routing protocols must be developed to support efficient and reliable communication, considering factors such as network topology, quality of service requirements, and energy efficiency [7]. The problem will be more complex and challenging, especially in large-scale 5G and 6G networks. To solve these problems, researchers used block optimization methods

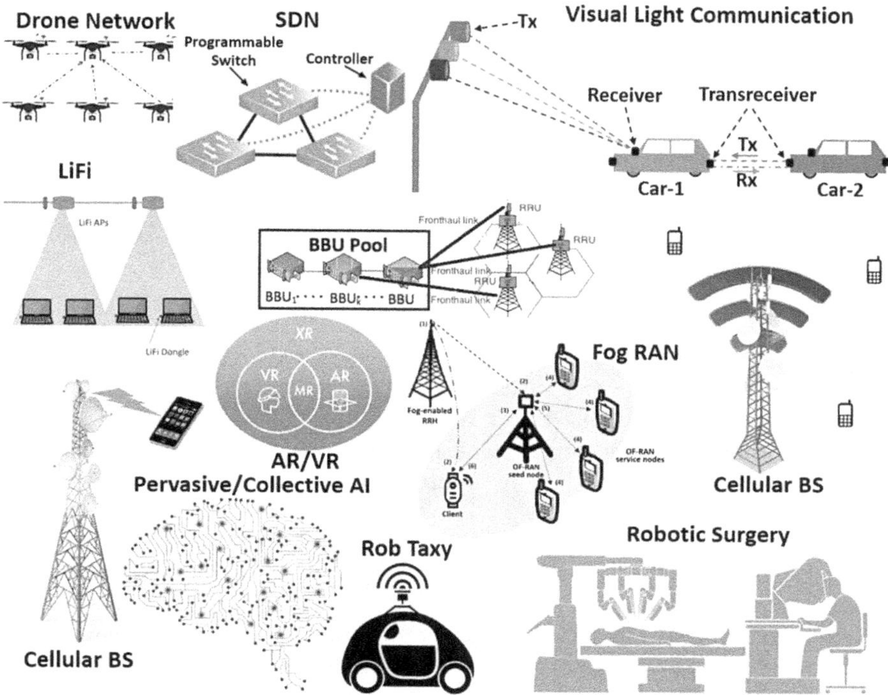

FIGURE 1.1 Next-generation (6G) wireless communication vision.

where they divided the communication system into blocks such as modulation, coding, channel, demodulation, etc., and tried to optimize each block independently without considering its effect on the other blocks and used mathematical and information theory concepts that only capture the approximate behavior of the system. These make it more challenging to perform end-to-end optimization of the communication system in practice and produce sub-optimal performance.

In recent years, machine learning (ML)-based approaches have emerged over traditional complex algorithms to support these challenges optimally and efficiently. ML-based methods work on data-driven approaches rather than model-based approaches. These methods take data from real hardware in a real-world environment and optimize the system for the best performance.

1.2 MACHINE LEARNING ALGORITHMS

Machine learning is a statistical method in which computers approximate complex system models with great accuracy [8]. Its application is crucial in several disciplines, including speech processing, image processing, and artificial intelligence. The phrase "learning" refers to forecasting the results from new observations. The learning algorithms themselves perform this action. Data features are utilized by algorithms to learn. These characteristics could either be continuous or discrete.

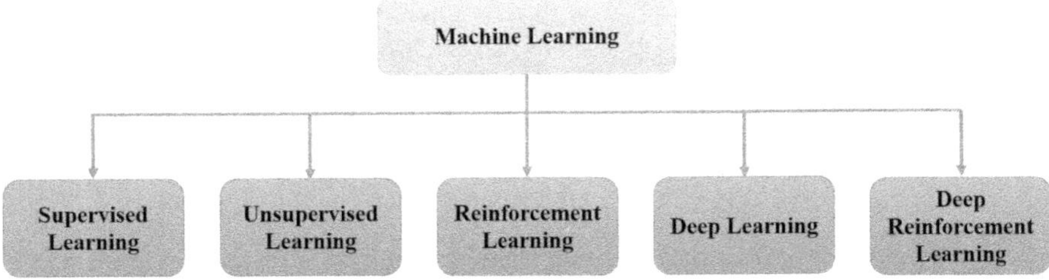

FIGURE 1.2 Various machine learning techniques.

Data is frequently organized in a data frame that is referred to as the "design matrix." In this format, each row represents a distinct item, and each column corresponds to the values associated with that object. Following this, the features undergo the feature engineering process in which new features are created and deleted based on the subject matter and the formulation of the problem.

A wide array of learning algorithms is included in ML. Figure 1.2 illustrates these learning methods for your convenience. Before giving results, each of these algorithms has to go through the training process. Following training, the taught algorithm becomes what is referred to as the "model." Deep learning (DL) is a subfield of machine learning (ML) that involves the use of a great number of processing layers to discover various data representations at various levels of abstraction [9]. The similarities it has with the human brain are used in deep learning [8]. The methods of deep learning discussed in my thesis are deep neural networks (DNN) and deep reinforcement learning (DRL), together with their respective training structures, and this part covers those topics.

1.2.1 Deep Neural Networks

MLPs are deep neural networks, but instead of having a single layer between the input and output nodes, they include multiple layers. The feed-forward deep network that maps values to their corresponding incoming and outgoing values is called an MLP. MLPs are regarded as "universal approximators" because the approximation features of feed-forward networks are quite generic [10].

In addition to being a universal approximator, convolutional neural networks (CNN) have recently gotten much attention because the size of training datasets has grown. This phenomenon is commonly referred to as "big data." CNNs currently have greater depth (in terms of the number of layers) and width (in terms of the number of neurons per layer) as a result of larger model sizes, higher computation speeds, and the ability to leverage distributed computing. Because of this rise in depth and width, CNNs are now capable of accurately performing complicated classification and regression tasks. CNN's designs have progressed to the point where they have what is essentially the same number of connections per neuron as a cat's brain [8]. Every 2.4 years, the number of neurons found in contemporary CNNs will double. If this trend continues, by the year 2050 a CNN will most likely have the same amount of neurons as a human brain [8]. Due to this excessive growth, CNN is now a very desirable option for solving AI problems that could be extremely challenging. Some examples of these problems include cascaded channel estimation for reconfigurable intelligent surfaces (RIS) and beamforming design for multi-user mmWave RIS-based communications, as well as RIS-assisted non-orthogonal multiple access networks [11] and other similar problems.

1.2.2 Deep Reinforcement Learning

Reinforcement learning (RL) is a class of ML in which the model training is based on rewarding desired behaviors and/or punishing undesired ones. An RL agent can typically perceive and comprehend its surroundings, act, and learn through trial and error. RL can be of two types: (**a**) deep RL and (**b**) vanilla RL. Because it records its experience in a table, the vanilla RL is also considered to be a "tabular" solution. At the same time, the deep reinforcement learning makes use of DNN to assess the amount of experience it has had and to predict the discounted rewards to select the most appropriate behaviors [12]. Therefore, deep reinforcement learning is comparable to supervised learning in certain respects. A dataset containing the actions of the agent, the current state of the environment, and the reward are all analyzed by the DNN to determine which behavior would most likely result in the highest reward in the future.

When CNN is used as an estimator in RL, it can cause problems with convergence in deep RL. Using DNNs as an estimator in RL requires the development of specialized algorithms and models that can effectively learn from the agent's experiences in the environment. This involves solving various technical challenges, such as dealing with high-dimensional input data, handling noisy or incomplete observations, and managing the trade-off between exploration and exploitation. RL with DNN as an estimator is an AI problem because it involves developing algorithms and models that enable an AI agent to learn to make decisions and take actions in complex environments [13].

1.2.3 Centralized Learning

The tremendous volume of data produced by mobile and Internet of Things (IoT) devices is the driving force behind the development of DL. The data is continuously uploaded to the cloud in a traditional DL system where it is analyzed, and further features are extracted. After that, the models are trained in a more effective manner on servers with a high level of performance. This centralized machine learning strategy is depicted in Figure 1.3a. It is possible to deploy and make use of models in a scalable manner thanks to ML-as-a-service providers such as Amazon Web Services, Google Cloud, and Microsoft Azure. When there are a lot of interactions with cloud services, a greater collection of training data occurs, and as a result, more intelligent ML-based apps are developed. On the other hand, users are increasingly concerned about the privacy of the data used for training. The machine learning model used in centralized learning is constructed in the cloud, where the data are also transferred. A user connects to the model by way of an API and submits a request to make advantage of one of the available services.

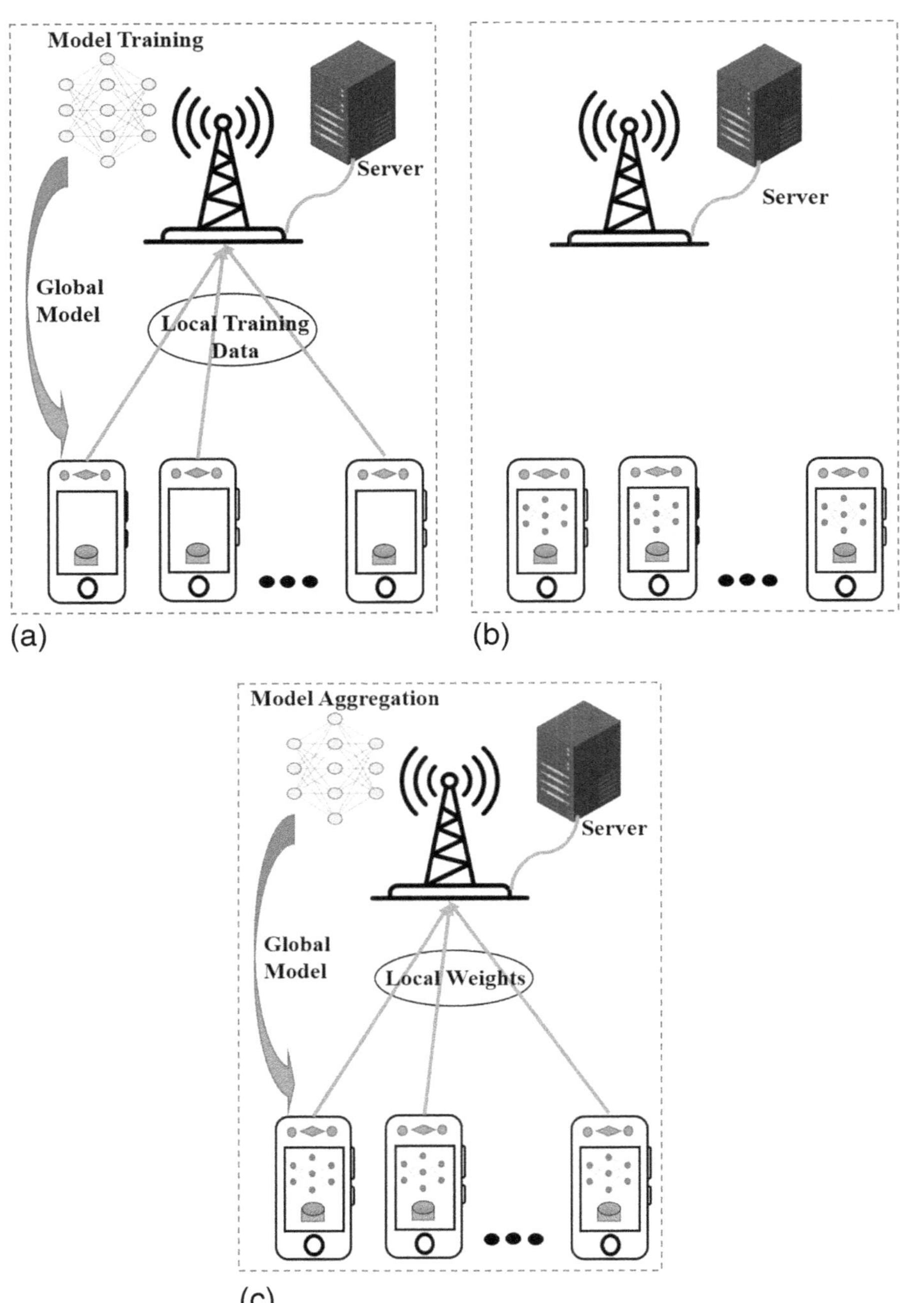

FIGURE 1.3 Centralized vs. Distributed vs. Federated Learning. (a) Centralized Learning, (b) Distributed On-Site Learning, (c) Federated Learning.

1.2.4 Distributed Learning

Researchers developed on-the-device distributed machine learning in response to the growing model privacy problems, as well as the rise in the volume of training datasets. On the local platform, the generation of the dataset, the training of the model, testing, and inference are all carried out. On-site machine learning asks the server to distribute a pre-trained or generic machine learning model to the devices, as shown in Figure 1.3b. This is done in lieu of sending a request to the cloud that contains the user's private data. After the model has been deployed by the cloud, the individual devices will next personalize it by testing it, deriving conclusions from it, and training it using their own data. In the process of distributed learning, once the generic model has been sent to the local devices by the cloud, there is no need for further communication between the cloud and the local device.

1.2.5 Federated Learning

Federated Learning (FL) [14] provides the ability to apply DL to devices with limited memory and processing resources while maintaining user security during model training. FL balances model performance and user privacy via three design elements. First, each device uses its raw data to train a local model. Second, after local training, the model weights are shared with the server for aggregation, making a global, more accurate model. Third, the global model is then shared with the individual users. During this FL approach, the individual user need not upload their data to the central server, thus providing more privacy and security. A wide range of mobile applications, such as those used for predicting health issues based on user data, can run DL models on cell phones employing FL, which allows a better model with privacy for the user data.

Local training, model aggregation, and model dissemination are the three steps of an FL process. The entire training process varies from a traditional mobile edge computing system in three ways. In mobile edge computing systems, a device can offload some of its work to the cloud while asynchronously doing its tasks. However, in FL each device must complete its local model training before performing weight uploading. Second, with FL, the cloud cannot aggregate the global model until each device offloads its local model to the cloud. This needs precise synchronous processing and imposes a latency constraint. This training procedure is frequently repeated numerous times. Third, the submitted model sizes in FL should be the same for all IoT devices. However, uploaded data sizes vary from device to device in generic mobile edge computing. Figure 1.3 depicts the three stages of FL.

1.2.6 Deep Learning in Resource Allocation

Numerous researchers have used RL approaches to address power management and resource allocation issues in 5G networks. In [56], a deep Q-learning algorithm enhances power management, beam shaping, and interference mitigation. [57] uses a Q-learning method to divide communication and computing resources among slice requests that arrive at random, each with a distinct weight and QoS requirement. However, these works often merge many network applications into a single agent for collaborative optimization. On the other hand, some recent studies investigated federated reinforcement learning's potential uses in wireless communication. To distribute heterogeneous resources and compute offloading in the 5G ultra-dense network scenarios, the authors of [58] used deep reinforcement learning (DRL). In this, decentralized FL training is done on the RL model.

1.3 HETEROGENEOUS NETWORK

Heterogeneous networks (HetNets) are communication networks that consist of multiple types of nodes with diverse capabilities, such as different radio access technologies (RATs) and varying transmission ranges. These networks are increasingly prevalent in modern telecommunications, encompassing a variety of devices such as mobile phones, Internet of Things (IoT) devices, and base stations.

Resource allocation in heterogeneous networks refers to the process of efficiently distributing network resources, such as bandwidth, power, and time, among the various nodes to optimize network performance. The goal is to achieve improved connectivity, reduced latency, enhanced throughput, and better overall user experience.

Reinforcement learning (RL) is a powerful machine learning technique that has gained significant attention for solving complex decision-making problems, especially in dynamic and uncertain environments. RL models learn from interactions with the environment and use trial-and-error learning to find optimal strategies to achieve specific goals.

When applied to resource allocation in heterogeneous networks, reinforcement learning allows autonomous decision-making agents to learn how to allocate resources efficiently based on the current network state and the desired performance objectives. The agents can be base stations, IoT devices, or any other network entities that need to make decisions regarding resource allocation.

The key components of applying reinforcement learning to resource allocation in heterogeneous networks are as follows:

1. **State Representation**: Defining the network state is crucial for RL agents to make informed decisions. The state may include information such as the number of connected devices, data traffic levels, channel conditions, and available resources.
2. **Action Space**: The action space corresponds to the possible decisions an RL agent can make in the network. For resource allocation, actions might include adjusting transmit power, allocating bandwidth, or choosing between different radio access technologies.
3. **Reward Design**: Rewards are used to provide feedback to the RL agent, guiding it towards making decisions that lead to desired outcomes. The reward function should be carefully designed to incentivize actions that align with network optimization objectives, such as maximizing throughput, minimizing latency, or conserving energy.
4. **Exploration vs. Exploitation**: RL agents balance exploration of new strategies with exploitation of learned policies. They try different actions to gather information about the environment initially, but over time, they exploit their learned knowledge to make better decisions.

5. **Learning Algorithm**: Various RL algorithms can be employed, such as Q-learning, Deep Q Networks (DQNs), Proximal Policy Optimization (PPO), or Deep Deterministic Policy Gradients (DDPG). The choice of algorithm depends on the complexity of the problem and the available data.
6. **Challenges**: Resource allocation in heterogeneous networks is a challenging problem due to the dynamic nature of wireless channels, varying user demands, and the scale of modern networks. Additionally, there are constraints like limited processing capabilities, communication overhead, and the need for real-time decision-making.

By using reinforcement learning for resource allocation in heterogeneous networks, operators can achieve more efficient and adaptive network management, leading to improved user experiences and network performance. As RL research advances and hardware capabilities increase, its applicability in diverse network scenarios will likely expand, offering promising solutions to complex resource allocation problems in dynamic and heterogeneous environments.

1.4 REINFORCEMENT LEARNING

Reinforcement learning (RL) is a type of machine learning that focuses on how an agent can learn to make decisions in an environment to achieve specific goals. It is inspired by the behavioral psychology concept of learning through rewards and punishments. In RL, an agent interacts with an environment, performs actions, and receives feedback in the form of rewards or penalties based on its actions. The agent's objective is to learn a strategy (policy) that maximizes the cumulative reward it receives over time.

Here's a detailed explanation of the key components and concepts in reinforcement learning:

1. **Agent and Environment**:
   - The RL framework consists of two main components: the agent and the environment.
   - The agent is the decision-making entity that takes actions in the environment to achieve its goals.
   - The environment is the external system with which the agent interacts. It provides the state of the environment to the agent, receives actions from the agent, and updates its state based on those actions.
2. **State (s), Action (a), and Reward (r)**:
   - At each time step t, the environment is in a certain state denoted by s(t). The state represents the current situation or configuration of the environment.
   - The agent selects an action a(t) based on the current state. Actions are the decisions or moves the agent can take in the environment.
   - After the agent performs an action, the environment transitions to a new state s(t+1) and provides feedback in the form of a reward r(t). The reward indicates how favorable the action was in the given state.

3. **Policy (π)**:
    - The policy is the strategy or decision-making function followed by the agent. It maps states to actions and guides the agent's behavior in the environment.
    - In RL, the goal is to learn the optimal policy (π*) that maximizes the expected cumulative reward over time.
4. **Reward Function**:
    - The reward function, denoted by R(s, a), quantifies the immediate feedback received by the agent after performing action a in state s.
    - The goal of the agent is to find a policy that maximizes the sum of expected rewards over time, also known as the return.
5. **Return (G)**:
    - The return is the cumulative sum of rewards obtained from time t to the end of the episode (or the infinite horizon in some cases).
    - It is defined as G(t) = r(t+1) + r(t+2) + r(t+3) + … + r(T), where T is the terminal time step.
6. **Value Function (V) and Action-Value Function (Q)**:
    - The value function represents the expected cumulative reward an agent can achieve from a given state onward, following a particular policy.
    - It is defined as V(s) = E[G(t) | s(t) = s], where E[.] denotes the expectation.
    - The action-value function (also known as the Q-function) is similar to the value function but takes into account the action taken in the given state.
    - It is defined as Q(s, a) = E[G(t) | s(t) = s, a(t) = a].
7. **Exploration and Exploitation**:
    - In RL, agents face a trade-off between exploration and exploitation.
    - Exploration refers to the agent's strategy of trying new actions to gather information about the environment.
    - Exploitation refers to the agent's strategy of using its current knowledge to make decisions that are expected to yield high rewards.
    - Striking the right balance between exploration and exploitation is critical for finding an optimal policy.

1.5 FEDERATED LEARNING

Federated Learning is a decentralized machine learning approach that enables training of models across multiple devices or edge devices while keeping the data on those devices. It was introduced to address the challenges of centralized machine learning systems where all the data is collected and stored on a central server. In centralized systems, data privacy, communication costs, and latency can become significant concerns.

In Federated Learning, the training process occurs directly on the edge devices, such as smartphones, IoT devices, or other user devices, without requiring the raw data to be sent to a central server. Instead, only model updates or gradients are exchanged between the devices and the central server. This approach allows for privacy preservation and reduces communication overhead since only model updates, which are usually much smaller in size, are transmitted.

The key components and workflow of Federated Learning are as follows:

1. **Client Devices**: These are the edge devices or user devices that participate in the Federated Learning process. Each device has its own local data, which is typically heterogeneous and private.
2. **Central Server**: The central server coordinates the Federated Learning process. It distributes the initial model to the client devices, receives model updates from them, and aggregates the updates to improve the global model.
3. **Global Model**: The global model is the model being trained and improved through the Federated Learning process. The initial version of the model is sent to the client devices by the central server.
4. **Local Training**: On each client device, the global model is used to train the model locally on the device's data. The training process usually involves stochastic gradient descent (SGD) or variations of it.
5. **Model Updates**: After local training, each client device computes the gradients of its local model and sends only these gradients (model updates) to the central server.
6. **Model Aggregation**: The central server receives the model updates from multiple client devices and aggregates them to create an improved global model. The aggregation process can be as simple as averaging the gradients or can involve more complex techniques like Federated Averaging or Federated Proximal algorithms.
7. **Iteration**: The process of local training, model updates, and model aggregation is repeated over multiple iterations to further refine the global model.

Federated Learning offers several advantages:

1. **Privacy Preservation**: Since data remains on the client devices and only model updates are shared, Federated Learning enhances data privacy and security.
2. **Reduced Communication**: Sending model updates instead of raw data significantly reduces communication costs and network latency.
3. **Decentralization**: The decentralized nature of Federated Learning makes it suitable for large-scale distributed systems without relying on a centralized server.
4. **Adaptability to Heterogeneous Data**: Federated Learning can accommodate diverse and heterogeneous data from different client devices.

However, there are challenges associated with Federated Learning, such as dealing with data heterogeneity, handling unreliable or slow clients, and ensuring the security and integrity of the model during the aggregation process. Researchers continue to explore and develop techniques to address these challenges and expand the applicability of Federated Learning to various domains, including healthcare, finance, and IoT applications. Figure 1.4 represents the various steps required in a Federated Learning-based system.

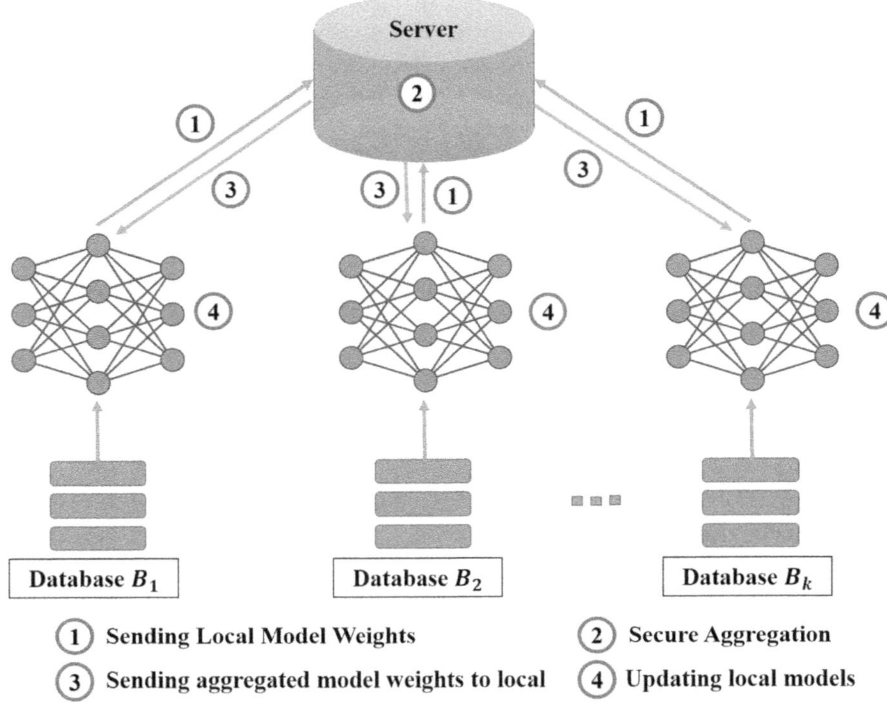

FIGURE 1.4 Architecture of a Federated Learning-based ML system.

1.6 FEDERATED DEEP REINFORCEMENT LEARNING IN RESOURCE ALLOCATION

Federated Deep Reinforcement Learning (FDRL) in Resource Allocation for Heterogeneous Networks is an advanced approach that combines the concepts of Federated Learning, Deep Reinforcement Learning, and resource allocation to optimize the performance of communication networks with diverse devices and technologies.

1.6.1 Overview of Federated Deep Reinforcement Learning (FDRL) in Resource Allocation

1. **Heterogeneous Networks**: Heterogeneous networks consist of various types of devices, such as mobile phones, IoT devices, and base stations, with different capabilities and communication technologies (e.g., 5G, Wi-Fi, etc.). Resource allocation in such networks aims to efficiently distribute network resources (e.g., bandwidth, power, and time) to enhance network performance and user experiences.

2. **Deep Reinforcement Learning (DRL)**: Deep Reinforcement Learning is an extension of reinforcement learning where artificial neural networks (deep learning) are used to approximate the value function or policy. DRL models, such as Deep Q Networks (DQNs) and Deep Deterministic Policy Gradients (DDPG), can handle high-dimensional state and action spaces, making them suitable for complex problems.
3. **Federated Learning**: Federated Learning enables the training of machine learning models across multiple devices or edge nodes while keeping the data on those devices. In the context of resource allocation for heterogeneous networks, this means that each device can locally optimize its resource allocation decisions based on its specific environment and constraints without sharing raw data with a central server.

1.6.2 Applying Federated Deep Reinforcement Learning in Resource Allocation

1. **Agent and Environment**: Each device in the heterogeneous network acts as an independent agent. The environment includes all the devices and their interactions, network states, available resources, and user demands.
2. **State Representation**: The state representation comprises information about the device's current state, such as network conditions, traffic load, user demands, available resources, and neighboring devices' states.
3. **Action Space**: The action space consists of the various resource allocation decisions the device can make, including adjusting transmit power, allocating bandwidth, and selecting radio access technologies.
4. **Reward Function**: The reward function provides feedback to each device based on the quality of the resource allocation decisions made. Rewards can be designed to optimize different network objectives, such as maximizing throughput, minimizing latency, or conserving energy.
5. **Federated Learning and Model Updates**: The central server initiates the training process by sending an initial DRL model to each device. Each device performs local training using its local environment and then sends model updates (gradients) to the central server without sharing raw data. The server aggregates the model updates from all devices and updates the global DRL model.
6. **Collaboration and Learning from Each Other**: During the training process, devices in the network can learn from each other's experiences without sharing raw data. This enables them to collectively improve the global model based on insights from different devices' environments.
7. **Decentralized Decision-Making**: Each device can independently make resource allocation decisions based on the global DRL model, which is updated through Federated Learning. This decentralized decision-making approach allows devices to adapt to their local conditions while benefiting from the knowledge of the global model.

1.6.2.1 Challenges and Advantages

Applying Federated Deep Reinforcement Learning in resource allocation for heterogeneous networks presents several challenges and advantages:

Challenges:

- Dealing with heterogeneity: Devices may have different capabilities and data distributions, requiring robust learning algorithms to handle varying environments.
- Communication overhead: Model updates need to be transmitted between devices and the central server, which can be challenging in large-scale networks.
- Privacy and security: Ensuring data privacy during the Federated Learning process is critical, especially when dealing with sensitive user data.

Federated Deep Reinforcement Learning (FDRL) in resource allocation for heterogeneous networks offers several advantages that make it a promising approach for optimizing network performance and addressing the challenges of resource allocation in diverse environments. Some of the key advantages include:

1. **Privacy Preservation**: One of the most significant advantages of FDRL is its ability to preserve user privacy. Since the training of models occurs locally on each device without sharing raw data, sensitive user information remains on the individual devices. This mitigates privacy concerns associated with centralized approaches where data is collected and stored on a central server.
2. **Communication Efficiency**: FDRL reduces communication overhead by transmitting only model updates or gradients between the devices and the central server. Compared to transmitting large volumes of raw data, model updates are typically much smaller in size, leading to significant communication cost savings and lower latency.
3. **Decentralized Decision-Making**: FDRL allows for decentralized decision-making, where each device can independently optimize its resource allocation based on its local environment and constraints. This autonomy enables devices to adapt quickly to changes in their surroundings, enhancing network efficiency and responsiveness.
4. **Scalability**: Federated Learning inherently supports scalability, making it suitable for large-scale heterogeneous networks with a diverse range of devices and technologies. FDRL can handle a substantial number of devices without requiring a massive central server or extensive computational resources.
5. **Adaptability to Heterogeneous Data**: Heterogeneous networks often consist of devices with different capabilities, communication technologies, and data distributions. FDRL can effectively handle such data heterogeneity by training models locally on each device, allowing the global model to be updated collaboratively from diverse sources.

6. **Global Knowledge Sharing**: While each device trains its model locally, FDRL enables the exchange of model updates among devices during the Federated Learning process. This knowledge sharing allows devices to learn from each other's experiences, leading to an improved global model that benefits from insights gathered across the network.
7. **Real-time Learning and Adaptation**: The decentralized and iterative nature of FDRL enables real-time learning and adaptation in dynamic environments. Devices can continuously update their models and make resource allocation decisions based on the latest information, improving overall network efficiency and performance.
8. **Robustness and Fault Tolerance**: FDRL's decentralized nature makes it more robust and fault-tolerant compared to centralized approaches. If some devices are unavailable or experience issues, the Federated Learning process can continue with the participation of other devices, ensuring continuity in resource allocation optimization.
9. **Energy Efficiency**: By optimizing resource allocation based on local conditions and constraints, FDRL can lead to improved energy efficiency in heterogeneous networks. Devices can allocate resources more intelligently, reducing unnecessary energy consumption and enhancing overall network sustainability.

In summary, Federated Deep Reinforcement Learning in resource allocation for heterogeneous networks addresses privacy, communication, and scalability challenges while facilitating decentralized decision-making and real-time adaptation. These advantages make FDRL a promising approach for optimizing network performance in diverse and dynamic environments, while ensuring data privacy and efficient resource utilization.

REFERENCES

1. M. Z. Chowdhury, M. Shahjalal, S. Ahmed, and Y. M. Jang, "6G wireless communication systems: Applications, requirements, technologies, challenges, and research directions," *IEEE Open Journal of the Communications Society*, vol. 1, pp. 957–975, 2020.
2. F. Tariq, M. R. A. Khandaker, K.-K. Wong, M. A. Imran, M. Bennis, and M. Debbah, "A speculative study on 6G," *IEEE Wireless Communications*, vol. 27, no. 4, pp. 118–125, 2020.
3. "5G Evolution and 6G," White Paper 38.211, 01 2020, version 16.3.0 Release 16. [Online]. Available: https://www.docomo.ne.jp/english/binary/pdf/corporate/technology/whitepaper_6g/DOCOMO_6G_White_PaperEN_20200124.pdf
4. A. Al-Fuqaha, M. Guizani, M. Mohammadi, M. Aledhari, and M. Ayyash, "Internet of things: A survey on enabling technologies, protocols, and applications," *IEEE Communications Surveys Tutorials*, vol. 17, no. 4, pp. 2347–2376, 2015.
5. J. Xie, Z. Song, Y. Li, Y. Zhang, H. Yu, J. Zhan, Z. Ma, Y. Qiao, J. Zhang, and J. Guo, "A survey on machine learning-based mobile big data analysis: Challenges and applications," *Wireless Communications and Mobile Computing*, vol. 2018, p. 8738613, 2018.
6. H. Tataria, M. Shafi, A. F. Molisch, M. Dohler, H. Sjoland, and F. Tufvesson, "6G wireless systems: Vision, requirements, challenges, insights, and opportunities," *Proceedings of the IEEE*, vol. 109, no. 7, pp. 1166–1199, 2021.

7. X. You, C.-X. Wang, J. Huang, X. Gao, Z. Zhang, M. Wang, Y. Huang, C. Zhang, Y. Jiang, J. Wang, M. Zhu, B. Sheng, D. Wang, Z. Pan, P. Zhu, Y. Yang, Z. Liu, P. Zhang, X. Tao, S. Li, Z. Chen, X. Ma, S. Han, K. Li, C. Pan, Z. Zheng, L. Hanzo, X. S. Shen, Y. J. Guo, Z. Ding, H. Haas, W. Tong, P. Zhu, G. Yang, J. Wang, E. G. Larsson, H. Q. Ngo, W. Hong, H. Wang, D. Hou, J. Chen, Z. Chen, Z. Hao, G. Y. Li, R. Tafazolli, Y. Gao, H. V. Poor, G. P. Fettweis, and Y.-C. Liang, "Towards 6G wireless communication networks: vision, enabling technologies, and new paradigm shifts," *Science China Information Sciences*, vol. 64, no. 1, p. 110301, 2020.
8. I. Goodfellow, Y. Bengio, and A. Courville, *Deep Learning*. Cambridge, MA: The MIT Press, 2016.
9. Y. LeCun, Y. Bengio, and G. Hinton, "Deep learning," *Nature*, vol. 521, p. 436–444, 2015.
10. C. M. Bishop, *Pattern Recognition and Machine Learning (Information Science and Statistics)*. Berlin, Heidelberg: Springer-Verlag, 2006.
11. Z. Ding, L. Lv, F. Fang, O. A. Dobre, G. K. Karagiannidis, N. Al-Dhahir, R. Schober, and H. V. Poor, "A state-of-the-art survey on reconfigurable intelligent surface-assisted non-orthogonal multiple access networks," *Proceedings of the IEEE*, vol. 110, no. 9, pp. 1358–1379, 2022.
12. V. Mnih, K. Kavukcuoglu, D. Silver, A. Graves, I. Antonoglou, D. Wierstra, and M. Riedmiller, "Playing atari with deep reinforcement learning," arXiv preprint arXiv:1312.5602, 2013.

2 Federated Deep Reinforcement Learning-based Resource Allocation in Heterogeneous Networks

Satish Kumar and Rajarshi Mahapatra
SPM IIIT-Naya Raipur, Raipur, Chhattisgarh, India

2.1 INTRODUCTION

Next-generation wireless communication networks need to address the increasing demand for higher data rates, greater capacity, broader coverage, and efficient utilization of wireless spectrum. New physical layer techniques, such as communication at mmWave frequencies, M-MIMO, and HetNets, have been introduced to achieve these objectives. However, these techniques also present challenges that make the temporal behavior of the wireless channel more irregular and complex. For example, communication at mmWave frequencies suffers from high path loss and susceptibility to blockage, which makes the channel highly variable over time and space. Similarly, M-MIMO introduces new types of interference, such as inter-user interference, making the channel behavior more dynamic and challenging to model—while with HetNets, the power, spectrum, and resource allocation among multiple BSs become more complex.

To address these challenges, new techniques have been developed to enhance the performance of next-generation wireless communication networks. For example, adaptive beamforming and beam tracking techniques can improve the reliability and robustness of mmWave communication by adapting to changes in the channel environment. Similarly, advanced signal processing and resources such as power and spectrum allocation and interference management techniques can mitigate the effects of interference in M-MIMO systems and HetNets, improving the system's capacity and performance. In the previous chapter, we discussed the challenges of M-MIMO and communication at mmWave frequencies and provided methods to overcome these problems.

In this chapter, we discuss the challenges related to HetNets. To address the challenges related to massive coverage requirements and higher data-rate in the fifth

DOI: 10.1201/9781003303114-2

generation of communication systems, the bigger cells are divided into many smaller ones. The capacity and coverage of these smaller cells depend upon the data rate and the user requirements. This use of smaller cells makes the existing network dense, complex, and relies on effectively sharing network resources among themselves while mitigating the interference [1]. However, unplanned deployments, individual user QoS requirements, and interference caused by other micro and femtocells make resource allocation with interference coordination (RAIC) in such HetNets very complex and computationally expensive [2]. Using these HetNets, a service provider divides users from macro BS (MBS) to micro BS (MiBS) and femto BS (FBS) according to user density, mobility, and data rate requirements. Table 2.1 represents the details of such different types of cells provided by 3GPP in its rel-16 [3]. These MiBSs and FBSs usually transmit on the same channels as MBSs to increase spectral efficiency. The above approach increases cell interference and is thus more prone to SER.

People use distributed resource allocation to manage interference among different MBSs, MiBSs, and FBSs. However, with the increasing density of the network, such resource allocation methods have become more complex and computationally inefficient. Moreover, unplanned user and cell deployment require self-organized and intelligent RAIC techniques to make HetNets intelligent and computationally efficient. Furthermore, the non-convexity and complexity make it challenging to obtain a globally optimal RAIC strategy for HetNets. Researchers have recently proposed many methods to address the optimal RAIC techniques. Such solutions are mainly based on linear programming, game and queuing theory, and Markov approximation rules.

These methods require accurate information regarding network placement, the number of users, required data rate, CSI, available resources, etc., to obtain the optimal RAIC among FBS/MiBS and MBS cells. However, such complete information is challenging to obtain in real-time 5G networks, making implementing optimal RAIC strategy challenging and impractical. Furthermore, increasing users could randomly create an FBS and connect to the core network through fiber or DSL link. Thus, the overall cellular network needs to be adaptable in real time. Researchers have proposed many schemes to optimize resource allocation for a long time while

TABLE 2.1
Details of Different Types of Cells in 5G HetNet [128]

	Types of Cell		
	Femto (InH)	**Micro (UMi-Street Canyon)**	**Macro (UMa)**
Indoor/Outdoor	Indoor	Indoor/Outdoor	Indoor/Outdoor
Bandwidth	20 MHz	20 MHz	20 MHz
Maximum Output Power	24 dBm	35 dBm	49 dBm
ISD	20 m	200 m	500 m
UE Mobility (Horizontal Plane)	3 km/h	3 km/h	3 km/h
Minimum BS-UE Distance (2D)	-	10 m	35 m
BS Antenna Height h_{BS}	3 m	10 m	25 m

minimizing the interference between MiBSs/FBSs and MBSs in cellular HetNets. To obtain the desired data rate and coverage in 5G networks, all these schemes rely on user fairness and reliability while being distributive, less complex, and computationally inexpensive [4, 5]. However, most such traditional systems lack the capacity for self-organization and flexibility to adjust to changing network conditions.

With the increasing number of smaller cells in 5G, RAIC has become too complex. Moreover, unplanned HetNet deployments require self-organizing and intelligent RAIC techniques. The non-convexity and complexity make it challenging to obtain an optimal RAIC strategy. Additionally, 5G networks are naturally dispersed systems, meaning they are designed to operate in a distributed and decentralized manner. There are several reasons for this. Firstly, 5G networks rely on high-frequency mmWave bands, which have a limited range and are easily blocked by obstacles such as buildings and trees. To provide reliable and consistent coverage, 5G networks need to be deployed in a dense network of small cells closer to the users. This distributed deployment enables the network to provide seamless coverage and capacity where needed, and it also helps reduce the interference caused by neighboring cells [6]. Secondly, 5G networks are designed to support a wide range of use cases, including massive IoT, mission-critical services, and high-bandwidth applications such as virtual and augmented reality. 5G networks need to be highly flexible and adaptable to support these diverse use cases. This flexibility is achieved by using a software-defined network (SDN) and network function virtualization (NFV) architecture, which allows network functions to be distributed and dynamically reconfigured based on the changing network conditions and service requirements [7]. Finally, 5G networks are also designed to support multi-access edge computing (MEC), which involves deploying computing resources closer to the network edge to enable low-latency and high-bandwidth applications. MEC requires a distributed architecture where the computing resources are distributed across multiple nodes in the network, including base stations and edge servers. All these requirements make HetNet an integral part of 5G wireless networks.

Deep Reinforcement Learning (DRL), which uses a DNN as a predictor, has recently emerged to perform many complex tasks efficiently. This work proposed a DRL-based RAIC technique to optimize power and spectrum allocation with interference coordination in downlink 5G HetNets while guaranteeing the individual user QoS goals. Centralized DRL techniques would require a lot of data interchange, whereas fully decentralized approaches would cause poor performance and slower convergence. To further address these issues, we suggest an FL method to DRL, in which BS cooperatively trains the central DRL network by just sharing model weights rather than training data. Simulation results show that the proposed federated DRL-based RAIC method performs better, takes less execution time under different coverage and data rate requirements, and converges quickly.

2.1.1 Supervised Learning-based Resource Allocation

DL can be used in a supervised setting to solve the problem of resource management by employing a traditional optimization strategy as the training ground truth. A deep neural network (DNN) can approximate the input-output relationship by being trained on a vast dataset. In this way, the DNN can take the place of many more conventional optimization strategies that need a high level of computing complexity. An NP-hard power allocation problem under interference channels is addressed in [8], which suggests a DNN-based strategy to solving the problem. The weighted minimal mean-squared error (WMMSE) algorithm [9] is used to create the required output, and it is applied to each sample individually. Despite the fact the DNN has a lower computational complexity than the WMMSE, simulations show its performance is fairly comparable to that of the WMMSE. The linear sum assignment programming (LSAP) problem often arises in wireless resource management, and [10] proposes a training approach identical to the one already in use. Previously, the Hungarian algorithm was successful in solving the LSAP problem; however, due to the Hungarian algorithm's high level of computing complexity, its application to the real-time system cannot be considered. In a similar vein, the article [11] suggests a DL-based resource allocation framework as a solution to the issue of energy-efficient power regulation in multi-cell networks. The DNN is taught to make an approximation of the solution of a branch-and-bound (B&B) algorithm that operates at its highest possible level of performance.

2.1.2 Unsupervised Learning-based Resource Allocation

The supervised learning-based system has two drawbacks, despite the fact its training paradigm is rather basic. On the other hand, traditional algorithms can only produce outcomes that are not ideal in various conditions. Because these inferior answers are treated as though they were the absolute truth while being trained, the DL-based technique cannot produce improved results. A significant amount of data is required to train a DNN; as a consequence, a significant amount of data for training that is labeled is required. Unsupervised learning approaches have been developed to optimize the objective directly through unsupervised training of the DNN's parameters; these approaches solve the drawbacks of the supervised learning paradigm. Many different loss functions have been devised for the DNN, and the DNN models can outperform traditional heuristic optimization strategies in a wide variety of settings. For instance, in [12], a fully connected neural network is proposed as a method for determining the power of users under the interference channel. The training target for this method is the sum rate of all users. In addition, spectrum and energy efficiency have been considered as possible training objectives in [13]. The results of [14] show that the effectiveness of both unsupervised learning approaches is superior to the WMMSE heuristic method currently being used.

2.1.3 Learning Assisted Optimization for Resource Allocation

The deep neural network (DNN) completely takes the place of the optimization technique in both the supervised and unsupervised paradigms. Recent developments in the DL-based methodology have resulted in the incorporation of prior knowledge regarding the optimization problem. This serves to improve sample efficiency. When discrete and continuous variables are considered, many problems associated with the administration of wireless resources can be rewritten as mixed integer nonlinear programming (MINLP) problems. Researchers investigated the resource management of cloud radio access networks (RANs) and device-to-device (D2D) systems in [15, 16], respectively; [15, 16] may be found here. The B&B method will be applied to solve both MINLP problems, which may be viewed as having the potential to be posed by either of these two different types of systems. By utilizing a DNN-learned pruning policy, it is feasible to speed up the process, which is possible using DNN. The task of learning about pruning policies can be further transformed into a binary classification problem that can be solved by DL [16]. This is achieved by picking problem-independent features that are constant, as well as problem-dependent traits suitable for the activity at hand.

2.1.4 Deep Reinforcement Learning for Resource Allocation

Deep Reinforcement Learning has also been used to solve a variety of issues pertaining to wireless communication systems. These issues include network slicing, integrated designs of caching, processing, software-defined and virtual vehicle network-based communication, and others. DRL has been utilized in the management of resources for dynamic spectrum access as well as power control. [17] suggests a DRL-based spectrum access method as a possible solution. In DRL, the current state is determined by the history of actions and observations, and the reward function is determined by successful transmission of information. Additionally, DRL has been utilized for the purpose of power allocation in [18]. In this, the centralized training method is utilized, and the greedy policy views each user to be an agent that investigates the surrounding environment.

To optimize the cumulative long-term benefits, DRL tries to create an agent that can discover the best course of action across various states through interaction with the environment. Since a DNN has a high representational ability to approximate the value function or the direct strategy, unlike traditional RL, DRL uses this DNN to express the policy. Figure 2.1 represents the block diagram of a DRL model. DRL

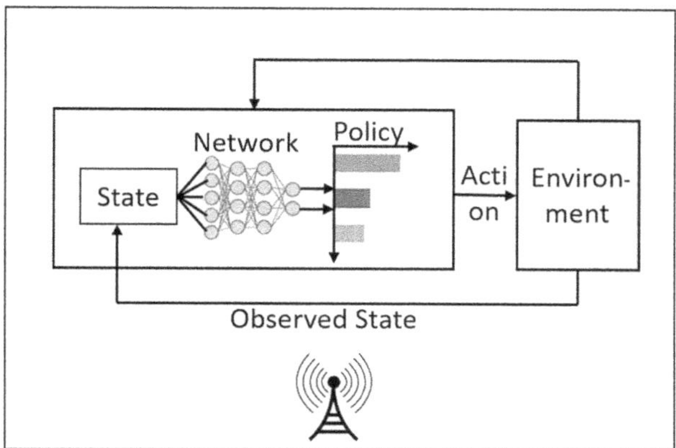

FIGURE 2.1 Block diagram representation of a DRL network.

techniques optimize any unknown system by interacting with it, which makes the DRL-based schemes a perfect solution for scenarios whose parameters are continuously changing.

In reinforcement learning, an agent learns to make a sequence of decisions, called actions, in an environment to maximize a numerical reward signal. The environment provides feedback to the agent in the form of rewards or penalties based on its actions. The agent's goal is to learn a policy, which is a mapping from states to actions that maximizes the cumulative reward it receives over time.

The key components of reinforcement learning are as follows:

1. **Agent**: The learner or decision-maker that interacts with the environment. It observes the state of the environment, selects actions, and receives rewards.
2. **Environment**: The external system or world in which the agent operates. It can be as simple as a computer simulation or as complex as a real-world scenario.
3. **State**: The current situation or configuration of the environment as perceived by the agent. It provides the necessary information for the agent to make decisions.
4. **Action**: The specific decision or choice made by the agent based on the current state. Actions can have short-term consequences and affect future states and rewards.
5. **Reward**: A numerical signal that indicates the desirability or quality of the agent's actions. The agent's objective is to maximize the cumulative reward it receives over time.

Reinforcement learning algorithms aim to discover an optimal policy through exploration and exploitation. Exploration involves trying out different actions to gather information about the environment and learn which actions lead to higher rewards. Exploitation involves exploiting the learned knowledge to make decisions that are likely to maximize rewards.

Some popular reinforcement learning algorithms include Q-learning, SARSA, and Deep Reinforcement Learning algorithms like Deep Q-Networks (DQN) and Proximal Policy Optimization (PPO). These algorithms use various techniques, such as value functions, policy gradients, and neural networks, to learn and improve decision-making policies.

2.1.4.1 Deep Q-Networks

A Deep Q-Network (DQN) is a Deep Reinforcement Learning algorithm that combines Q-learning, a popular value-based reinforcement learning method, with deep neural networks. DQN was introduced by researchers at DeepMind in 2013 and has since become a fundamental algorithm in the field of deep reinforcement learning. The key idea behind DQN is to use a deep neural network to approximate the Q-value function, which represents the expected cumulative reward an agent will receive by taking a particular action in a given state. By approximating this function with a deep neural network, DQN can handle high-dimensional state spaces, such as images, and learn complex decision-making policies.

Here's a high-level overview of how DQN works:

1. **Experience Replay**: DQN utilizes an experience replay buffer to store and randomly sample experiences encountered during interactions with the environment. Each experience consists of a state, action, reward, and next state. Experience replay breaks the sequential correlation of experiences and allows for more efficient learning.
2. **Deep Neural Network**: DQN employs a deep neural network, often a convolutional neural network (CNN), as a function approximator for the Q-value function. The network takes the state as input and outputs the Q-values for all possible actions.
3. **Q-Learning Update**: The network is trained using a variant of the Q-learning algorithm called the Bellman equation. During training, the network adjusts its weights to minimize the difference between the predicted Q-values and the target Q-values. The target Q-values are calculated using a target network, which is a copy of the main network with fixed weights. This helps stabilize the learning process.
4. **Exploration and Exploitation**: To balance exploration and exploitation, DQN typically uses an epsilon-greedy policy. It selects the action with the highest predicted Q-value most of the time (exploitation), but occasionally chooses a random action (exploration) to discover new states and actions.

DQN has been successfully applied to various tasks, including playing Atari games, controlling robots, and optimizing complex systems. It has demonstrated the ability to learn directly from raw sensory inputs, making it particularly suitable for domains where high-dimensional and continuous observations are involved.

Since the introduction of DQN, numerous extensions and improvements have been proposed, such as Double DQN, Dueling DQN, and Rainbow, which incorporate additional enhancements to improve stability, sample efficiency, and performance.

Overall, DQN has played a pivotal role in advancing the field of deep reinforcement learning, showcasing the potential of combining deep neural networks with reinforcement learning to solve complex decision-making problems.

2.1.4.2 Deep Deterministic Policy Gradient

DDPG, which stands for Deep Deterministic Policy Gradient, is a Deep Reinforcement Learning algorithm that combines ideas from both policy-based and value-based methods. It is designed to handle continuous action spaces in reinforcement learning problems. DDPG was introduced by researchers at DeepMind in 2015 and has been successful in various applications, including robotics control and continuous control tasks.

Here's an overview of how DDPG works:

1. **Actor-Critic Architecture**: DDPG follows an actor-critic framework, which consists of two neural networks: an actor network and a critic network. The actor network learns a deterministic policy that directly maps states to actions, while the critic network approximates the action-value function (Q-value) to evaluate the quality of the actor's chosen actions.

2. **Replay Buffer**: Similar to DQN, DDPG also uses an experience replay buffer. Experiences consisting of states, actions, rewards, and next states are stored in the buffer and randomly sampled during training. This allows for more efficient learning by breaking the temporal correlation of experiences.
3. **Actor Network**: The actor network takes the current state as input and outputs the corresponding action. It aims to learn the optimal policy by maximizing the expected cumulative reward. The actor network is trained using the gradient ascent on the expected Q-value computed by the critic network.
4. **Critic Network**: The critic network approximates the action-value function (Q-value) and estimates the expected cumulative reward for a given state-action pair. It is trained using the Bellman equation and temporal difference learning. The critic network provides feedback to the actor network by computing the gradient of the Q-value with respect to the actor's chosen actions.
5. **Exploration and Exploitation**: To balance exploration and exploitation, DDPG uses a technique called "exploration noise." Gaussian noise is added to the actor's chosen actions during training to encourage exploration of the action space. As training progresses, the exploration noise is typically decayed to focus more on exploitation.

DDPG has several advantages, including its ability to handle high-dimensional continuous action spaces and its stable learning behavior. The deterministic policy provided by the actor network allows for efficient optimization and learning of complex policies.

Extensions and enhancements to DDPG have also been proposed, such as prioritized experience replay and Hindsight Experience Replay (HER), to further improve its performance and sample efficiency in various domains.

In summary, DDPG is a powerful algorithm for solving reinforcement learning problems with continuous action spaces. By combining policy-based and value-based approaches, it enables the learning of optimal deterministic policies and has shown promising results in a wide range of applications.

2.1.5 Federated Learning-based Resource Allocation

Only a few contributions have utilized FL to train distributed control and DRL models. Until now, FL has mostly been applied to supervised learning issues in disciplines like natural language processing and computer vision. A federated control technique has been suggested in this domain in [19] to address coordination issues involving a large number of agents, edge caching issues [20, 21], mobile edge computing strategies and offloading [22], and IoT applications [23]. The authors of [24] took underlay mode D2D-enabled wireless networks into account to optimize spectrum utilization and a DRL-based FL-aided decentralized resource allocation technique to maximize aggregate capacity and reduce overall power consumption while maintaining the required QoS for both cellular customers and D2D users. Through independent simulations for the 5G mm-wave and 6G terahertz scenarios, the authors have assessed the performance of the suggested systems.

FedAverage and FedProx are popular FL algorithms for training ML models in distributed environments. [25] provides a detailed explanation of the FedAvg algorithm and its variants, such as FedProx and FedNova. One of the main advantages of using FedAvg in FL is its simplicity and ease of implementation, which makes it suitable for resource-constrained IoT devices. In [25], the FedAvg has been shown to perform well in various applications, including image classification and natural language processing. Similarly, the FedProx algorithm has been applied for resource allocation in HetNets in [26]. The authors propose a FedProx-based algorithm incorporating a proximal term to regularize resource allocation decisions and ensure device fairness. The algorithm optimizes energy efficiency and system performance by considering the trade-off between power consumption and data rate.

2.2 REINFORCEMENT LEARNING IN RESOURCE ALLOCATION

Resource allocation in heterogeneous networks refers to the process of efficiently assigning and managing network resources, such as bandwidth, power, and computing capacity, in a network composed of diverse and heterogeneous devices or components. Reinforcement learning (RL) can be employed as a technique to optimize resource allocation in such networks.

In the context of heterogeneous networks, RL can be used to learn dynamic resource allocation policies that adapt to changing network conditions, varying traffic demands, and the characteristics of different devices. Here's how RL can be applied to resource allocation:

1. **State Representation**: The first step is to define the state space. The state typically includes information about network conditions, such as network load, traffic patterns, device capabilities, and channel conditions. The state should capture the relevant aspects of the network that affect resource allocation decisions.
2. **Action Selection**: The RL agent selects actions based on the current state. In the context of resource allocation, actions represent the allocation decisions, such as assigning bandwidth to different devices, adjusting power levels, or scheduling tasks. The action space depends on the specific resources being allocated and their associated constraints.
3. **Reward Design**: The agent receives rewards based on the quality of its resource allocation decisions. The reward can be defined based on various performance metrics, such as throughput, delay, energy efficiency, fairness, or a combination of multiple objectives. The reward function guides the RL agent towards learning resource allocation policies that maximize network performance.
4. **Exploration and Exploitation**: To learn an effective resource allocation policy, the RL agent needs to balance exploration and exploitation. It explores different allocation strategies to gather information about the network and exploit the learned knowledge to make better resource allocation decisions over time. Exploration can be achieved through stochastic action selection or by introducing exploration noise.

5. **Training and Learning**: The RL agent learns from interactions with the environment. It collects experiences by selecting actions, observing the resulting states, and receiving rewards. These experiences are used to update the agent's policy through the RL training algorithm. Common RL algorithms used for resource allocation include Q-learning, Deep Q-Networks (DQN), Proximal Policy Optimization (PPO), or actor-critic methods like Advantage Actor-Critic (A2C) or Proximal Policy Optimization.
6. **Dynamic Adaptation**: Heterogeneous networks are subject to changing conditions and evolving traffic demands. RL enables the resource allocation policy to adapt dynamically to such variations. The agent can continuously learn and update its allocation strategy based on the real-time network state and performance feedback.

By applying RL to resource allocation in heterogeneous networks, it becomes possible to optimize resource utilization, improve network performance, and achieve efficient allocation of resources based on specific objectives and constraints. RL-based resource allocation has been investigated in various domains, including wireless networks, cloud computing, and Internet of Things (IoT) environments.

Further, if we use DRL-based methods in a distributed manner, then in many scenarios, we can achieve better results [27]. Recently, these DRL-based techniques have been used explicitly in wireless communication systems such as resource allocation [28, 29], dynamic spectrum access [30], and energy harvesting [31].

This work uses a DRL-based scheme to optimally allocate resources such as power and bandwidth to MiBSs, FBSs, and MBSs in a 5G cellular HetNet while mitigating interference and guaranteeing the QoS requirements of the individual user in the network. DRL is an online learning method in which prior knowledge of the system/scenario is not required. The long-term output rather than the immediate short-term output is considered in DRL-based systems. Q-learning is one of the widely used DRL-based methods. [32] proposes an autonomous Q-learning algorithm that obtains optimal transmit power and channel requirements in HetNets. An online DRL has been used to solve the user association in vehicular networks [33]. To get rate adaptation in cellular networks, authors used a DRL-based technique [34]; however, the joint optimization problem's large action and state spaces make optimal RAIC solutions with Q-learning more challenging.

Earlier publications in this field suggested fully centralized or distributed Deep Reinforcement Learning to address the resource allocation with interference coordination issue. Training along with decision-making are centralized in the former situation, and all BSs communicate all state data to the server at each time step. In the second scenario, each agent (base station) trains and operates the deep learning model independently without knowing its effect on the neighboring cell and without coordination with the other BSs. This chapter presents a federated DRL algorithm implementation that incorporates the most advantageous aspects of both distributed and centralized methods. A class of ML problems known as FL involves many clients (such as mobile devices or entire scenarios) working together to train a DNN model while maintaining a decentralized training process.

2.3 AIM OF THE CHAPTER

The goal of this chapter is to discuss a federated DRL-based Deep-Q network approach to jointly optimize the resource allocation and interference coordination in 5G cellular HetNets. This joint optimization method is utilized to get the maximum long-term downlink reward while guaranteeing the QoS of all the users. Moreover, since the above optimization problem is non-convex and distributed, we have investigated the multi-user DRL method, and optimal policy has been obtained using Deep Q-learning Network (DQN). Here, we focus on maximizing the downlink sum rate in a mobile multi-cell 5G network. By exchanging their model weights in an FL manner and gaining access to local data, the BS collaborates with the central server to efficiently train the DNN model available at the central server that serves as the brain of the DRL controller. After implementation and software simulation using Python and MATLAB environment, the proposed federated DRL-based DQN method converges quickly and performs better than other classical RAIC schemes. The main points of this chapter are listed below:

- A federated DRL-based network architecture optimizes the resource allocation with interference coordination in a HetNet environment.
- The network hyper parameters are optimized for a HetNet with a 3GPP compliant 5G channel model.
- In the proposed F-DRL scheme, each base station uses the DRL algorithm to generate the model weights locally. These intermediate model weights are then sent to the central server regularly for aggregation and central model training. After training, the central model's weight is shared with the individual base station. Using these weights, the model at individual base stations then generates optimal resource allocation.
- The F-DRL scheme reduces the communication overhead to 0.2% while showing superior performance and converges faster than the centralized approach.

2.4 SYSTEM MODEL

HetNets are mainly used to satisfy the increased data rate and coverage requirements in 5G and beyond networks. To fulfill these requirements, many smaller cells are used to offload the users from MBS to MiBSs and FBSs while maintaining the QoS of all the users. Figure 2.2 represents a three-tier 5G HetNet scenario we have considered in this work.

N_m, N_{mi}, and N_f are the number of available macro, micro, and femtocells, respectively in the scenario, whereas K_m, K_{mi}, and K_f are the number of users per unit macro, micro, and femtocells, respectively. Thus, the total number of users available in the scenario is $N = (N_m \times K_m) + (N_{mi} \times K_{mi}) + (N_f \times K_f)$. If a particular user i selects BS_l at time t, user association $b^l(t)$ will become 1; otherwise, 0. Moreover, the entire spectrum available to a particular BS is divided into K channels and entirely used by its users. Now, the binary channel allocation vector $c^k(t)$ of a particular user i will be '1'

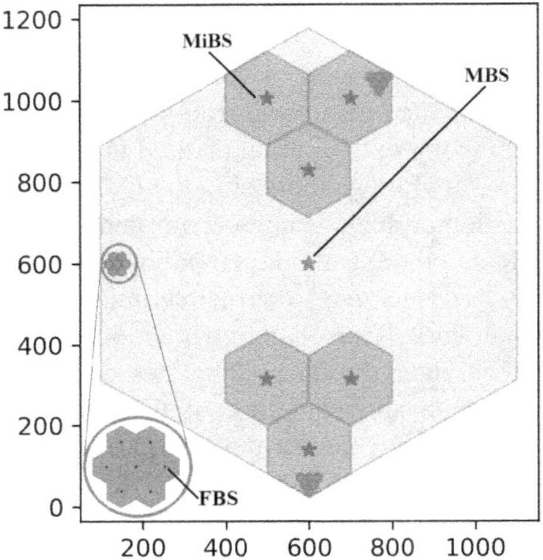

FIGURE 2.2 Three tiers 5G HetNet Scenario (all dimensions are in meters).

when the user utilizes the channel; otherwise, 0. At the same time, the different users of a particular BS are multiplexed with OFDMA. Since all the micro and the femtocells are located in the macro cell coverage, some femtocell users can also be located inside the microcell coverage and all the femto and microcell users are located in macrocell coverage. It generates interference among all the cells, and now it is required to take co-channel interference into the system design. The transmit power should be a finite numerical value in a practical application. Assume that the transmit power between i^{th} user and l^{th} BS is given by $p^k(t) = (p^1(t) +, \ldots, + p^K(t))$, $l \in \mathrm{B}$, $i \in \mathrm{N}$, $k \in \mathrm{K}$. Here, B, N and K represent the set containing all the BSs, users, and channels, respectively. Each user can measure the CSI $h^l(t)$ for all BSs $l \in \mathrm{B}$. Then, the instantaneous signal-to-interference-plus-noise ratio (SINR) at i^{th} user from BS_l using C_k is

$$\Gamma_{li}^{k}(t) = \frac{h_i^l(t)c_i^k(t)(t)p_{li}^k(t)}{\sum_{j \in \mathrm{B}-\{l\}} h_i^l(t)c_i^k(t)(t)p_{li}^k(t) + WN_0} \tag{2.1}$$

where N_0 is the noise power and W is the channel bandwidth. Now from the Shannon formula, the downlink data rate that could be achieved by i^{th} user from a particular BS on the channel C_k is given here:

$$r_{li}^{k}(t) = W\log_2\left(1 + \Gamma_{li}^{k}(t)\right) \tag{2.2}$$

Finally, the data rate that a particular user can achieve is given as:

$$r_i(t) = \sum_{l=0}^{L-1} b_i^l(t) \sum_{k=1}^{K} W\log_2\left(1 + \Gamma_{li}^{k}(t)\right) \tag{2.3}$$

Now the sum data rate of the entire HetNet system can be expressed as

$$R(t) = \sum_{i=0}^{N-1} R_i(t) \tag{2.4}$$

The main goal of this HetNet is to achieve the maximum sum rate while maintaining the QoS requirement of each user. To ensure the minimum QoS requirements Ω_i of all the users from their associated BSs, each user must receive the minimum SINR $\Gamma_i(t)$ required to maintain Ω_i. Now the optimization problem for the power and channel allocation in the HetNet can be formulated as follows:

$$\text{and}\left\{\hat{p}_{li}^{k}(t)\right\} = \arg\max R(t)$$

$$\text{Subject to: } 0 \le p_{li}^{k}(t) \le P_{\max} \forall l, i$$

$$r_i(t) = \sum_{l=0}^{L-1} b_i^l(t) \sum_{k=1}^{K} \Gamma_{li}^{k}(t) \ge \Omega_i \tag{2.5}$$

Moreover, the transmit power of BS_l to user i over channel k is assumed to be $p^k(t)$. Hence, the total cost of ith user can be defined as:

$$\varphi_i(t) = \sum_{l=0}^{L-1} \varphi_i^l(t) = \sum_{l=0}^{L-1} \lambda_l b_i^l(t) \sum_{k=1}^{K} c_i^k(t)(t)\, p_{li}^{k}(t) \tag{2.6}$$

The BS_l's transmit power's unit price is represented by λ_l. Then the ith UE's utility is defined as $w_i(t)$ as the difference between the cost and the achievable profit:

$$w_i(t) = \rho_i r_i - \varphi_i^l(t) = \sum_{l=0}^{L-1} b_i^l(t)\left[\sum_{k=1}^{K}\left[\rho_i p_{li}^{k}(t) - \lambda_l c_i^k(t)(t)\, p_{li}^{k}(t)\right]\right] \tag{2.7}$$

Here, $\rho_i > 0$ denotes the amount of profit made from each channel's capacity. Now the RAIC goal could be accomplished by increasing its long-term reward to its maximum potential. The value of the long-term reward denoted by $\Theta_i(t)$ is calculated as the sum of the instantaneous benefits accrued over an infinite time period and is expressed as follows:

$$\Theta_i(t) = \sum_{\tau=t}^{+\infty} \gamma^{\tau-t} w_i(\tau) \tag{2.8}$$

where the discounting factor is represented by $\gamma \in [0, 1)$. For DRL, we have used discrete power levels that take values between P_{min} and P_{max} as action space. These discrete power levels have been defined as:

$$\mathbb{A} = \left\{ P_{min}, \frac{P_{max} - P_{min}}{M_P - 1}, \frac{2\left(P_{max} - P_{min}\right)}{M_P - 1}, \ldots, P_{max} \right\} \tag{2.9}$$

Where M_p is the number of power levels for a different type of BSs, the P_{max} is chosen according to the BS type as provided in 3GPP TR 38.901 (Rel-16) Table-7.8-1 [3]. All agents of a similar BS type have the same action space, whereas different BS types have different action spaces.

2.5 PROPOSED FEDERATED-DRL ALGORITHM

The suggested RAIC algorithm is covered in more detail in this section. At first, the problem 2.5 has been recast in a DRL scenario, wherein each base station acts as an agent to maximize the sum data rate of all its associated user equipment while minimizing interference to the neighboring cells. Each base station has unique control procedures that allot the optimal power levels to the base station based on the observable state of its own and neighboring cells. The base stations must train DNN models in the context of DRL with either the Q-values or the control action as their outputs. To train the DNN models quickly enough to adapt to the current network conditions, one of the crucial difficulties is a huge amount of training data from the BSs that increases the communication overhead of the system and might hinder user security. Therefore, to address these problems, researchers proposed F-DRL architecture, which enables the federated central server and the base stations as agents to develop a resource allocation model by uploading their intermediate model weights with the central server while maintaining privacy of their user's data. Since this work considers the resource allocation among different BSs, all BSs have almost equal information about the communication scenario and know about their own and neighboring cell users (Figure 2.3).

2.6 SIMULATION RESULTS

In this simulation, we consider a 5G HetNet that has 1 UMa [3] as MBSs, 6 UMi-Street Canyon [3] as interfering MiBSs, 19 Indoor-office [3] as FBSs with 20 users per MBS, 5 users per MiBS and four users per FBS. The total number of users in this system is 126. These cells' radius, bandwidth, and other properties can be found in 3GPP TR 38.901 (Rel-16) [3]. The considered 5G HetNet with all the users are represented in Figure 2.4. The maximum transmit powers of MBSs, Interfering MiBSs, and FBSs are chosen according to the 3GPP TR 38.901 (Rel-16) Table-7.8-1 [3] that are 49 dBm, 44 dBm, and 24 dBm, respectively. The total number of resource blocks = 51 with subcarrier spacing = $30KHz$, the downlink center frequency at which transmission happens $f = 5.5GHz$ *and* channel bandwidth $W = 20MHz$. The channel

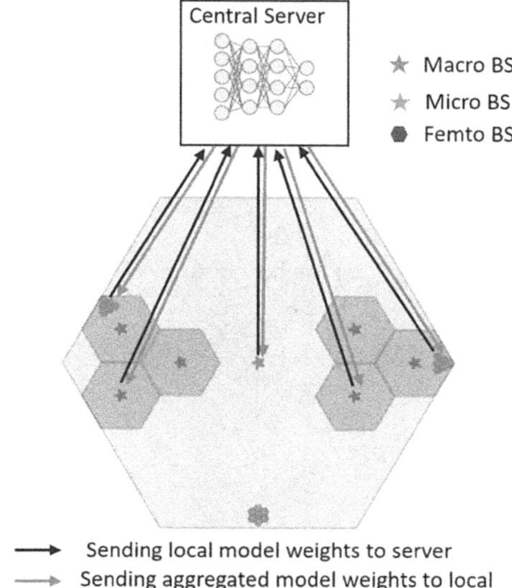

FIGURE 2.3 Block diagram of the F-DRL scheme.

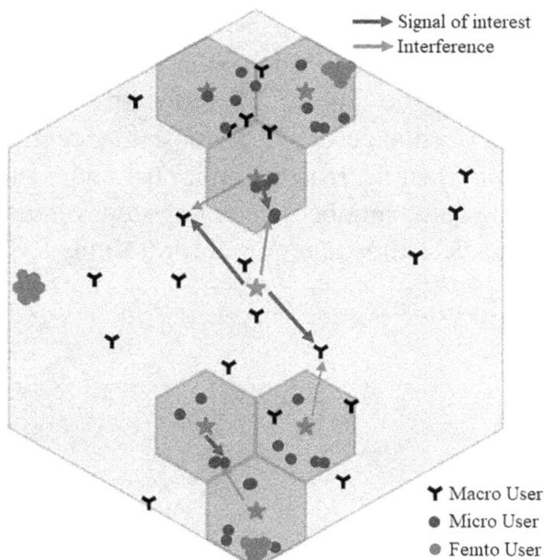

FIGURE 2.4 Three tiers 5G HetNet with users.

model between BSs and the users as tapped delay line (TDL-A) and path loss models for MBSs, MiBS, and FBSs from 3GPP TR 38.901 (Rel-16) Table-7.4.1-1. [3] has been used. The UE noise figure and the noise temperature are 9dB and 290*K*, respectively. The equivalent profit ρ_i for the reinforcement learning is U(0, 1). In this simulation for federated operation, the aggregation period is assumed to be 500 episodes ($A_g = 500$).

ALGORITHM 1: FEDERATED DRL STRATEGY FOR RAIC

Input: Learning Rate = l_r, num_Episodes = EP, Agg_period = A_g
Initialize $Q(s, a)$ and $Q^*(s, a)$ arbitrary
for *n: = 1 to EP episodes* **do**
  Initialize initial state s
  for *all steps of each episodes* **do**
    define $a = \max_a Q(S_t, a)$ for best action
    calculate $Q_{estimated} = Q*(S_{t+1}, a)$
    $Q(S_t, a_t) \leftarrow Q(S_t, a_t) + \alpha\left(R_{t+1} + \gamma\left(Q*(S_{t+1}, a)\right)\right) - Q(S_t, a_t)$
    $s \leftarrow s'$
  end
  if n%Ag == 0 **then**
    Send s to the central server for aggregation and central model training
    Get aggregated model weight from the server to the individual BSs
  end
end

In this work, the joint optimization of the RAIC problem is modeled as a Markov decision process. In any DRL-based system, the learning rate severely affects the training efficiency of the scheme. Figure 2.5 shows the effect of different learning rates δ on the training performance of the proposed system. In the beginning, the training steps are huge and then decrease continuously with increasing training episodes. With the increase in δ, the number of training steps required to achieve optimal RAIC is reduced, and thus less time is required for training.

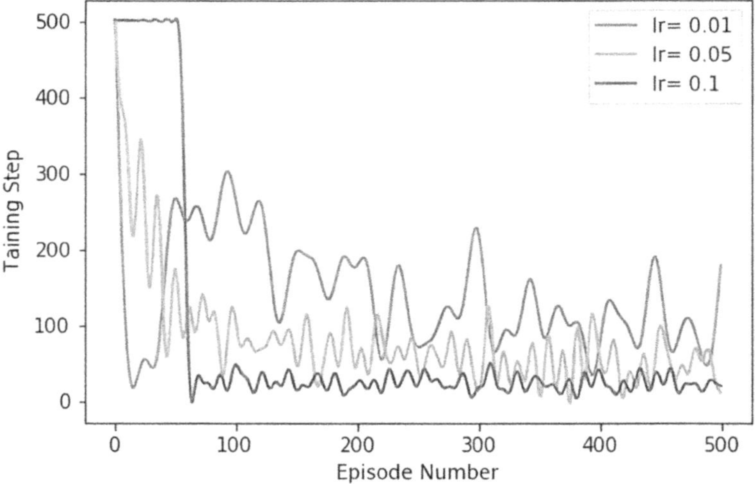

FIGURE 2.5 Training efficiency for different learning rate δ.

2.6.1 Spectral Efficiency Performance

Figure 2.6 represents the spectral efficiency of the fully centralized, fully decentralized, and F-DRL-based proposed scheme. From this simulation result, we can infer that the fully centralized scheme's spectral efficiency is superior to specific training episodes. After 6800 training episodes, the proposed F-DRL-based RAIC scheme performs similarly to the centralized scheme. Here, in the centralized scheme, the communication overhead is higher because, at each BS, it is required to update the data at each time instant to the central server. In the proposed scheme, the network overhead is far less. In all the models, the fully decentralized schemes show poor performance.

The F-DRL-based proposed scheme outperforms the decentralized scheme. Figure 2.7 represents the sum capacity of the HetNets with the proposed scheme for 20 MHz bandwidths of the frequency spectrum. From this graph following observations were made:

1. Initially, the Sum capacity increases with increasing training episodes and then saturates after specific training episodes.
2. The system achieves an average sum rate of over 105 Mbps after 8000 training episodes.
3. After 6800 training episodes, the proposed F-DRL scheme performs better than the fully centralized and fully decentralized scheme.
4. The fully decentralized approach provides the SR that is the minimum of all three schemes and also saturates at the lowest SR.

This graph shows that the system achieves an average sum rate of over 105 Mbps after 8000 training episodes. After 6800 training episodes, the proposed F-DRL scheme performs better than the fully centralized scheme.

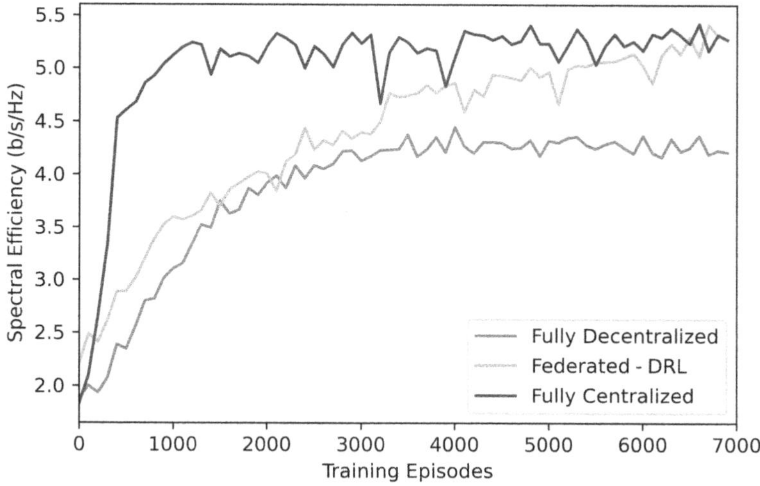

FIGURE 2.6 Spectral efficiency of different RAIC schemes.

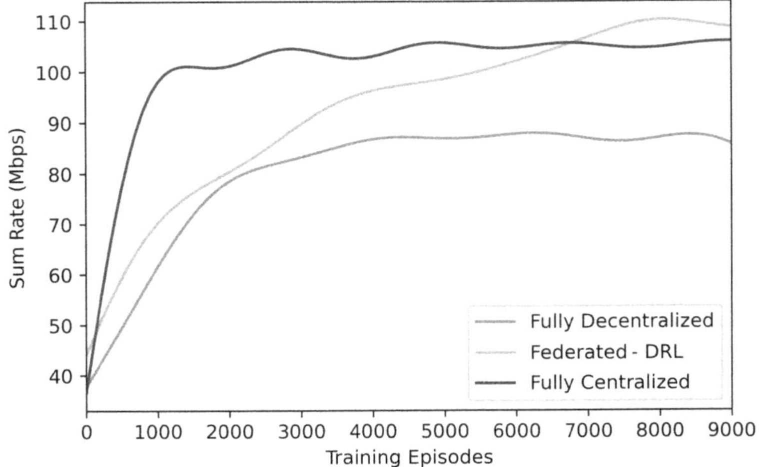

FIGURE 2.7 Sum capacity of the HetNet for 20 MHz frequency bandwidth.

TABLE 2.2
Comparision between Fully Centralized, Fully Decentralized, and the Proposed F-DRL Scheme

	F-DRL	Fully Centralized	Fully Decentralized
Mean SE (bit/S/Hz)	5.60	5.61	4.319
Avg. Execution Time (μS)	250	740	240
Communication Overhead	0.002	1	0

2.6.2 Timing and Complexity Analysis

Table 2.2 compares the fully centralized, fully decentralized, and proposed F-DRL scheme. The proposed scheme only updates the central server at $A_g = 500$ episodes, whereas in the fully centralized approach, it is required to update the central server after each episode. Thus, the communication overhead in the proposed scheme is 500 times less than the fully centralized scheme. In a fully decentralized approach, there is no need to update the central server; thus, the communication overhead is zero. In the fully centralized scheme, a large volume of data must be processed at the central server in each episode; thus, it requires approximately 3 times higher execution time than the F-DRL.

2.7 SUMMARY

In this chapter, an F-DRL-based joint optimization of resource allocation with interference coordination strategy has been proposed and successfully implemented for three tiers 5G HetNets. The joint optimization problem has been designed to obtain long-term rewards while maintaining the QoS requirements of all the users in the system. Since the joint optimization of the RAIC problem is non-convex and combinatorial, thus we have used DRL and Deep Q-Learning approaches for optimal resource allocation between all the users and their preferred BS. To further use the advantage of distributed RL while mitigating the effect of centralized RL, federated learning with DRL has been used. This proposed scheme provides spectral efficiency superior to the fully centralized scheme while executing 3 times faster, having 0.2% communication overhead and converging faster than the fully centralized scheme.

To maximize the return, here in this work we have used DNN with RL algorithms for generating the Q-values or executing action probabilities. In the FL approach, a centralized statistical model must be used to generate the optimal resource allocation strategy for learning from the actions generated by various BSs in the network. This work uses the FL approach to assume that the CNN-based RL process saves each BS state locally. After a specific time interval, it only updates the intermediate model weights to the central server. After getting these intermediate model weights from each BSs, the central model is trained using the local weights. After central model training, the weights are again shared with individual BSs to update their local models. After that, the model at the individual BSs generates the optimal power and resource allocation strategy for them. The following objective function is to be solved as a result of the FL training:

$$\min_{\delta} P(\delta) = \sum_{l=1}^{L} w_i L_i(\delta_i) \tag{2.10}$$

where δ and $P(\delta)$ represent the global model weights and global function that need to be optimized, δ_l is the local model weights, and L_l is the local loss function at BS_l. The contribution of the lth cell in data size is represented by w_l.

Formally, value-based RL algorithms estimate an expected return beginning from state s when taking action a using an action-value function $Q(s, a)$:

$$Q_\pi(S_t, a) = E_\pi \left\{ \sum_{k=1}^{\infty} \gamma^{k-1} r_{t+k-1} \middle| s_t, a \right\} \tag{2.11}$$

$$= E_{S_{t+1},a} \left\{ r_t + \gamma Q_\pi(S_{t+1}, a) \middle| S_t, a_t \right\} \tag{2.12}$$

The RL agent seeks the best action-value function, $Q * (S_t, a)$, which is the highest expectation of the cumulative discounted return from stage S_t:

$$Q^*(S_t, a) = E_{S_{t+1}} \left\{ r_t + \gamma \max_a Q^*(S_t, a) \middle| S_t, a \right\} \tag{2.13}$$

Figure 2.3 represents the basic block diagram of the F-DRL scheme. Each BSs uses DRL (block diagram in 2.1) to generate the model weights locally. These intermediate model weights are then sent to the central server at regular intervals. After aggregating the local model weights, the central server uses these aggregated weights to train the central modal. After training, the server will update the central modal weights to the individual local BSs. Then BSs use the local model to generate the optimal resource allocation. The interval after which the central server is updated with individual BSs intermediate model weights is known as the aggregation period (A_g). The overall learning procedure for F-DRL-based proposed RAIC scheme is summarized in algorithm 3.

REFERENCES

1. H. Mehrpouyan, M. Matthaiou, R. Wang, G. K. Karagiannidis, and Y. Hua, "Hybrid millimeter-wave systems: A novel paradigm for hetnets," *IEEE Communications Magazine*, vol. 53, no. 1, pp. 216–221, 2015.
2. P. Tehrani, F. Restuccia, and M. Levorato, "Federated deep reinforcement learning for the distributed control of nextg wireless networks," in *2021 IEEE International Symposium on Dynamic Spectrum Access Networks (DySPAN)*, 2021, pp. 248–253.
3. 3GPP, "Study on channel model for frequencies from 0.5 to 100 GHz," 3rd Generation Partnership Project (3GPP), Technical Report (TR) 138.901, 11 2020, version 16.1.0 Release 16. [Online]. Available: https://www.etsi.org/deliver/etsi_tr/138900_138999/138901/16.01.00_60/tr_138901v160100p.pdf
4. C. Niu, Y. Li, R. Q. Hu, and F. Ye, "Fast and efficient radio resource allocation in dynamic ultra-dense heterogeneous networks," *IEEE Access*, vol. 5, pp. 1911–1924, 2017.
5. M. A. Kamal, H. W. Raza, M. M. Alam, M. M. Su'ud, and A. B. A. B. Sajak, "Resource allocation schemes for 5G network: A systematic review," *Sensors (Basel)*, vol. 21, no. 19, 2021.
6. H. Shokri-Ghadikolaei, C. Fischione, G. Fodor, P. Popovski, and M. Zorzi, "Millimeter wave cellular networks: A mac layer perspective," *IEEE Transactions on Communications*, vol. 63, no. 10, pp. 3437–3458, 2015.
7. P. K. Agyapong, M. Iwamura, D. Staehle, W. Kiess, and A. Benjebbour, "Design considerations for a 5g network architecture," *IEEE Communications Magazine*, vol. 52, no. 11, pp. 65–75, 2014.
8. H. Sun, X. Chen, Q. Shi, M. Hong, X. Fu, and N. D. Sidiropoulos, "Learning to optimize: Training deep neural networks for interference management," *IEEE Transactions on Signal Processing*, vol. 66, no. 20, pp. 5438–5453, 2018.
9. Q. Mao, F. Hu, and Q. Hao, "Deep learning for intelligent wireless networks: A comprehensive survey," *IEEE Communications Surveys Tutorials*, vol. 20, no. 4, pp. 2595–2621, 2018.
10. M. Lee, Y. Xiong, G. Yu, and G. Y. Li, "Deep neural networks for linear sum assignment problems," *IEEE Wireless Communications Letters*, vol. 7, no. 6, pp. 962–965, 2018.
11. B. Matthiesen, A. Zappone, K.-L. Besser, E. A. Jorswieck, and M. Debbah, "A globally optimal energy-efficient power control framework and its efficient implementation in wireless interference networks," *IEEE Transactions on Signal Processing*, vol. 68, pp. 3887–3902, 2020.
12. F. Liang, C. Shen, W. Yu, and F. Wu, "Towards optimal power control via ensembling deep neural networks," *IEEE Transactions on Communications*, vol. 68, no. 3, pp. 1760–1776, 2020.
13. W. Lee, M. Kim, and D.-H. Cho, "Deep power control: Transmit power control scheme based on convolutional neural network," *IEEE Communications Letters*, vol. 22, no. 6, pp. 1276–1279, 2018.
14. Q. Shi, M. Razaviyayn, Z.-Q. Luo, and C. He, "An iteratively weighted mmse approach to distributed sum-utility maximization for a mimo interfering broadcast channel," *IEEE Transactions on Signal Processing*, vol. 59, no. 9, pp. 4331–4340, 2011.

15. Y. Shen, Y. Shi, J. Zhang, and K. B. Letaief, "Lorm: Learning to optimize for resource management in wireless networks with few training samples," *IEEE Transactions on Wireless Communications*, vol. 19, no. 1, pp. 665–679, 2020.
16. M. Lee, G. Yu, and G. Y. Li, "Learning to branch: Accelerating resource allocation in wireless networks," *IEEE Transactions on Vehicular Technology*, vol. 69, no. 1, pp. 958–970, 2020.
17. S. Wang, H. Liu, P. H. Gomes, and B. Krishnamachari, "Deep reinforcement learning for dynamic multichannel access in wireless networks," *IEEE Transactions on Cognitive Communications and Networking*, vol. 4, no. 2, pp. 257–265, 2018.
18. Y. S. Nasir and D. Guo, "Multi-agent deep reinforcement learning for dynamic power allocation in wireless networks," *IEEE Journal on Selected Areas in Communications*, vol. 37, no. 10, pp. 2239–2250, 2019.
19. S. Kumar, P. Shah, D. Z. Hakkani-Tur, and L. Heck, "Federated control with hierarchical multi-agent deep reinforcement learning," ArXiv, vol. abs/1712.08266, 2017.
20. X. Wang, R. Li, C. Wang, X. Li, T. Taleb, and V. C. M. Leung, "Attentionweighted federated deep reinforcement learning for device-to-device assisted heterogeneous collaborative edge caching," *IEEE Journal on Selected Areas in Communications*, vol. 39, no. 1, pp. 154–169, 2021.
21. X. Wang, C. Wang, X. Li, V. C. Leung, and T. Taleb, "Federated deep reinforcement learning for internet of things with decentralized cooperative edge caching," *IEEE Internet of Things Journal*, vol. 7, no. 10, p. 15, 2020. [Online]. Available: http://urn.fi/URN:NBN:fi:aalto-202011066354
22. J. Ren, H. Wang, T. Hou, S. Zheng, and C. Tang, "Federated learning-based computation offloading optimization in edge computing-supported internet of things," *IEEE Access*, vol. 7, pp. 69 194–69 201, 2019.
23. D. C. Nguyen, M. Ding, P. N. Pathirana, A. Seneviratne, J. Li, D. Niyato, and H. V. Poor, "Federated learning for industrial internet of things in future industries," *IEEE Wireless Communications*, vol. 28, no. 6, pp. 192–199, 2021.
24. Q. Guo, F. Tang, and N. Kato, "Federated reinforcement learning-based resource allocation in d2d-enabled 6g," *IEEE Network*, pp. 1–7, 2022.
25. I. Kholod, E. Yanaki, D. Fomichev, E. Shalugin, E. Novikova, E. Filippov, and M. Nordlund, "Open-source federated learning frameworks for iot: A comparative review and analysis," *Sensors*, vol. 21, no. 1, 2021.
26. P. Zheng, Y. Zhu, Z. Zhang, Y. Hu, and A. Schmeink, "Federated learning in heterogeneous networks with unreliable communication," in *2021 IEEE Globecom Workshops (GC Wkshps)*, 2021, pp. 1–6.
27. S. D. Whitehead, "A complexity analysis of cooperative mechanisms in reinforcement learning," in *Proceedings of the Ninth National Conference on Artificial Intelligence - Volume 2, ser. AAAI'91*. AAAI Press, 1991, pp. 607–613.
28. Y. Xu, G. Gui, H. Gacanin, and F. Adachi, "A survey on resource allocation for 5g heterogeneous networks: Current research, future trends, and challenges," *IEEE Communications Surveys Tutorials*, vol. 23, no. 2, pp. 668–695, 2021.
29. B. Wen, Z. Gao, L. Huang, Y. Tang, and H. Cai, "A q-learning-based downlink resource scheduling method for capacity optimization in lte femtocells," in *2014 9th International Conference on Computer Science Education*, 2014, pp. 625–628.
30. B. Hamdaoui, P. Venkatraman, and M. Guizani, "Opportunistic exploitation of bandwidth resources through reinforcement learning," in *GLOBECOM 2009 – 2009 IEEE Global Telecommunications Conference*, 2009, pp. 1–6.
31. W. Sun, W. Fan, M. Zhao, W. Song, X. Cai, T. Liu, and Z. Jia, "Deep reinforcement-learning-guided backup for energy harvesting powered systems," *IEEE Transactions on Computer-Aided Design of Integrated Circuits and Systems*, vol. 41, no. 2, pp. 346–358, 2022.

32. A. Asheralieva and Y. Miyanaga, "An autonomous learning-based algorithm for joint channel and power level selection by d2d pairs in heterogeneous cellular networks," *IEEE Transactions on Communications*, vol. 64, no. 9, pp. 3996–4012, 2016.
33. Z. Li, C. Wang, and C. Jiang, "User association for load balancing in vehicular networks: An online reinforcement learning approach," *IEEE Transactions on Intelligent Transportation Systems*, vol. 18, pp. 2217–2228, 2017.
34. E. Ghadimi, F. Davide Calabrese, G. Peters, and P. Soldati, "A reinforcement learning approach to power control and rate adaptation in cellular networks," in *2017 IEEE International Conference on Communications (ICC)*, 2017, pp. 1–7.

3 A Comprehensive Overview of Internet of Nano-Things (IoNT) in the Next-Generation Heterogeneous Networks

Deployment Aspects, Applications, and Challenges

Hadi Zahmatkesh
OsloMet - Oslo Metropolitan University, Oslo, Norway

3.1 INTRODUCTION

The Internet of Things (IoT) concept has radically changed how people utilize the Internet over the last few years. The devices and objects involved have the ability to perceive and gather data from the surrounding environment. Subsequently, this data can be transmitted over the Internet, where it undergoes processing to serve diverse objectives. For example, in the medical field, Body Area Networks (BANs) can gather and send important information of people receiving medical treatment from a doctor to service providers' computing systems to carefully watch and check a large number of patients more efficiently and precisely. In addition, sensors deployed in the environment around us can provide valuable information in terms of real-time monitoring and rehabilitation regarding elderly and physically disabled people [1, 2].

Together with the development in the Internet and sensing/actuation technologies, the latest advancement in nanotechnology and design of nano-scale constituent such as nano-sensors and nano-antennas provides a new class of services and applications in different industries including healthcare [3, 4] and agriculture [5]. This has brought a new challenging paradigm called Internet of Nano-Things (IoNT), a system of nano-scale connected devices to transfer data to the cloud [6]. Unlike IoT, which is

DOI: 10.1201/9781003303114-3

only deployed in a way that it can be seen by the naked human eye, IoNT deployments are at scales not visible by the naked human eye [7].

In IoNT, operational tasks such as sensing and actuation are performed by nanomachines having dimensions ranging from 1 to 100 nm [8]. Moreover, there are great potentials for the IoNT in terms of application areas, especially in the field of healthcare [9]. For example, in the IoNT paradigm utilizing molecular communication, digital information can be attached in chemical molecules to spread via absorbing and complex media [10]. According to researchers from the center for High-rate Nanomanufacturing in the University of Massachusetts, nano-products are likely to fuel the next economic boom [11]. It would work with governments and industries to find practical manufacturing solutions for commercial products utilizing technology developments and nano-science. For example, researchers are investigating how to design polymer patterns from small nano-particles or nano-elements to utilize them in a broad range of applications. In the healthcare domain, nanotechnology was first used in 2001 when a medical technology company called Given Imaging [12] introduced a capsule called PillCam, which contains a camera and light that a patient can swallow [13]. These methods of treatment are not altogether satisfactory, and scientists are working to come up with new precise and convenient treatment methods using nanotechnology [13].

The aforementioned instances represent only a fraction of the domains in which IoNT and nano-sensing technology have made noteworthy advancements. In addition to the studies already mentioned, it is essential to look into the challenges and opportunities the IoNT paradigm presents to make efficient use of the new spectrum introduced for the IoNT. Therefore, this chapter provides an overview of the IoNT and focuses on strategies to consider while addressing the IoNT challenges.

3.2 COMPARISON TO OTHER STUDIES

Several studies in the literature cover various aspects of the IoNT. For instance, in [1], the authors discuss applications and various challenges such as data collection and middleware challenges to realizing the concept of IoNT. They also propose their solutions for the mentioned challenges. A similar study in [4] discusses the applications of IoNT in healthcare and investigates its requirements to support various application categories. Moreover, the authors discuss IoNT challenges in the healthcare domain. The study in [6] discusses the highest level of general development in electromagnetic communications among nano-devices from the communication and information theoretic perspective. Moreover, this study highlights the main research challenges in the IoNT and nano-networks in terms of protocols, channel modeling, and information encoding. A concise overview of IoNT applications in different fields is provided in [14]. The authors also discuss the general architecture of the IoNT in healthcare applications that can be improved based on specific applications. The survey in [7] reviews the current status and future promises of the IoNT and discusses its challenges for possible enhancement of their applications. Similarly, an overview of the application areas of the IoNT and its architecture is provided in [15]. Moreover, the authors discuss various challenges of the IoNT that need to be addressed so that it can be an essential part of people in the near future.

Although several survey studies have been published regarding the IoNT, none of them presents a comprehensive review that considers aspects such as application areas, network architecture and standards, and deployment issues, as well as their future challenges in the 5G era. Table 3.1 shows a comparison between our study and similar papers in terms of various aspects of the IoNT.

Our main contributions in this work relative to the recent literature in the field are as follows:

- This chapter provides a comprehensive overview of the IoNT paradigm considering the main application areas, architecture, and their limitations.
- We present potential IoNT-enabling technologies that can be used to support several breakthrough innovations in services, products, and processes in various environments and fields.
- We provide deployment aspects of the IoNT that allow researchers to have precise planning approaches in the design of the IoNT paradigm.
- We also discuss the role of cloud computing, fog computing, and big data analytics in support of the IoNT.
- Finally, we present some open research issues and discuss future research challenges.

The rest of this chapter is organized as follows. Section 3.2.1 provides general discussions about 5G and its role in realizing the IoNT paradigm. Section 3.2.2 presents the characteristics of the IoNT and its main components. IoNT-enabling technologies and their main application areas are discussed in Sections 3.2.3 and 3.2.4, respectively. Section 3.2.5 discusses the architecture of the IoNT from the perspective of a computer network using TCP/IP reference model. Section 3.2.6 discusses the role of cloud computing, fog computing, and big data analytics in support of the IoNT. Deployment aspects of the IoNT is presented in Section 3.2.7. Section 3.2.8 provides future research directions and discusses some open research issues. Finally, Section 3.3 concludes this chapter.

3.2.1 5G and IoNT

The architecture of 5G wireless communication technology includes various Radio Access Technologies (RATs), an IP network, an aggregator, nano-core, etc. 5G gives the ability to remotely manage the users and provides better and faster solutions for them. The concept of a flat IP can be used by a 5G network so that various RANs can utilize the same nano-core for the communication. Various RANs are supported by the architecture of the 5G network; these RANs include Universal Mobile Telecommunication System (UMTS), Global System for Mobile Communication (GSM), Long-Term Evolution (LTE), LTE-Advanced (LTE-A), and Wireless Fidelity (Wi-Fi), to just name a few. The hierarchical architecture of 5G identifies devices using normal IP addresses, while the flat IP architecture uses symbolic names for the identification of devices. The flat IP architecture reduces the overall cost and may help minimize the latency, as various RANs are connected to the same single nano-core. It also decreases the number of elements in the network in the data path, which

TABLE 3.1
Comparison of Similar Studies Related to IoNT

Ref.	Main Theme	Architecture	Enabling Technologies	Applications	Standards	Deployment	Challenges
[1]	IoNT applications and challenges	X	-	X	-	-	X
[4]	Applications of IoNT in healthcare	X	-	X	-	-	X
[6]	Electromagnetic communications among nano-devices	X	-	-	-	-	X
[7]	Current status and future promises of the IoNT	-	-	X	X	-	X
[14]	General architecture of the IoNT in healthcare domain	X	-	X	-	-	-
[15]	IoNT main challenges	X	-	X	-	-	X
***	Deployment aspects, applications, and IoNT Challenges	X	X	X	X	X	X

- = Not Considered, X = Considered
***= Our study

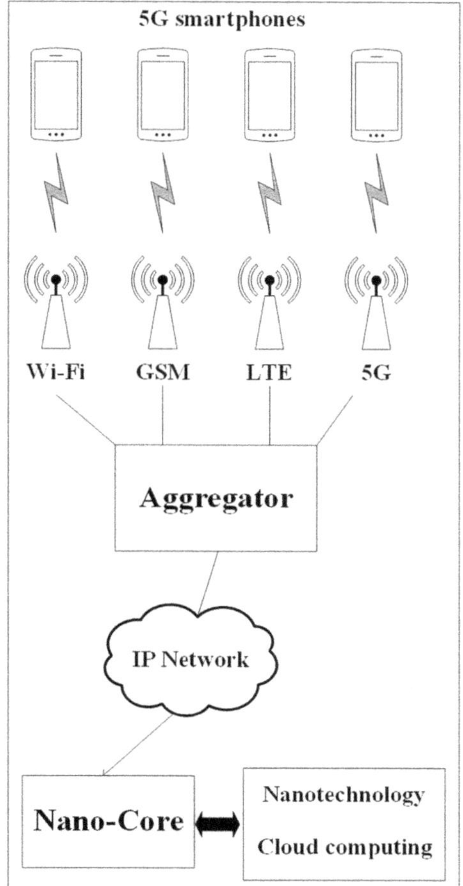

FIGURE 3.1 An architecture of 5G network.

in turn reduces operational costs and capital expenditure [16]. This architecture is shown in Figure 3.1.

In this architecture, the mobile unit communicates with all the RANs that transfer the signals towards the aggregator, which aggregates all the traffic coming from various RANs and forwards it towards the gateway. In addition, nanotechnology in the nano-core section of the architecture works on nano-materials between 0.1 and 100 nm. This is a revolutionary technology that will completely change the mobile communication industry towards the concept of IoNT by influencing sensors and actuators and making the devices more secure and intelligent. In the 5G nano-core, mobiles can be considered nano-equipment that is flexible and transparent and able to sense the environment. Cloud computing is another technology in the 5G nano-core section of the architecture. It uses the Internet and centralized remote servers for matching between the data and the users' applications. The remote sensor in the 5G network allows users to utilize applications without installing them; therefore, they can have access to their data from any computer using the Internet.

3.2.2 IoNT Characteristics and Architecture

3.2.2.1 IoNT Characteristics

The main characteristics of an IoNT paradigm can be summarized in threefold: 1) Instrumented, 2) Interconnected, and 3) Intelligent.

***Instrumented*:** The IoNT has emerged as a fully instrumented system with a network of thousands of devices such as wearable sensors, cameras, actuators, etc. Through these devices the IoNT can have access to real and reliable data in practice.

***Interconnected*:** The IoNT must have a vast set of subsystems that are interconnected and collaborating to provide continuous access for the targeted data that can be based on various resources. The combination of these accurate interconnected and instrumented systems creates a network of regular and nano-scaled items in the physical world to tackle tasks considered to be almost impossible.

***Intelligent*:** The abovementioned IoNT characteristics (instrumented and interconnected) make the best use of information collected from sensors and systems at various sizes if an intelligent approach has been utilized. Any IoNT project will often use AI, and in particular, Machine Learning (ML) techniques, to provide the optimum quality of life for people while maximizing network resources. Apart from employing ML techniques to enhance the performance and functionality of the IoNT network, they can also be applied to comprehend and gain insights from acquired data. For example, in medical scenarios, ML can be harnessed to grasp the progression of a patient's disease, enhance overall health monitoring, identify health-related concerns, and predict potential health disruptions in advance. As an example, the study in [17] utilizes ML to anticipate an impending diabetes crisis using a network of sensors known as the Internet of Bio-Nano Things.

3.2.2.2 IoNT Architecture

IoNT is a distributed network of billions of nano-sensors and nanodevices that should be able to communicate with one another across the Internet. The interconnection of nano-sensors and nano-devices in the IoNT needs improvement in the networking components to carry out the communication process in nano-networks. Regardless of the application of the IoNT (e.g., healthcare, smart office, etc.), the architectures of the IoNT contain the following components [1, 6]:

***Nano-nodes*:** These are the endpoints (e.g., nano-actuators and nano-sensors) that can carry out basic operations such as processing and computation. They can only send out signals over extremely short distances because of their limited memory, energy, and communication capabilities. For instance, a nano-machine with communication capabilities embedded in diverse objects, such as books and keys, is an example of a nano-node. These devices can also be employed as biological nano-sensors inside the human body.

***Nano-routers*:** They have more computing resources than nano-nodes and are suitable for gathering data from limited nano-machines such as nano-sensors. In addition, nano-routers, by transmitting extremely basic control commands (e.g., on/off and sleep commands), can monitor nano-nodes' behaviors.

***Nano-micro interface devices*:** These devices aggregate the information forwarded by nano-routers. They have the ability to send data not only to micro-scale devices but also from micro-scale to nano-scale. These devices can be considered as hybrid devices that can communicate in nano-scale using nano-communication technologies.

***Gateways*:** They are utilized to allow for remote system monitoring over the Internet. Gateways are capable of receiving data from nano-micro interface devices and transmitting it to the relevant service provider. For instance, in a medical context, all sensor data from the human body can be transmitted to the healthcare-related service provider over the Internet.

Please note that the network architecture described in Figure 3.2 can be divided into two communication segments. The first segment is the communication between nano-node and the gateway, which can be a Local Area Network (LAN), and the second one is the communication between the gateway and the cloud. This segment can be a 5G network, which can provide connection for LAN and public networks. Moreover, according to the specific applications of the IoNT, the network segments presented in Figure 3.2 can be either static or dynamic. For example, the topology of the IoNT can be static in industrial applications, while in the healthcare scenarios, it is a dynamic approach since nano-sensors should be mobile for specific health applications and move around the human body. Detailed discussions related to deployment models for the IoNT are presented in Section 3.2.7.

3.2.3 IoNT-Enabling Technologies

IoNT has the potential to support several breakthrough innovations in services, products, and processes in different environments and fields. It can be utilized in different areas such as agriculture, health, computers and communication, and energy production to help address major challenges we currently face. In this section, we discuss some of the most widely used examples of these technologies.

Nano-medicine can benefit from the implementation of IoNT in different applications such as diagnostic and imaging, drug delivery systems, and nano-dentistry. Nano-devices can also be utilized in cancer therapy [18, 19]. Moreover, smart nano-devices can help detect and terminate tumor cells inside the body [20]. Nano-sensors are another important type of infrastructure that facilitates a wide range of applications. Nano-sensors can be utilized in agriculture to monitor the health of crops and

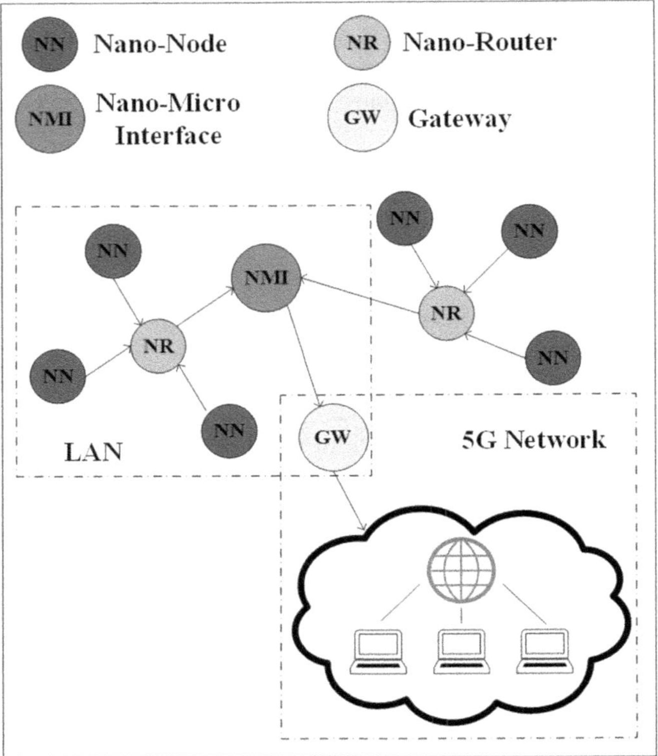

FIGURE 3.2 IoNT network architecture and its main components.

manage water resources by adjusting input to crop and animal needs. They can also be applied in the food industry to control the level of bacteria and other important indicators. Moreover, they can be used in healthcare to monitor patients' health-related issues. Plus, certain cutting-edge nano-sensors have been created by employing synthetic biology techniques to alter organisms, including bacteria [21]. The goal is to develop bio-computers utilizing DNA and proteins to recognize specific chemical targets, store limited information, and communicate their status through color changes or other detectable signals.

In the realm of IoNT, numerous nano-sensors crafted from non-biological substances like carbon nano-tubes can function as wireless nano-antennas, enabling them to both sense and transmit signals [21]. Nano-sensors are tiny and can gather information from millions of points. Smart devices can then integrate the data to generate comprehensive maps revealing even slight modifications in various conditions such as vibration, light, magnetic fields, electrical currents, and other environmental conditions. Today, traffic lights, wearables, and surveillance cameras are connected to the Internet—with IoNT, billions of nano-sensors can collect vast amounts of real-time information and send it to the cloud to provide thorough, up-to-date, and inexpensive information about our cities, factories, and even our bodies.

Commonly used wireless communication technologies such as enhanced Machine Type Communication (eMTC), Narrow-Band IoT (NB-IoT), LoRa, and SigFox can be utilized for the integration of nano-devices with IoT to realize the IoNT paradigm. eMTC [22] was released in 2014 in LTE-Advanced (LTE-A) pro release 12, and additional features such as extended coverage and low cost, as well as narrow-band and simplified operation, were added to release 13 with the intent to decrease cost and power consumption [23]. NB-IoT [24] is another wireless communication technology specifically designed for IoT. Compared to the eMTC, NB-IoT has a reduced bandwidth that causes reduction in device complexity and peak data rate [22]. NB-IoT has been utilized in many different applications and scenarios such as intelligent parking and smart hospitals [25]. NB-IoT faces competition from two other technologies, namely LoRa (Long Range) and SigFox [26].

LoRa serves as a physical layer Low Power Wide Area Network (LPWAN) solution, providing coverage up to 10–15 kilometers in rural areas and 3–5 kilometers in urban settings. Its data rates range from 0.3 to 37.5 kb/s [27]. LoRa's versatility allows it to be applied in various scenarios and applications, such as smart agriculture [28, 29], enabling real-time analytics and connectivity in agricultural processes. Additionally, LoRa plays a crucial role in providing real-time connectivity for other IoT-based applications, including healthcare and intelligent buildings [28]. On the other hand, SigFox was the first LPWAN technology designed specifically for IoT in 2009, operating on a frequency band of 868 MHz [27]. It has an impressive coverage range of 30–50 kilometers in rural areas and 3–10 kilometers in urban areas. With a data transmission rate of 100 bps, SigFox stands as a balanced solution, striking a middle ground between power consumption and coverage.

An additional alternative technology for facilitating communication among IoT devices is Wi-Fi, which can support a range of up to 100 meters [30]. With the escalating demand for wireless data and the substantial increase in communication devices, cellular technologies are undergoing a transformation towards a heterogeneous network architecture, incorporating small cells [31, 32] to meet the demand for higher capacity. However, due to the limited licensed spectrum available for cellular networks, efforts to enhance system capacity through network densification result in inter-cell interference [33]. Consequently, cellular technologies can opt to utilize the unlicensed spectrum bands at 2.4 and 5 GHz, which are currently allocated for Wi-Fi usage. The ZigBee technology is also suitable for communication between IoT devices since it was designed to provide low power consumption and low data rate communication [34]. It is based on the IEEE 802.15.4 standard and supports a range of typically 50 meters for the communication between smart IoT devices.

3.2.4 IoNT Applications

IoNT represents the integration of nano-scale devices into the current communication networks and the Internet, offering diverse applications across various fields in the era of IoT. The subsequent subsections present and describe some typical

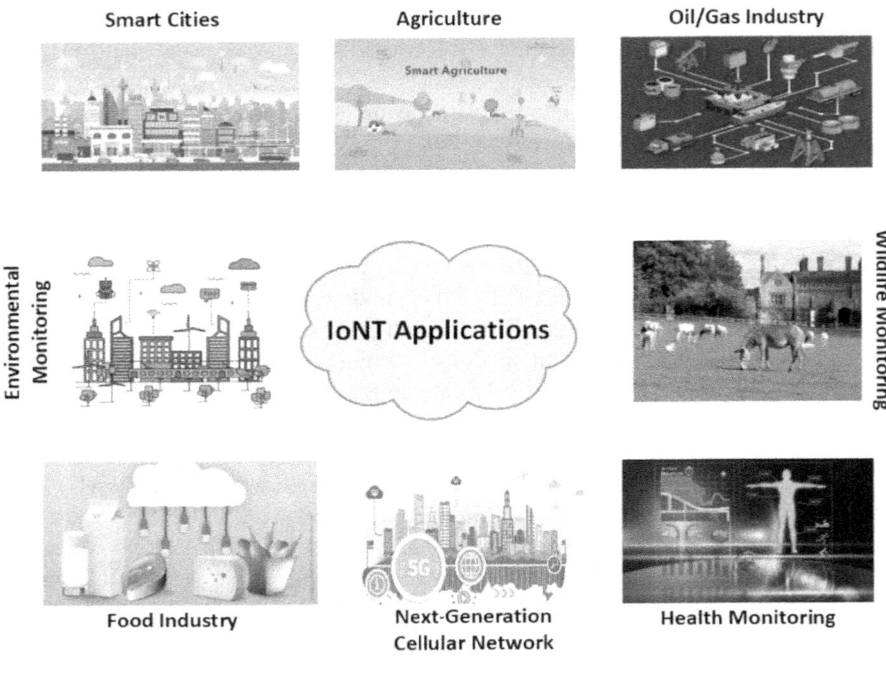

FIGURE 3.3 IoNT applications.

applications of IoNT. Figure 3.3 visually depicts different applications of IoNT in various domains:

***Smart Cities*:** The main goal of smart cities [35] is to make use of the IoT to improve the quality of life of people by monitoring the availability of parking spaces, managing and controlling traffic in the city, controlling air quality, and even activating the alarm when the trash container is full [36]. The implementation of smart cities provides communication and interaction with sensors, actuators, surveillance cameras, vehicles, and home appliances. With IoNT, nano-sensors can be utilized to identify the location of pollution in the air and clean up that specific pollution [37]. They can also be utilized to collect enormous amounts of information in real time to enhance the quality of life and provide better services and applications [38].

***Health Monitoring*:** Nano-sensors have several applications in the field of healthcare monitoring [39]. They can monitor different health-related factors including cholesterol, sodium levels, and glucose. Furthermore, these sensors have the potential to identify cancer-causing tumors [40, 41]. Additionally, by implanting nano-sensors within patients' bodies, it becomes possible to track and monitor various health indicators through a smartwatch that connects to the Internet. Moreover, they can be used to repair demyelinated neurons by recognizing the affected area and locating a myelin sheath there [42]. Although it is difficult to identify the precise location to place the sheath, with IoNT the nano-sensors can pass on the nerve impulse signal to the destination or toward the other end of the nerve [14].

***Agriculture*:** Nanotechnology and nano-scale sciences have important impacts on various aspects of the agriculture industry. For example, they can provide a field sensing system to track crop health and environmental conditions [43]. With IoNT, networks of wireless sensors can be deployed to control field conditions such as temperature, moisture level, plant diseases, etc., to provide necessary information to minimize resource input and maximize output [44]. IoNT can also help to control and improve agricultural productivity. Various types of nano-particles can be embedded into a nano-sensor to manage and control pest/fungi in plants [45]. In other words, IoNT has the capability to revolutionize precision agriculture by leveraging wireless communication and remote sensing technologies, leading to enhanced efficiency and productivity in agricultural practices [14].

***Environmental Monitoring*:** IoNT is becoming increasingly important in the domain of environmental monitoring. Parameters such as temperature and real-time traffic can be more efficiently monitored by deploying nano-sensors in public places such as bus stops, airports, hotels, and railway stations [15]. Moreover, IoNT can develop a pollutant sensing system to detect air, water, and soil pollution and help in the enhancement of remediation technologies. For instance, nano-sensors can be utilized to detect various toxic compounds in different environmental systems [46]. This information may be forwarded to a central location for further assessment, and quick reaction can be provided as a response to this dangerous issue.

***Wildlife Monitoring*:** The use of nano-sensors and IoNT can have significant effects on wildlife monitoring. These devices can aid in the early detection of diseases in animals and reduce economic losses if implemented accurately and used correctly [47]. For example, nano-sensors can be implemented on some insects to detect diseases or abnormal conditions in the environment [48]. Through a gateway in the vicinity, this information is sent to the control room and required actions are provided. Moreover, nano-sensors and wearable technologies can be mounted on animals to detect and monitor various health-related parameters such as sweat constituents [49], body temperature [50], and stress [51], as well as the presence of viruses and pathogens [52, 53].

***Next-Generation Cellular Networks*:** IoNT uses the Terahertz (THz) band for communication among sensors. The THz band communication (0.1 THz – 10 THz) provides high bandwidth and high data rate communications. It can therefore be used in the next generation of cellular systems (e.g., small cells) as a part of Heterogeneous Networks (HetNets) [54]. The THz band communication will provide ultra-high speed data communication within the coverage area of a few tens of meters in indoor and outdoor environments for both static and mobile users [55]. They can, for example, be used to provide ultra-high definition multimedia streaming to smart devices or ultra-high definition video conferencing required by the next generation of cellular networks. Moreover, directional THz band links can deliver an incredibly fast wireless backhaul to the small cells [55].

***Oil and Gas Industry*:** The utilization of nano-sensors can boost oil recovery rates by traversing through small holes on the surface of rocks and aiding in the identification of oil trapped within the rocks. Currently, various tools are available to investigate the field, but the offered resolution is not high enough such that the location of the oil is recognized by using a large magnetic source and receiver to map the nano-particles, which are injected using recycled water [56]. IoNT allows nano-sensors to interact and communicate with each other using molecular communication, and the obtained information can be transmitted using a gateway located in the vicinity. This, in turn, allows the location of oil to be effectively mapped without the need for a particular magnetic source and receiver [14].

***Food Industry*:** Another application of the IoNT is in the food industry. Many food products such as fruits, meats, eggs, fresh vegetables, and dairy products are either perishable or semi-perishable. The enhancement of shelf life is one of most important areas of nanotechnology to improve the ability to preserve the quality and safety of the products [43]. With IoNT, nano-sensors can provide the status of the food freshness in real time, which in turn helps determine the exact expiry date of the product [14]. Employing nano-sensors in food packaging offers advantages in detecting the proliferation of microorganisms. Moreover, nano-sensors can play a role in online process control by regulating storage conditions to prevent incidents of food poisoning [57].

3.2.5 IoNT Network Architecture and Standards

In IoNT, nano-networks connect nano-devices that can sense, collect, process, and store information. Due to the use of nano-scale devices in nano-networks, the process of transferring data and information between various components of a network should be suitable for nano-devices for communication in nano-scale networks. Communication in nano-scale networks can be divided into molecular communication and electromagnetic communication. The architecture of a network is defined as a communication approach to cope with the complexity of the system [58]. In computer networks, a stack of layers is used to present the functionality of a system. Each layer provides a service for the layer above by using services implemented by the layer below with the addition of the required functionality. With the help of the layered approach, system designers can concentrate on the certain part of a system by expressing the details of other layers. In addition, a change in the performance of a service in a layer does not have any effect on other layers' design if the same service is provided by that layer. The architecture for the IoNT may be discussed from the perspective of a computer network using the TCP/IP reference model [59]. In subsequent subsections we discuss a network architecture of the IoNT, starting with the physical layer, which is responsible for transmission of data over physical media. We then proceed to the link layer, which ensures reliable data transmission between nano-devices over communication links—and finally, to the network layer, which facilitates data transmission within a network of interconnected nano-devices via communication links.

3.2.5.1 Physical Layer

The physical layer of the IoNT serves as the foundation for addressing various aspects of transmission, reception, and information propagation. This layer tackles challenges such as hardware selection and interfaces appropriate for nano-devices within the IoNT context. Other issues are related to signal modulation, signal propagation, and signal amplification [58], which are briefly discussed in this section. The physical layer provides the hardware and interfaces of nano-devices. To fulfill the diverse application-specific requirements of the IoNT, multiple implementations will be necessary. These implementations should meet criteria such as performance levels and compatibility with the surrounding environment. For example, in molecular communication, sender and receiver bio-nano-machines are two common hardware devices that have the capability to encode and decode messages onto or from information molecules. The physical layer may provide an interface to communicate between sender and receiver. Moreover, it might act as an interface for optical or electrical systems such as Wireless Body Sensor Networks (WBSNs) [40, 60], which work outside the communication environment.

Additionally, the physical layer offers methods for encoding signals using information molecules. For instance, in the field of molecular communication, a bio-nano-machine distinguishes between different types of information molecules through a range of receptors, where each molecule type corresponds to a unique signal [61, 62]. Furthermore, within molecular communication, the physical layer encompasses diverse mechanisms for the transmission of information molecules throughout the environment. Information molecules may passively or actively propagate in the environment without the use of chemical energy [63, 64] or with the use of transport bio-nano-machines, respectively [65, 66]. The physical layer can provide a mechanism for signal amplification as well. The distance from the sender nano-device results in the reduction of signal strength. For instance, in molecular communication, information molecules that are spread at longer distances may not be detected by the receiver bio-nano-machines [67]. Therefore, positive feedback loops are utilized by biological systems to increase the signal strength over a long distance [58].

3.2.5.2 Link Layer

In the IoNT architecture, the link layer provides mechanisms for a group of nano-devices to communicate in a reliable manner within a specific range. This range can be described as an environment where information can spread from a source to a recipient in a reasonable amount of time. This layer can also provide mechanisms for handling errors that may happen when information propagates from a source to a recipient. In this regard, error handling is an important task since information may arrive with a substantial amount of jitter, which may result in errors in the receiver side. For instance, in biological systems, a redundant amount of information may be transmitted by a sender bio-nano-machine, and the receiver responds when it discovers a threshold amount of such information. A huge amount of information increases the signal-to-noise ratio and decreases the effect of noise or fluctuation [64]. Moreover, the link layer of the IoNT architecture provides mechanisms for monitoring the transmission rate (e.g., flow control). In scenarios of molecular communication, the transmitting bio-nano-machine might need to adjust the transmission rate if the receiver's chemical kinetics are rapid, extremely slow, or subject to variability under different conditions. In the absence of flow control, the receiver could transmit information molecules at a rate higher than the reaction rate at the receiver, leading to eventual loss of molecules [68]. Moreover, this layer can serve as a mechanism to allocate the shared medium among a group of nano-devices for media access control. This helps prevent transmission errors that occur when multiple devices attempt to transmit information simultaneously. For instance, a set of bio-nano-machines can employ media access control by transmitting and synchronizing information at different intervals [68] or by sending information of the same type with distinct characteristics to facilitate demultiplexing at the receiver [69].

3.2.5.3 Network Layer

The network layer of the IoNT provides a way for a set of nano-devices to communicate within a range greater than the one achieved by the link layer. This layer can form a particular group of nano-devices connected through communication links to meet application-specific goals such as delay and energy efficiency in spreading information over a network. The network layer can also provide mechanisms for routing in transmission of information from a sender to a receiver. For example, in molecular communication, without network routing, the range of molecular communication would be limited to a range obtained by a single communication link. Network routing would increase the complexity and scale of the IoNT that can meet the requirements of the applications. For instance, if a broadcast message is transmitted by a source to a destination, if the destination is in the local network, the router does not carry out any action but if the destination is outside the local network, the router chooses the best communication link to the next router on the way to the destination and transmits the information though the communication link [70, 71].

The network layer also provides a system to process information in a network of nano-devices. The network may contain a set of nano-devices placed close to each other and therefore can sense similar information. In cases like this, information can be combined and transmitted from a particular nano-device to an application that in turn enhances the performance and energy efficiency, as well as the accuracy of the information obtained in the communication system [72]. Congestion control is another task that can be provided by the network layer. Without congestion control, a network of nano-devices can face a huge amount of transmitted information, which in turn decreases the speed of transmission and increases communication interference and the rate of collision.

3.2.5.4 Upper Layers

Architecture of the IoNT may also include transport and application layers according to the TCP/IP model. The transport layer can provide an option for reliable and end-to-end transmission of information. Furthermore, it could potentially offer a means for managing errors and ensuring the sequential transmission of information from a

TABLE 3.2
A TCP/IP-based Architecture for the IoNT

TCP/IP Layers	Main Tasks
Application Layer	– Implementation of application-specific functionalities
Transport Layer	– Reliable end-to-end transmission – Error handling – In-sequence transmission of information
Network Layer	– Network formation – Network routing – Network processing – Congestion control
Link Layer	– Error handling – Flow control – Media access control
Physical Layer	– Hardware and interfaces – Signal modulation – Signal propagation – Signal amplification

source to a destination. The application layer of the IoNT can implement application-specific functionalities that will be more obvious in the future as the applications are being developed. A summary of the IoNT network architecture discussed in this section is presented in Table 3.2.

3.2.5.5 Standards

Standards, in computer networks, permit communication of various computers developed by various manufacturers that will be important in facilitating the development of the IoNT. For instance, in synthetic biology, a standard library for bio-nano-machines was developed to engineer synthetic living organisms from a group of DNA sequences [73]. The IEEE P1906.1 standard was established by Standards Working Group for Nano-networking in 2011 to develop standardization of molecular communication, which is one of the main application areas of the IoNT [74, 75]. The main goals of the working group are to provide a conceptual framework and a common terminology for molecular communication, as well as to investigate a practical architecture and a group of protocols for molecular communication by designing, implementing, and evaluating the molecular communication systems.

3.2.6 Big Data Analytics, Cloud Computing, and Fog Computing in Support of the IoNT

The rapid development and emergence of nanotechnology and nano-sensors would finally result in a network of millions of devices interconnected at nano-scale [76]. Such networks will ultimately generate huge amounts of data too complex to be monitored by traditional data processing methods. In this section, we discuss different sources that contribute to the big data concept in the IoNT era. In addition, we discuss various important properties of the IoNT that can be provided by the concept of cloud and fog computing.

3.2.6.1 Big Data Analytics in Support of the IoNT

The term "big data" is defined as a set of huge amounts of data that are too complex to be monitored and handled by traditional data processing methods. Over the past few years, great progress has occurred for the applications of nanotechnology in various fields according to the new nano-scale devices and materials. These applications include medicine, agriculture, electronics, food, energy, transportation, and communications, to name just a few [77]. A tremendous impact has occurred in the field of nano-electronics where quicker nano-transistors have been designed for communication technologies [78]. In medical industry, nano-medicine has been utilized in different healthcare applications such as biological imaging and diagnosis tools. Moreover, nano-particles have played a significant role in stem cell research for the differentiation between regenerative medicine, tissue engineering, and treatment of different diseases [77]. Therefore, nanotechnology-based IoT has created the IoNT paradigm, where communication networks and Internet-connected devices would play a significant role in the IoT era to analyze the big data related to nano things [79]. Different attributes characterize the big data concept and are discussed as follows:

***Volume*:** The term "big" was referred to approximately 1.1 zettabytes of data in 2016 [80]. In addition, the Compound Annual Growth Rate (CAGR) has changed from 6% for IP WAN to 47% for mobile Internet in 2020. For instance, in a medical context with a network of a large number of nano-sensors in and off body, it is expected that the sensors have frequent communication with the sink node or routers, which may eventually generate a vast amount of data. The existence of bio-nano sensors would cause a transition from non-continuous to a continuous source of information, considering that on-body sensors transmit information continuously. The integration of such data in the IoNT era would increase the data size significantly. Moreover, the available health databases consist of a large quantity of data that will need big data processing techniques to monitor and deal with such volume of information.

***Variety*:** It is the proliferation of various data formats such as pictures, audio files, webpages, etc., which make it impossible to reveal it in a structural way. Therefore, it is not possible to handle big data using the traditional Relational Data Base Management Systems (RDBMS). Availability of various communication technologies along with heterogeneous smart nano-sensors will produce a variety of data types. Moreover, variety can happen by integration of heterogeneous data from different sources such as sensors and medical lab reports. This data can be organized in different ways, such as structured (e.g., health databases), semi-structured (e.g., digital logs of sensor communication), and unstructured (e.g., visual aids and text reports) approach.

***Velocity*:** It is referred to the frequency in which data is generated and processed. For example, in medical applications, smart devices generate a continuous set of information that requires specific technologies to collect, merge, and process information in real time while maintaining the integrity of the data. In the IoNT, nano-sensors are limited in terms of energy and data storage capacity. Thus, to transfer information, data can be transmitted in several small packets with high speed to maintain the data integrity. Therefore, frequent communication is a requirement for a bio-nano-sensor implementation design [81]. The data generation speed as a result of frequent communication and the requirement for data analysis in real time regarding accurate and timely decision-making are two important factors that contribute toward the velocity attribute. These factors make the problem a "big data" analytic problem.

***Value*:** It is the detailed approach of obtaining extra benefits from pure data. The volume of big data leads to an increase in the cost of poor data. Therefore, there is a need for effective and thorough analysis of information and the value that can be obtained by the data. The value of the data obtained from nano-devices is dependent on many factors, such as the age of data and the ultimate goal. Fresh data are more valuable than data produced a long time ago. Similarly, if the data obtained from nano-sensors are enough for the decision-making process, the value of the data is much more than the cases where it only plays a partial role. However, the aggregation of data from nano-sensors with the data obtained from other sources can improve the value of data. Usually, the value of single data from a specific period of time reduces over time, while the value of the aggregated data increases, especially in the healthcare domain [82].

3.2.6.2 Cloud Computing in Support of the IoNT

Data integration is a challenging and important task when it comes to nano-sensors-based data analytic solutions. This is performed by gathering information from conventional and non-conventional sources and then storing this information for the future to extract additional information. In the healthcare context, for instance, it is not possible for all the objects to form such facilities in-house to handle the huge amount of data. Moreover, it is also challenging to extend the storage and computational capacity according to demands. In this regard, cloud technology can enable the development of a patient-centered model where the data can be sent from the patient's body to the cloud. This, in turn, allows doctors to have access and study the patient's information. Cloud computing can reduce healthcare-related costs by providing early diagnosis; in addition, it allows medical practitioners to share treatments with their colleagues. Such clinical projects utilizing cloud computing facilities include Health Cloud Exchange [83], Husky Healthcare Social Cloud [84], e-Health Cloud [85, 86], and informatics for integrating biology and bedside (i2b2) [87], which is adopted by more than 100 medical centers around the world.

There are popular companies that provide cloud services commercially, which include Google Cloud, Microsoft Azure, Amazon Web Services, IBM Cloud, Cloudera, MapR, and HortonWorks. They also support batch processing with Hadoop and provide facilities for stream processing (e.g., Spark), which are suitable for conventional data processing. The authors in [88] also provide a cloud-based framework for big data analytics in healthcare applications by implementing Hadoop cluster.

3.2.6.3 Fog Computing in Support of the IoNT

Fog computing [89] is an emerging concept that can help in data storage and processing for immediate decision-making. By introducing a fog layer between the cloud and data source, data can be processed swiftly, enabling rapid decision-making. Consequently, necessary actions can be executed with minimal latency. Fog computing can help reduce latency, preserve network bandwidth, and move the most sensitive data to the best place for analyzing [90]. In addition, it supports real-time interactions and mobility where applications require direct communication with mobile devices. In the IoNT era, data aggregation and processing of the most time-sensitive data are provided by fog computing, as it processes data at the edge of the network close to where the data is generated instead of transferring the huge amount of data to the cloud.

Another aspect of fog computing's role can pertain to the addressing scheme of nano-devices, which may involve an "object name service" server responsible for storing identification tags linked to these nano-devices. This server conducts the lookup process based on an electronic product code, providing the nano-device ID, along with other pertinent information about the device's location [90]. Within intra-body networks, nano-sensors and nano-actuators can be introduced into the human body and monitored remotely through the Internet by healthcare providers. For example, when nano-devices like nano-capsules containing essential drugs are injected into a patient's body, they receive a signal to move to the location where the drugs need to be delivered to specific cells. Upon reaching the designated location, these nano-capsules transmit a message containing their ID and location, confirming their successful arrival. The fog computing system, utilizing the "object name service" server, carries out the process of checking the location for these nano-devices.

3.2.7 Deployment Aspects of the IoNT

IoNT has come across several challenges such as node deployment, Quality of Service (QoS), power consumption reduction, and so on. Nano-device deployment is an important issue due to its impact on the cost and network capability of the IoNT. In the literature, deployment strategies of the IoNT can be classified into the following categories: *static* vs. *dynamic* and *random* vs. *deterministic* deployments. We discuss these categories in detail in the following subsections.

3.2.7.1 Random vs. Deterministic

In random deployments, nano-devices can be located randomly to reduce the deployment cost of the IoNT. Due to the miniature characteristic of nano-devices such as nano-sensors, they are usually deployed randomly [91]. For instance, in [92], a framework for data delivery in nano-scale is proposed where a number of nano-sensors are randomly deployed in small areas to help in disaster management. In this framework, data is transmitted from different subsystems using a nano-router towards a gateway connected to the Internet. The system also considers various IoNT-related limitations such as the hop count and the level of remaining energy. In [93], the authors propose a peer-to-peer routing protocol for nano-networks. In this work, they assumed 2D uniform random topology where identical nano-sensors are randomly deployed. They also considered packet collisions and redundant retransmission to optimize the proposed protocol. The authors in [94] proposed an energy-efficient wireless sensor network MAC protocol consisting of several nano-nodes, nano-routers, and one nano-micro interface placed at the center of a rectangular field. In this study, nano-nodes are deployed randomly and move around the field with constant velocity.

To assess the performance of the proposed protocol, the authors compared it with a similar nano-MAC protocol called Smart-MAC [95] and found out that the proposed protocol enhances the performance of the network in terms of packet loss rate, energy consumption, and scalability. A wireless sensor network at nano-scale is proposed in [96], which is useful for detection of intra-body disease as an important application of the IoNT. The model assumes hexagonal cell-based nano-sensors, which are randomly deployed at various body organs to communicate with their exact neighbor cell based on the combinations of transmission methods. The authors in [97] proposed a multi gateway probabilistic polling technique for electromagnetic-based wireless nano-sensor networks (WNSNs). In the network model, nano-sinks and gateway are deployed in a random manner following uniform distribution. To obtain minimal latency and full network connectivity, they are also equipped with THz transceivers with maximal transmission power. The authors revealed that the proposed scheme performs better than the multi gateway autonomous polling scheme in terms of bandwidth efficiency. Similarly, the maximum achievable throughput and the network capacity in the terahertz band is thoroughly investigated in [98] by considering random deployment of nano-devices.

In deterministic deployments of IoNT, nano-devices can be placed on predefined locations to enhance the performance of the entire network in terms of network lifetime, connectivity, and energy consumption. For instance, beside considering random deployment of nano-sensors, the authors in [93] analyzed their proposal for a peer-to-peer routing protocol for nano-networks using a deterministic deployment strategy called grid layout. The grid layout is investigated because of its direct application on software-defined metamaterials [99]. In the study presented in [94], the authors considered a scenario where multiple nano-routers are deterministically deployed throughout the area to investigate energy-efficient wireless nano-sensor network MAC protocol for communication in the terahertz band. In this study, the nano-routers are deployed in such a way that at least one nano-router exists within the radio range of any of them. The authors revealed that the proposed MAC protocol outperforms the Smart-MAC protocol presented in [95] in terms of energy efficiency and packet loss rate.

3.2.7.2 Static vs. Dynamic

In the IoNT, nano-nodes and nano-router, as well as nano-micro interface, can be either static or dynamic according to the specific applications. For instance, in the industrial applications, it can be assumed that the topology of the IoNT is static, while in the healthcare scenarios it is a dynamic approach since nano-sensors should be mobile for certain health applications and move around the human body. In [6], the authors discussed some of the most important use cases of the IoNT using static deployment of nano-devices in dairy farming including grass monitoring, animal health and feed management, and monitoring field condition. In [100], a research investigation introduced a data dissemination scheme based on flooding, utilizing static node deployment. This scheme aimed to enable scalable communication at the nano-scale, providing efficient networking and application services concurrently. Similarly, [101] explored the static deployment of nano-devices on the Internet of Nano-Things (IoNT). The authors proposed a channel-aware forwarding technique specifically designed for terahertz band wireless nano-sensor networks operating on electromagnetic principles. They revealed that the proposed scheme performs better than traditional forwarding schemes in terms of channel capacity. The work in [102] proposed a routing algorithm for nano-networks considering static node deployment based on physical layer network coding. The authors showed the advantages of the proposed routing algorithm in terms of energy efficiency and throughput compared to the point-to-point routing in nano-networks. Moreover, the static deployment of nano-devices in the network model of an electromagnetic-based WNSN is considered in [97]. The authors presented a multi gateway probabilistic polling scheme and showed the advantages of the proposed scheme over the autonomous polling technique in terms of bandwidth efficiency.

In [96], a dynamic WNSN for intra-body applications is proposed where nano-sensors are deployed within each cell inside the human body and communicate with their neighbor cell based on the combination of transmission methods. In this study, a hexagonal-shaped cell is used to model each cell in the body, which in turn provides a 3D structure for WNSN. The authors present a comparative analysis related to the energy efficiency of the proposed method with a different combination of transmission methods. Similarly, various applications of the IoNT in a medical context and the requirements necessary to support various application scenarios are thoroughly discussed in [4]. The authors also outline the most important challenges for realizing healthcare application in the IoNT. The work in [3] presents a function-centric nano-networking concept to address various nano-machines in the IoNT by considering location information and functional capabilities of the machines. The authors finally show the advantages of the proposed method with similar studies available in the literature. An energy-efficient data delivery framework in nano-scale is proposed in [103] where several nano-sensors are dynamically deployed inside the human body to address the challenges of data delivery in IoNT in terms of energy consumption and hop count. The authors finally revealed that the proposed algorithm improves the network lifetime. A summary of different deployment strategies of the IoNT discussed in this section is presented in Table 3.3.

3.2.8 Open Research Issues and Future Challenges

Despite different research activities and fast developments that have been achieved recently with regard to the IoNT paradigm, there are still several research issues and challenges waiting to be considered regarding the performance improvements of the IoNT.

TABLE 3.3
A Summary of Different Deployment Strategies of the IoNT

Reference	Main Theme	Random	Deterministic	Static	Dynamic
[3]	Discussion on the combination of in-body nano-communication with the IoT as the IoNT	X	-	-	X
[4]	Applications of IoNT in healthcare	X	-	-	X
[6]	Electromagnetic communications among nano-devices	-	-	X	-
[93]	Proposes a peer-to-peer routing protocol for nano-networks	X	X	X	-
[91]	Focuses on designing MAC protocols for WNSNs that prioritize energy efficiency and are aware of spectrum utilization	X	-	-	-
[92]	Proposes a data delivery framework in nano-scale with randomly deployed nano-sensors	X	-	-	-
[94]	Proposes an energy-efficient wireless sensor network MAC protocol consisting of several nano-nodes, nano-routers, and one nano-micro interface	X	X	-	X
[95]	Introduces a novel NS-3 module called Nano-Sim, which serves as a simulation model for Wireless Nano-Sensor Networks (WNSNs) utilizing electromagnetic communications within the Terahertz frequency range	X	X	-	-
[96]	Proposes a WSN at nano-scale, useful for detection of intra-body diseases	X	-	-	X
[97]	Proposes a multi gateway probabilistic polling technique for electromagnetic-based wireless nano-sensor networks	X	-	X	-
[98]	Investigates maximum achievable throughput and the network capacity of nano-networks in the terahertz band	X	-	-	-
[99]	Presents an innovative category of systems that allows for programmatic control over the electromagnetic behavior	-	X	-	-
[100]	Puts forward a data dissemination scheme based on flooding, employing static node deployment to achieve scalable communication at the nano-scale	-	-	X	-
[101]	Proposes a channel-aware forwarding technique for electromagnetic-based wireless nano-sensor networks in the terahertz band for static deployment of nano-devices in the IoNT	X	-	X	-
[102]	Proposes a routing algorithm for nano-networks considering static node deployment based on physical layer network coding	-	-	X	-
[103]	Proposes an energy-efficient framework for data delivery in nano-scale networks where several nano-sensors are dynamically deployed inside the human body	-	-	-	X
[104]	Introduces the concept of Internet of Bio-Nano Things, which merges synthetic biology and nano-technology to form an integrated concept	-	-	-	X

- = Not Considered, X = Considered

3.2.8.1 Architecture and Protocols

Because of the limitations of nano-devices in processing and power management, they cannot deal with complex communication protocols. The performance of nano-networks can be improved by providing additional technologies such as fog computing to coordinate and improve the behavior of nano-devices. For example, Fog Nano Datacenters have advantageous characteristics that improve the ability of researchers and developers to properly design service delivery for IoT real-time applications [105]. Therefore, research into architecture and communication protocols of the IoNT would be helpful to manage nano-devices and perform the complex computations. For instance, existing approaches to nano-scale communications, such as employing terahertz-band electromagnetic waves or molecules as carriers of information, do not offer a feasible solution for implementing networks of molecular-scale devices due to their overly complex communication mechanisms [106]. Moreover, it is required to investigate new service-oriented architectures to make nano-devices (e.g., nano-sensors and nano-actuators) compatible to cope with the very large amount of data in the IoNT.

3.2.8.2 Reliability

Reliability in the IoNT is a crucial issue that needs to be guaranteed for both the data going to the nano-nodes from a remote command center and for the data coming from the nano-machines to a sink. Various aspects can affect the reliability of the nano-network, such as failure of the nano-machine and temporary molecular interference in the channel. Actually, if molecules suddenly burst, it can cause unexpected errors in nano-nodes and create disconnections of the network at various points. If this affects only some of the nano-machines in the network, an alternative path can be determined by the routing protocol, while if this affects the whole network, little actions can be performed based on the specific applications. For example, the number of nano-machines can be increased to cover the same area. However, this can challenge routing of the information as well as the channel access in the network. Regarding molecular interference, more comprehensive approaches are required to deal with this issue. For instance, peaks of attenuation can be achieved by absorbing molecules, but still some transmission channels containing path-loss may be usable [6]. Therefore, it is important to investigate protocols in which nano-machines can sense available channels with the help of chemical nano-sensors.

Considering the hardware perspective, reliability can be improved by including more links and components to overcome the faulty components probability in the deployed nano-network, which increases the complexity and installation costs. A few studies have tried to enhance the reliability of the network by considering error reduction, fault tolerance, and/or failure probabilities [107, 108]; however, other studies have examined the Packet Delivery Ratio (PDR) as an indicator of the reliable in-network communication [109–111]. In addition, the reliability of the link between two nodes in IoNT is evaluated and improved by considering reliability routing decision algorithms. According to the literature, it is obvious that reliability improvement is extremely important and should be further investigated in IoNT networks since data congestion is growing mainly around the gateways.

3.2.8.3 Throughput

Unlike PDR, throughput deals with the total number of data packets successfully transmitted to the destination over time, which can be interpreted as the number of bits processed over time. Throughput is considered a key factor in IoNT studies and has been addressed using several routing algorithms. Several attempts have been proposed to enhance the throughput by data aggregation points [76, 112–114]. However, more research should be conducted in this area for more throughput enhancement in the IoNT where energy can be depleted quickly without a wise resource utilization.

3.2.8.4 Lifetime

Nano-sensor nodes in the IoNT are subject to death and stop functioning because of their limited energy capacities, which can severely degrade the overall nano-network lifetime. In this regard, several attempts have been proposed to prolong the network lifetime via fair energy consumption over the networks [92, 115, 116], whereas other attempts show the advantage of using Maximum Likelihood Routing Algorithm (MLRA) [117–119] over random methods in extending the network lifetime. However, this is still an open research issue in the literature and necessitates further investigation.

3.2.8.5 Data Collection and Routing Technology

Collecting a large amount of data generated by nano-devices is an important issue in realizing the IoNT paradigm. Traditional architectures of the wireless sensor networks use a number of sinks to gather data from the sensors, but this may not be possible for nano-networks due to their limitations and constraints. In this regard, routing is an important factor in transmission of data through communication networks. Sensor networks usually use routing algorithms that focus on enhancing the scalability and energy efficiency of the entire networks. However, some significant differences between traditional sensors and nano-sensors have a noticeable effect on the IoNT algorithm design [1]. For instance, nano-sensors use less energy than traditional micro-sensors, which need energy harvesting to supply power to the devices. Moreover, nano-sensors have limited computational processing capabilities and memory storage, and therefore, little knowledge of the architecture of the communication environment. Thus, it is important to investigate new routing algorithms of the IoNT to efficiently deal with the substantial volume of data generated by the network's nano-devices.

3.2.8.6 Security and Privacy

One of the most important challenges in the IoNT paradigm is security. As mentioned in the previous sections, applications that are based on IoNT are used with mobile phones, vehicles, sensors, household appliances, and various other large-scale infrastructure systems. In other words, if not today, in the future they will be a part of our everyday life information exchange processes quite often. Since these devices have specific control and monitoring procedures and will heavily rely on an existing connection to the Internet, they have the potential to raise various types of security- and privacy-related issues [120]. Furthermore, the popularity of terahertz networks as enabling technology for wearables [121] and e-health applications [4] [122] introduces a higher level of sensitivity for the data to be exchanged. Particularly for e-health applications, potential intruders can in fact cause security breaches such as exploiting private biological data gathered by either in-body or wearable sensors, disruption of medical applications that may indeed cause serious damage, for example, in case there is a dedicated drug delivery application controlled by a wearable device, or modification of communication links at the nano-communication level of the Body Area Networks (BAN) [123].

Conducting research to ensure data integrity and user privacy within the IoNT, as well as developing novel authentication schemes for nano-things, holds the utmost importance to prevent the theft of individuals' personal information. Additionally, crucial research is needed to explore new neighbor discovery mechanisms that leverage the high directivity of terahertz antennas, enabling the determination of relative location and orientation among various nano-objects. For these purposes, the mechanisms that require strict key management mechanisms still suffer from issues related to key revoking for IoNT applications similar to many other sensor base networks. Furthermore, again similar to other architectures that involve restricted energy resources, while new security-related protocols are recommended, the energy efficiency- and performance-related issues should also be carefully considered. The absolute positioning of nano-scale networks is more difficult compared to classical sensor networks, which may introduce some challenges for applications in need of localizations. However, it is possible to introduce mechanisms to allow only nearby nano-machines to communicate and eliminate the remote attackers from interfering [124]. Finally, for nano-level potential Denial of Service attacks (DoS), it is also necessary to develop state-of-the-art intrusion detection systems in nano levels.

3.2.8.7 Service Discovery

In the IoNT paradigm, it is expected that every nano-node connects to the network in a seamless manner and its presence should be known by other devices in the network simultaneously. However, for several applications, it would be enough to inform only the nano-router or the nano-micro interface in the vicinity about the presence of the new nano-node in the network. Therefore, network association and new service discovery solutions are required to deal with this issue by considering the fact that a large number of nano-things may be included in such a network. Moreover, in medical applications of the IoNT, the challenge would be related to the large amount of information shared in the network. Therefore, new service discovery solutions would be needed to investigate biological environments and to communicate with biological objects to obtain information.

3.2.8.8 Context Awareness

Context awareness poses a significant challenge in the era of IoNT. In medical applications, networks deployed outside the human body can interact with the external environment to define and update their context. On the other hand, nano-networks positioned inside the body necessitate a non-invasive system to discern their context, especially in scenarios where node mobility plays a crucial role. In addition, some applications organize and coordinate several contexts for particular services that can considerably take advantage of the integration of on-body, intra-body, and external networks. For instance, a patient's change of context can be detected by an on-body network, and the external network can be alerted about the presence of an allergen in the patient's body. Then, the appropriate drug is released in the body by notifying the intra-body network to control the allergic reaction.

3.3 CONCLUSION

Presently, the IoNT paradigm is experiencing rapid growth with the overarching goal of enriching people's quality of life. These advancements in IoNT and nano-technology will profoundly impact diverse sectors, including next-generation heterogeneous networks, healthcare, environmental monitoring, agriculture, and more. In this chapter, we conducted a comprehensive examination of various studies from the literature to provide an overview of IoNT, encompassing its architecture and principal application areas. We also highlighted potential IoNT-enabling technologies that can foster groundbreaking innovations in services, products, and processes across different environments and fields. Furthermore, we delved into deployment aspects of IoNT, offering valuable insights for researchers to plan and design the IoNT paradigm with precision. Additionally, we explored the roles of cloud computing, fog computing, and big data analytics in supporting IoNT. Lastly, we addressed open research issues and discussed forthcoming challenges that demand careful consideration to enhance the performance of IoNT.

REFERENCES

[1] S. Balasubramaniam and J. Kangasharju, "Realizing the internet of nano things: challenges, solutions, and applications," *Computer*, vol. 46, no. 2, pp. 62–68, 2013.

[2] H. K. Bharadwaj, A. Agarwal, V. Chamola, N. R. Lakkaniga, V. Hassija, M. Guizani, and B. Sikdar, "A Review on the Role of Machine Learning in Enabling IoT Based Healthcare Applications," *IEEE Access*, vol. 9, pp. 38859–38890, 2021.

[3] M. Stelzner, F. Dressler, and S. Fischer, "Function Centric Nano Networking: Addressing nano machines in a medical application scenario," *Nano Communication Networks*, vol. 14, pp. 29–39, 2017.

[4] N. A. Ali and M. Abu-Elkheir, "Internet of nano-things healthcare applications: Requirements, opportunities, and challenges," in *IEEE 11th International Conference on Wireless and Mobile Computing, Networking and Communications (WiMob)*, 2015, pp. 9–14.

[5] K. Bhargava, S. Ivanov, and W. Donnelly, "Internet of nano things for dairy farming," in *Proceedings of the Second Annual International Conference on Nanoscale Computing and Communication*. ACM, 2015, p. 24.

[6] I. F. Akyildiz and J. M. Jornet, "The internet of nano-things," *IEEE Wireless Communications*, vol. 17, no. 6, pp. 58–63, 2010.

[7] M. Miraz, M. Ali, P. Excell, and R. Picking, "Internet of nano-things, things and everything: future growth trends," *Future Internet*, vol. 10, no. 8, p. 68, 2018.

[8] J. Chaudhry, U. Qidwai, M. H. Miraz, A. Ibrahim, and C. Valli, *Data security among ISO/IEEE 11073 compliant personal healthcare devices through statistical fingerprinting*. Institute of Electrical and Electronics Engineers, 2017.

[9] N. Kumar, N. Verma, I. Malhotra et al., "A technical review on application oriented comparative study of IoT, IoNT, and IoBNT," in *6th IEEE International Conference on Communication and Electronics Systems (ICCES)*, 2021, pp. 631–636.
[10] S. Liu, Z. Wei, B. Li, W. Guo, and C. Zhao, "Metric combinations in non-coherent signal detection for molecular communication," *Nano Communication Networks*, vol. 20, pp. 1–10, 2019.
[11] "Center for High-rate Nanomanufacturing, University of Massachusetts," https://www.uml.edu/
[12] "Given Imaging Ltd," https://www.givenimaging.com
[13] J. Kainic, "Microbots: Using Nanotechnology in Medicine," http://www.yalescientific.org, 2013.
[14] P. Kethineni, "Applications of internet of nano things: A survey," in *2nd IEEE International Conference for Convergence in Technology (I2CT)*, 2017, pp. 371–375.
[15] A. Nayyar, V. Puri, and D.-N. Le, "Internet of nano things (IoNT): Next evolutionary step in nanotechnology," *Nanoscience and Nanotechnology*, vol. 7, no. 1, pp. 4–8, 2017.
[16] A. Bagwari, G. S. Tomar, and J. Bagwari, *Advanced Wireless Sensing Techniques for 5G Networks*. CRC Press, 2018.
[17] H. Nieto-Chaupis, "Using Machine Learning for Anticipating a Diabetes Crisis Through a Sensors-based Internet of Bio-nano Things Network," in *IEEE/ITU International Conference on Artificial Intelligence for Good (AI4G)*, 2020, pp. 147–151.
[18] S. Findlay, "Drug delivery – exploring the nano options," http://www.frost.com/, 2008.
[19] A. Sinibaldi, "Cancer biomarker detection with photonic crystals-based biosensors: an Overview," *Journal of Lightwave Technology*, vol. **39**, pp. 3871–3881, 2021.
[20] C. Barton, *Nanotechnology, revolutionizing R&D to develop smarter diagnostics and therapeutics*, London: Business Insights Ltd, 2007.
[21] "Global Agenda - Top 10 Emerging Technologies of 2016," https://www.weforum.org/, 2016.
[22] S. K. Goudos, P. I. Dallas, S. Chatziefthymiou, and S. Kyriazakos, "A survey of IoT key enabling and future technologies: 5G, mobile IoT, sematic web and applications," *Wireless Personal Communications*, vol. 97, no. 2, pp. 1645–1675, 2017.
[23] A. Rico-Alvarino, M. Vajapeyam, H. Xu, X. Wang, Y. Blankenship, J. Bergman, T. Tirronen, and E. Yavuz, "An overview of 3GPP enhancements on machine to machine communications," *IEEE Communications Magazine*, vol. 54, no. 6, pp. 14–21, 2016.
[24] Y. Liu, Y. Deng, N. Jiang, M. Elkashlan, and A. Nallanathan, "Analysis of Random Access in NB-IoT Networks with Three Coverage Enhancement Groups: A Stochastic Geometry Approach," *IEEE Transactions on Wireless Communications*, 2020.
[25] H. Zhang, J. Li, B. Wen, Y. Xun, and J. Liu, "Connecting Intelligent Things in Smart Hospitals using NB-IoT," *IEEE Internet of Things Journal*, vol. 5, no. 3, pp. 1550–1560, 2018.
[26] M. Lauridsen, H. Nguyen, B. Vejlgaard, I. Z. Kovács, P. Mogensen, and M. Sorensen, "Coverage Comparison of GPRS, NB-IoT, LoRa, and SigFox in a 7800 km2 Area," in *85th IEEE Vehicular Technology Conference (VTC Spring)*, 2017, pp. 1–5.
[27] M. Centenaro, L. Vangelista, A. Zanella, and M. Zorzi, "Long-range communications in unlicensed bands: The rising stars in the IoT and smart city scenarios," *IEEE Wireless Communications*, vol. 23, no. 5, pp. 60–67, 2016.
[28] R. S. Sinha, Y. Wei, and S.-H. Hwang, "A survey on LPWA technology: LoRa and NB-IoT," *Ict Express*, vol. 3, no. 1, pp. 14–21, 2017.
[29] P. Boonyopakom and T. Thongna, "Environment Monitoring System through LoRaWAN for Smart Agriculture," in *5th IEEE International Conference on Information Technology (InCIT)*, 2020, pp. 12–16.

[30] E. Ferro and F. Potorti, "Bluetooth and Wi-Fi wireless protocols: A survey and a comparison," *IEEE Wireless Communications*, vol. 12, no. 1, pp. 12–26, 2005.
[31] F. Al-Turjman, E. Ever, and H. Zahmatkesh, "Small cells in the forthcoming 5G/IoT: Traffic modelling and deployment overview," *IEEE Communications Surveys & Tutorials*, vol. 21, no. 1, pp. 28–65.
[32] X. Yang, X. Wang, Y. Wu, L. P. Qian, W. Lu, and H. Zhou, "Small-cell assisted secure traffic offloading for narrowband Internet of Thing (NB-IoT) systems," *IEEE Internet of Things Journal*, vol. 5, no. 3, pp. 1516–1526, 2018.
[33] H. Zhang, X. Chu, W. Guo, and S. Wang, "Coexistence of Wi-Fi and heterogeneous small cell networks sharing unlicensed spectrum," *IEEE Communications Magazine*, vol. 53, no. 3, pp. 158–164, 2015.
[34] P. Kinney et al., "Zigbee technology: Wireless control that simply works," *Communications Design Conference*, vol. 2, 2003, pp. 1–7.
[35] E. Ever, F. M. Al-Turjman, H. Zahmatkesh, and M. Riza, "Modelling green HetNets in dynamic ultra large-scale applications: A case-study for femtocells in smart-cities," *Computer Networks*, vol. 128, pp. 78–93, 2017.
[36] A. Whitmore, A. Agarwal, and L. Da Xu, "The Internet of Things—A survey of topics and trends," *Information Systems Frontiers*, vol. 17, no. 2, pp. 261–274, 2015.
[37] M. N. K. Boulos and N. M. Al-Shorbaji, "On the Internet of Things, smart cities and the WHO Healthy Cities," 2014.
[38] H. F. Atlam, R. J. Walters, and G. B. Wills, "Internet of Nano Things: Security Issues and Applications," in *Proceedings of the 2nd International Conference on Cloud and Big Data Computing*. ACM, 2018, pp. 71–77.
[39] P. K. D. Pramanik, A. Solanki, A. Debnath, A. Nayyar, S. El-Sappagh, and K.-S. Kwak, "Advancing modern healthcare with nanotechnology, nanobiosensors, and internet of nano things: Taxonomies, applications, architecture, and challenges," *IEEE Access*, vol. 8, pp. 65 230–65 266, 2020.
[40] I. F. Akyildiz and J. M. Jornet, "Electromagnetic wireless nanosensor networks," *Nano Communication Networks*, vol. 1, no. 1, pp. 3–19, 2010.
[41] N. Rikhtegar and M. Keshtgary, "A brief survey on molecular and electromagnetic communications in nano-networks," *International Journal of Computer Applications*, vol. 79, no. 3, 2013.
[42] S. M. R. Al-Arif, N. Quader, A. M. Shaon, and K. K. Islam, "Sensor based autonomous medical nanorobots: A cure to demyelination," *Journal of Selected Areas in Nanotechnology*, 2011.
[43] H. Chen and R. Yada, "Nanotechnologies in agriculture: new tools for sustainable development," *Trends in Food Science & Technology*, vol. 22, no. 11, pp. 585–594, 2011.
[44] N. Scott and H. Chen, "Nanoscale science and engineering for agriculture and food systems," *Industrial Biotechnology*, vol. 9, no. 1, pp. 17–18, 2013.
[45] H. T. Gul, S. Saeed, F. Z. A. Khan, and S. A. Manzoor, "Potential of Nanotechnology in Agriculture and Crop Protection: A," *Applied Sciences and Business Economics*, vol. 1, no. 2, pp. 23–28, 2014.
[46] E. F. Mohamed, "Nanotechnology: Future of Environmental Air Pollution Control," *Environmental Management and Sustainable Development*, vol. 6, no. 2, pp. 429–454, 2017.
[47] S. Neethirajan, "Recent advances in wearable sensors for animal health management," *Sensing and Bio-Sensing Research*, vol. 12, pp. 15–29, 2017.
[48] V. Upadhayay and S. Agarwal, "Application of wireless nano sensor networks for wild lives," *International Journal of Distributed and Parallel systems*, vol. 3, no. 4, p. 173, 2012.

[49] T. Glennon, C. O'Quigley, M. McCaul, G. Matzeu, S. Beirne, G. G. Wallace, F. Stroiescu, N. O'Mahoney, P. White, and D. Diamond, "'SWEATCH': A wearable platform for harvesting and analysing sweat sodium content," *Electroanalysis*, vol. 28, no. 6, pp. 1283–1289, 2016.

[50] N. Sellier, E. Guettier, and C. Staub, "A review of methods to measure animal body temperature in precision farming," *American Journal of Agricultural Science and Technology*, vol. 2, no. 2, pp. 74–99, 2014.

[51] J. Lee, B. Noh, S. Jang, D. Park, Y. Chung, and H.-H. Chang, "Stress detection and classification of laying hens by sound analysis," *Asian-Australasian Journal of Animal Sciences*, vol. 28, no. 4, p. 592, 2015.

[52] N. A. Mungroo, G. Oliveira, and S. Neethirajan, "SERS based point-of-care detection of food-borne pathogens," *Microchimica Acta*, vol. 183, no. 2, pp. 697–707, 2016.

[53] N. Mungroo and S. Neethirajan, "Biosensors for the detection of antibiotics in poultry industry—a review," *Biosensors*, vol. 4, no. 4, pp. 472–493, 2014.

[54] I. F. Akyildiz, D. M. Gutierrez-Estevez, R. Balakrishnan, and E. Chavarria-Reyes, "LTE-Advanced and the evolution to Beyond 4G (B4G) systems," *Physical Communication*, vol. 10, pp. 31–60, 2014.

[55] I. F. Akyildiz, J. M. Jornet, and C. Han, "Terahertz band: Next frontier for wireless communications," *Physical Communication*, vol. 12, pp. 16–32, 2014.

[56] B. E. Usibe, A. I. Menkiti, M. U. Onuu, and J. C. Ogbulezie, "Development and analysis of a potential nanosensor communication network using carbon nanotubes," *International Journal of Materials Engineering*, vol. 3, no. 1, pp. 4–10, 2013.

[57] M. A. Augustin and P. Sanguansri, "Nanostructured materials in the food industry," *Advances in Food and Nutrition Research*, vol. 58, pp. 183–213, 2009.

[58] T. Nakano, A. W. Eckford, and T. Haraguchi, *Molecular Communication*. Cambridge University Press, 2013.

[59] T. Nakano, M. J. Moore, F. Wei, A. V. Vasilakos, and J. Shuai, "Molecular communication and networking: Opportunities and challenges," *IEEE Transactions on Nanobioscience*, vol. 11, no. 2, pp. 135–148, 2012.

[60] M. Chen, S. Gonzalez, A. Vasilakos, H. Cao, and V. C. Leung, "Body area networks: A survey," *Mobile Networks and Applications*, vol. 16, no. 2, pp. 171–193, 2011.

[61] M. S. Kuran, H. B. Yilmaz, T. Tugcu, and I. F. Akyildiz, "Interference effects on modulation techniques in diffusion based nanonetworks," *Nano Communication Networks*, vol. 3, no. 1, pp. 65–73, 2012.

[62] T. Nakano and M. Moore, "In-sequence molecule delivery over an aqueous medium," *Nano Communication Networks*, vol. 1, no. 3, pp. 181–188, 2010.

[63] M. Pierobon, I. F. Akyildiz et al., "Diffusion-based noise analysis for molecular communication in nanonetworks," *IEEE Transactions on Signal Processing*, vol. 59, no. 6, pp. 2532–2547, 2011.

[64] T. Nakano, Y. Okaie, and J.-Q. Liu, "Channel model and capacity analysis of molecular communication with Brownian motion," *IEEE Communications Letters*, vol. 16, no. 6, pp. 797–800, 2012.

[65] N. Farsad, A. W. Eckford, S. Hiyama, and Y. Moritani, "A simple mathematical model for information rate of active transport molecular communication," in *IEEE Conference on Computer Communications Workshops (INFO-COM WKSHPS)*, 2011, pp. 473–478.

[66] A. W. Eckford, N. Farsad, S. Hiyama, and Y. Moritani, "Microchannel molecular communication with nanoscale carriers: Brownian motion versus active transport," in *10th IEEE Conference on Nanotechnology (IEEE-NANO)*, 2010, pp. 854–858.

[67] T. Nakano and J. Shuai, "Repeater design and modeling for molecular communication networks," in *IEEE Conference on Computer Communications Workshops (INFOCOM WKSHPS)*, 2011, pp. 501–506.

[68] S. Balasubramaniam, N. T. Boyle, A. Della-Chiesa, F. Walsh, A. Mardinoglu, D. Botvich, and A. Prina-Mello, "Development of artificial neuronal networks for molecular communication," *Nano Communication Networks*, vol. 2, no. 2–3, pp. 150–160, 2011.

[69] M. J. Moore and T. Nakano, "Multiplexing over molecular communication channels from nanomachines to a micro-scale sensor device," in *Global Communications Conference (GLOBECOM), 2012 IEEE*. IEEE, 2012, pp. 4302–4307.

[70] L. C. Cobo and I. F. Akyildiz, "Bacteria-based communication in nanonetworks," *Nano Communication Networks*, vol. 1, no. 4, pp. 244–256, 2010.

[71] S. Balasubramaniam et al., "Opportunistic routing through conjugation in bacteria communication nanonetwork," *Nano Communication Networks*, vol. 3, no. 1, pp. 36–45, 2012.

[72] A. Einolghozati, M. Sardari, A. Beirami, and F. Fekri, "Consensus problem under diffusion-based molecular communication," in *45th IEEE Annual Conference on Information Sciences and Systems (CISS)*, 2011, pp. 1–6.

[73] A. Arkin, "Setting the standard in synthetic biology," *Nature Biotechnology*, vol. 26, no. 7, p. 771, 2008.

[74] S. F. Bush, "Wireless ad hoc nanoscale networking," *IEEE Wireless Communications*, vol. 16, no. 5, 2009.

[75] "IEEE 1906.1-2015 - IEEE Recommended Practice for Nanoscale and Molecular Communication Framework," https://standards.ieee.org/, 2015.

[76] A. Rizwan, A. Zoha, R. Zhang, W. Ahmad, K. Arshad, N. A. Ali, A. Alomainy, M. A. Imran, and Q. H. Abbasi, "A review on the role of nano-communication in future healthcare systems: A big data analytics perspective," *IEEE Access*, vol. 6, pp. 41 903–41 920, 2018.

[77] R. Singh, E. Singh, and H. S. Nalwa, "Inkjet printed nanomaterial based flexible radio frequency identification (RFID) tag sensors for the internet of nano things," *RSC Advances*, vol. 7, no. 77, pp. 48 597–48 630, 2017.

[78] S. E. Lyshevski, *Nano and molecular electronics handbook*. CRC Press, 2016.

[79] X. Zhang, J. Wang, and H. V. Poor, "Joint Resource Allocation Optimization Over Energy Harvesting Based 6G THz-Band Big-Data-Driven Nano-Networks," in *IEEE Global Communications Conference (GLOBECOM)*, 2020, pp. 1–6.

[80] C. V. N. I. Cisco, "The zettabyte era—trends and analysis, 2015–2020. white paper," 2016.

[81] R. Yu, T. W. Mak, R. Zhang, S. H. Wong, Y. Zheng, J. Y. Lau, and C. C. Poon, "Smart healthcare: cloud-enabled body sensor networks," in *14th IEEE International Conference on Wearable and Implantable Body Sensor Networks (BSN)*, 2017, pp. 99–102.

[82] I. D. Dinov, "Volume and value of big healthcare data," *Journal of Medical Statistics and Informatics*, vol. 4, 2016.

[83] S. Mohammed, D. Servos, and J. Fiaidhi, "HCX: a distributed OSGi based web interaction system for sharing health records in the cloud," in *IEEE/WIC/ACM International Conference on Web Intelligence and Intelligent Agent Technology (WI-IAT)*, vol. 3, 2010, pp. 102–107.

[84] R. Wooten, R. Klink, F. Sinek, Y. Bai, and M. Sharma, "Design and implementation of a secure healthcare social cloud system," in *Proceedings of the 12th IEEE/ACM International Symposium on Cluster, Cloud and Grid Computing (CCGrid)*. IEEE Computer Society, 2012, pp. 805–810.

[85] H. Löhr, A.-R. Sadeghi, and M. Winandy, "Securing the e-health cloud," in *Proceedings of the 1st ACM International Health Informatics Symposium*, 2010, pp. 220–229.

[86] A. Benharref and M. A. Serhani, "Novel cloud and SOA-based framework for E-Health monitoring using wireless biosensors," *IEEE Journal of Biomedical and Health Informatics*, vol. 18, no. 1, pp. 46–55, 2014.

[87] S. N. Murphy, G. Weber, M. Mendis, V. Gainer, H. C. Chueh, S. Churchill, and I. Kohane, "Serving the enterprise and beyond with informatics for integrating biology and the bedside (i2b2)," *Journal of the American Medical Informatics Association*, vol. 17, no. 2, pp. 124–130, 2010.
[88] S. Rallapalli, R. Gondkar, and U. P. K. Ketavarapu, "Impact of processing and analyzing healthcare big data on cloud computing environment by implementing hadoop cluster," *Procedia Computer Science*, vol. 85, pp. 16–22, 2016.
[89] H. Zahmatkesh and F. Al-Turjman, "Fog computing for sustainable smart cities in the IoT era: Caching techniques and enabling technologies-an overview," *Sustainable Cities and Society*, vol. 59, p. 102139, 2020.
[90] A. Galal and X. Hesselbach, "Nano-networks communication architecture: Modeling and functions," *Nano Communication Networks*, vol. 17, pp. 45–62, 2018.
[91] P. Wang, J. M. Jornet, M. A. Malik, N. Akkari, and I. F. Akyildiz, "Energy and spectrum-aware MAC protocol for perpetual wireless nanosensor networks in the Terahertz Band," *Ad Hoc Networks*, vol. 11, no. 8, pp. 2541–2555, 2013.
[92] F. Al-Turjman, "A rational data delivery framework for disaster-inspired internet of nano-things (IoNT) in practice," *Cluster Computing*, pp. 1–13, 2017.
[93] C. Liaskos, A. Tsioliaridou, S. Ioannidis, N. Kantartzis, and A. Pitsillides, "A deployable routing system for nanonetworks," in *IEEE International Conference on Communications (ICC)*, 2016, pp. 1–6.
[94] N. Rikhtegar, M. Keshtgari, and Z. Ronaghi, "EEWNSN: Energy Efficient Wireless Nano Sensor Network MAC Protocol for Communications in the Terahertz Band," *Wireless Personal Communications*, vol. 97, no. 1, pp. 521–537, 2017.
[95] G. Piro, L. A. Grieco, G. Boggia, and P. Camarda, "Nano-Sim: simulating electromagnetic-based nanonetworks in the network simulator 3," in *Proceedings of the 6th International Conference on Simulation Tools and Techniques (ICST)*, 2013, pp. 203–210.
[96] S. J. Lee, C. Jung, K. Choi, and S. Kim, "Design of wireless nanosensor networks for intrabody application," *International Journal of Distributed Sensor Networks*, vol. 11, no. 7, p. 176761, 2015.
[97] H. Yu, B. Ng, and W. Seah, "Multi-Gateway Polling for Nanonetworks under Dynamic IoT Backhaul Bandwidth," in *IEEE International Conference on Sensing, Communication and Networking (SECON Workshops)*, 2018, pp. 1–4.
[98] X.-W. Yao, C.-C. Wang, W.-L. Wang, and J. M. Jornet, "On the Achievable Throughput of Energy-Harvesting Nanonetworks in the Terahertz Band," *IEEE Sensors Journal*, vol. 18, no. 2, pp. 902–912, 2018.
[99] C. Liaskos, A. Tsioliaridou, A. Pitsillides, I. F. Akyildiz, N. V. Kantartzis, A. X. Lalas, X. Dimitropoulos, S. Ioannidis, M. Kafesaki, and C. Soukoulis, "Design and development of software defined meta-materials for nanonetworks," *IEEE Circuits and Systems Magazine*, vol. 15, no. 4, pp. 12–25, 2015.
[100] C. Liaskos and A. Tsioliaridou, "A promise of realizable, ultra-scalable communications at nano-scale: A multi-modal nano-machine architecture," *IEEE Transactions on Computers*, vol. 64, no. 5, pp. 1282–1295, 2015.
[101] H. Yu, B. Ng, and W. K. Seah, "Forwarding schemes for em-based wireless nanosensor networks in the terahertz band," in *Proceedings of the Second Annual International Conference on Nanoscale Computing and Communication*. ACM, 2015, p. 17.
[102] R. Zhou, Z. Li, C. Wu, and C. Williamson, "Buddy routing: a routing paradigm for nanonets based on physical layer network coding," in *21st IEEE International Conference on Computer Communications and Networks (ICCCN)*, 2012, pp. 1–7.
[103] F. Al-Turjman, "A Cognitive Routing Protocol for Bio-Inspired Networking in the Internet of Nano-Things (IoNT)," *Mobile Networks and Applications*, pp. 1–15, 2017.

[104] I. F. Akyildiz, M. Pierobon, S. Balasubramaniam, and Y. Koucheryavy, "The internet of bio-nano things," *IEEE Communications Magazine*, vol. 53, no. 3, pp. 32–40, 2015.

[105] A. Farahzadi, P. S. Farahsary, & J. Rezazadeh, "A survey on IoT fog nano datacenters," *Wireless Networks*, 1–35, 2022.

[106] M. Kuscu, & O. B. Akan, "The Internet of molecular things based on FRET," *IEEE Internet of Things Journal*, 3(1), 4–17, 2015.

[107] T. Lehtonen, P. Liljeberg, and J. Plosila, "Analysis of forward error correction methods for nanoscale networks-on-chip," in *Proceedings of the 2nd international conference on Nano-Networks (ICST)*, 2007, p. 3.

[108] S. D'Oro, L. Galluccio, G. Morabito, and S. Palazzo, "A timing channel-based MAC protocol for energy-efficient nanonetworks," *Nano Communication Networks*, vol. 6, no. 2, pp. 39–50, 2015.

[109] H. Yu, B. Ng, and W. K. Seah, "Pulse Arrival Scheduling for Nanonetworks under Limited IoT Access Bandwidth," in *IEEE 42nd Conference on Local Computer Networks (LCN)*, 2017, pp. 18–26.

[110] N. Islam and S. Misra, ""Catch the Pendulum": The Problem of Asymmetric Data Delivery in Electromagnetic Nanonetworks," *IEEE Transactions on Nanobioscience*, vol. 15, no. 6, pp. 576–584, 2016.

[111] T. Furubayashi, T. Nakano, A. Eckford, Y. Okaie, and T. Yomo, "Packet fragmentation and reassembly in molecular communication," *IEEE Transactions on Nanobioscience*, vol. 15, no. 3, pp. 284–288, 2016.

[112] Q. H. Abbasi, A. A. Nasir, K. Yang, K. A. Qaraqe, and A. Alomainy, "Cooperative in-vivo nano-network communication at terahertz frequencies," *IEEE Access*, vol. 5, pp. 8642–8647, 2017.

[113] F. Alam, R. Mehmood, I. Katib, N. N. Albogami, and A. Albeshri, "Data fusion and IoT for smart ubiquitous environments: A survey," *IEEE Access*, vol. 5, pp. 9533–9554, 2017.

[114] H.-H. Yen, "Energy aware and signal quality aware data aggregation touting in wireless nanosensor networks," in *Consumer Communications & Networking Conference (CCNC), 2017 14th IEEE Annual*. IEEE, 2017, pp. 879–884.

[115] S. Mohrehkesh and M. C. Weigle, "Optimizing energy consumption in terahertz band nanonetworks," *IEEE Journal on Selected Areas in Communications*, vol. 32, no. 12, pp. 2432–2441, 2014.

[116] R.-S. Liu, K.-W. Fan, Z. Zheng, and P. Sinha, "Perpetual and fair data collection for environmental energy harvesting sensor networks," *IEEE/ACM Transactions on Networking (TON)*, vol. 19, no. 4, pp. 947–960, 2011.

[117] A. Gunathillake, "Maximum Likelihood Coordinate Systems for Wireless Sensor Networks: from physical coordinates to topology coordinates," arXiv preprint arXiv:1805.00004, 2018.

[118] L. Lin, C. Yang, and M. Ma, "Maximum-likelihood estimator of clock offset between nanomachines in bionanosensor networks," *Sensors*, vol. 15, no. 12, pp. 30 827–30 838, 2015.

[119] Y. Fang, A. Noel, N. Yang, A. W. Eckford, and R. A. Kennedy, "Maximum likelihood detection for cooperative molecular communication," in *IEEE International Conference on Communications (ICC)*, 2018, pp. 1–7.

[120] H. F. Atlam, A. Alenezi, R. J. Walters, G. B. Wills, and J. Daniel, "Developing an adaptive Risk-based access control model for the Internet of Things," in *Internet of Things (iThings) and IEEE Green Computing and Communications (GreenCom) and IEEE Cyber, Physical and Social Computing (CPSCom) and IEEE Smart Data (SmartData)*, 2017, pp. 655–661.

[121] F. Afsana, M. Asif-Ur-Rahman, M. R. Ahmed, M. Mahmud, and M. S. Kaiser, "An Energy Conserving Routing Scheme for Wireless Body Sensor Nanonetwork Communication," *IEEE Access*, vol. 6, pp. 9186–9200, 2018.

[122] N. A. Ali, W. Aleyadeh, and M. AbuElkhair, "Internet of nano-things network models and medical applications," in *IEEE International Wireless Communications and Mobile Computing Conference (IWCMC)*, 2016, pp. 211–215.

[123] F. Dressler and S. Fischer, "Connecting in-body nano communication with body area networks: Challenges and opportunities of the internet of nano things," *Nano Communication Networks*, vol. 6, no. 2, pp. 29–38, 2015.

[124] G. Avoine, M. A. Bingöl, I. Boureanu, G. Hancke, S. Kardas, C. H. Kim, C. Lauradoux, B. Martin, J. Munilla, A. Peinado et al., "Security of distance-bounding: A survey," *ACM Computing Surveys (CSUR)*, vol. 51, no. 5, p. 94, 2018.

4 Emerging World of the Metaverse
An Indian Perspective

Aasita Bali and Arishmita Aditya
Christ University, Bangalore, India

4.1 INTRODUCTION

'Metaverse' is a compound word of 'transcendence' and 'universe.' It refers to a computer-generated world where both the real and the unreal coexist. In addition, the metaverse refers to a fully immersive three-dimensional digital environment that reflects the totality of shared online space across all dimensions of representation. The way we connect with the physical environment around us is rapidly changing as a result of recent technological advancements. Our digital footprint is therefore continuously recorded and monitored. In the metaverse, avatars are used as a means of self-representation, communication, and community-building. In the metaverse, digital currency is used, among other things, to buy goods like armor, weapons, and shields in video games. Users may also take leisurely aimless trips around the metaverse using a virtual reality headset and controllers.

The journey of the metaverse has been rapid, recent, and revolutionary. Previous technologies like Augmented Reality and Virtual Reality had already proved the potential of digital technology and how it can be used for promotion of a brand and gaming. In 2014, Google debuted its first Cardboard gadget and Google AR glasses. Sony and Samsung also stated that they were developing their own VR headsets. In 2016, HoloLens headsets from Microsoft were released. It produced holographic pictures that could be manipulated through augmented reality.

There was a momentous breakthrough when the entire world ran around their neighborhood trying to capture Pokémon using the augmented reality game Pokémon GO in 2016. IKEA, the Swedish furniture giant, entered the metaverse in 2017 with their Place app, which lets users choose a piece of furniture and see how it would appear in their home or workplace setting. Similarly, Burger King launched an augmented reality ad campaign in 2019 where customers could "burn" ads from its competitors in exchange for a free burger. The augmented reality display showed a rival's advertisement on fire. The sensory engagement of the users created a demand and impact to bring the brand and the audiences even closer.

In 2021, Microsoft launched a blended-reality platform, Mesh, as a significant step to integrate HoloLens and mixed reality. It is a collaborative platform that enables people from different remote locations to share holographic experiences on multiple devices. In the same year, the parent company of the social networking

DOI: 10.1201/9781003303114-4

site Facebook changed its name from Facebook to Meta, indicating a paradigm shift in their technological vision, i.e., from being a networking platform to engaging multiple technologies under one umbrella within their 'metaverse.' The Metaverse has the potential to be either a digital paradise or an addictive monster. With technological improvisation it will be closer to human intelligence, replicating human form, emotions, and settings, posing a challenge to human-to-human interaction (Press, n.d).

Though the chapter deals in details about the various possibilities of the metaverse and how it has changed the business and networking scenario in India, it also discusses how challenges faced by other digital communications have posed and might have a similar effect with the metaverse. Data collection, cybersecurity, privacy, and poor regulations are just a few issues policymakers and content providers might have to look out for while upgrading their digital architecture. The chapter begins with Internet of Things, followed by Emergence of the Metaverse, Network and Communication in the Metaverse, Socialization in the Metaverse, The Metaverse in India, technological limitations, Data Safety and Security, and finally, it concludes discussing the Dilemma and Conundrum of the Metaverse.

4.2 INTERNET OF THINGS AND VIRTUAL REALITY

In this digitally connected universe, it has become next to impossible to not come across the term 'Internet of Things.' From academics to medical science, the Internet of Things (IoT) can be found in an array of sectors, industries, and applications. The past 10 years have seen a tremendous surge of IoT platforms across different frameworks. Companies have already introduced numerous IoT-based products and services. Although it is frequently used, there is no one agreed-upon definition of what the Internet of Things truly entails. The International Telecommunication Union (ITU) defined it as a global infrastructure for the Information Society that enables improved services by linking objects (both real and virtual) based on current and developing interoperable information and communication technologies. The IoT's central concept of interconnection is the capacity for machines, devices, sensors, and people to connect to and interact with one another through the Internet.

IoT technologies are progressively being applied in our daily life. The most common applications are in relation to the smart industry and now, the metaverse. The 1992 science fiction book *Snow Crash* by Neal Stephenson featured the first mention of the metaverse—suggesting a digital planet that uses technologies like smartphones and the Internet for its communication and networks. Later, several developments lead to the expansion of the phrase 'metaverse' with more technological and networking advancements. Ernest Cline's *Ready Player One* is another book that helped the metaverse become more known in popular culture. Kye et al. (2021), in their article 'Educational applications of metaverse: possibilities and limitations,' described four categories of the metaverse as: Augmented reality, Lifelogging, Mirror worlds, and Virtual worlds. Augmented reality uses location-aware technology to expand the physical world. In the realm of Lifelogging, people employ cutting-edge technology to document and share their daily activities. The mirror world is a metaverse where the physical world's appearance, content, and infrastructure are translated into virtual reality. Dionisio et al. (2013) skillfully explain the concept of metaverse across four areas—immersive realism, the ubiquity of access and identity, interoperability, and scalability—in their article, '3D virtual worlds and the metaverse.' They also describe factors that support as well as constrain building a viable metaverse.

Sir Charles Wheatstone, a scientist, first described the idea of "binocular vision" in 1838. This idea states that humans may merge two separate pictures to create a single 3D view. Virtual Reality is described as a 3D online environment that allows multiple people to view and interact simultaneously. In this virtual world, the avatars and their social surroundings are designed based on artificial intelligence and algorithms. This provides a compelling alternative reality for interpersonal and sociocultural interaction. Advanced 3D visuals, avatars, and instant communication tools are all part of virtual reality technology. The booming rise of Zepeto and Roblox Corp, South Korea's successful gaming providers, have proved that the metaverse is here to stay in the online gaming industry.

4.3 EMERGENCE OF THE METAVERSE

The development of the metaverse is based on the convergence of technologies that allows multimodal interactions with digital objects and virtual environments. It is a persistent multiuser platform with a networked web of socially immersive surroundings. Modern VR headsets have also been developed as a result of the concept of 3D and virtual reality.

In 1935 American science fiction writer Stanley Weinbaum published the book, *Pygmalion's Spectacles*, in which the protagonist explores a fictional world using a pair of sensory-enhancing spectacles that provide sight, sound, taste, smell, and touch. In 1956, Morton Heilig developed the first VR device called 'Sensorama Machine,' which immersed the viewer by combining 3D film with sensory stimulations to recreate the feeling of riding a motorbike in Brooklyn. In 1960, he also secured a patent for the first head-mounted display.

The metaverse has been a popular topic in science fiction and literature for several decades. While there may be more recent works exploring it, some notable literary works have discussed or envisioned this concept. While not directly using the term 'metaverse,' this science fiction novel, *True Names* examines the concept of a vast computer network where users communicate with one another in a virtual environment. *Neuromancer* is considered a seminal work of cyberpunk fiction.

The World Wide Web grew throughout the 1990s and 2000s, which encouraged the development of online communities and social connections. Online games like "Ultima Online" (1997) and "EverQuest" (1999) gained popularity as they offered persistent virtual environments where players can communicate and collaborate. Linden Lab released "Second Life" in 2003, a virtual world where users can create content, interact with each other, and conduct business.

The emergence of virtual reality headsets like Oculus Rift and HTC Vive in early 2010 brought immersive VR experiences to a wider audience. 'VRChat' (2014) gained quick popularity as a social VR platform where users can explore user-generated worlds and interact with each other. As the world chased 'Pokémon Go' (2016), it demonstrated the potential of blending digital content with the real world.

Several companies are now attempting to replace traditional in-person employee onboarding with digital or VR-based alternatives. Rather than physically going to an office to learn processes, the notion is that virtual, metaverse-like settings offered via VR might act as a substitute with the added benefit of flexibility. Online skill platform SkillSoft is planning to use the metaverse to develop a new experiential learning setup. For example, Strivr, based in Palo Alto, California, is developing software that allows users to create virtual reality training programs focused on health and safety and customer service coaching. In 2018 Walmart proposed to partner with Strivr to purchase Oculus Go VR headsets to train personnel at its retail locations on how to use a new online order pickup station before it went live in shops.

4.4 NETWORK AND COMMUNICATION IN THE METAVERSE

The metaverse is a dynamic orchestrator of the next-generation Internet architecture that creates an immersive and self-adapting virtual environment where people engage in activities like those in the real world, such as socializing, working, and playing sports. Extensive reality, artificial intelligence, and blockchain are among the cutting-edge technologies helping the metaverse become the next alternative reality. Recent studies provide a detailed analysis and discussion about the multiple ways through which the metaverse applies and contributes in developing the existing world of network and communication (Figure 4.1).

In addition to other virtual and online venues, the metaverse was crucial in assisting individuals in coping with the pandemic and fostering social interaction despite physical distance. VR platforms such as VRChat, Rec Room, and AltspaceVR enabled people to create avatars and interact with each other in virtual worlds. Multiplayer online games like "Animal Crossing: New Horizons," "Fortnite," and "World of Warcraft" provided players with avenues for social interaction, collaboration, and shared experiences, fostering a sense of community. Platforms like Spatial and Mozilla Hubs allowed users to interact in virtual workspaces and classrooms. This makes it clearer how crucial a role the metaverse will play in creating smart communication platforms in the post-COVID 19 urban environment. However, this new paradigm shift in technology will create novel privacy and security threats in the digital metaverse ecosystem.

4.4.1 Building Smart Cities with the Metaverse

The metaverse has the potential to significantly contribute to the development of smart cities and businesses by offering enhanced connectivity, communication, data analysis, and virtual simulations. It enables businesses, government agencies, and residents to interact, collaborate, and share information in the smart city ecosystem. These experiences are predicted to further encourage people to migrate to the metaverse in the long term. The new conceptualization of place attachment is a dynamically evolving phenomenon impacted by how people experience different locations.

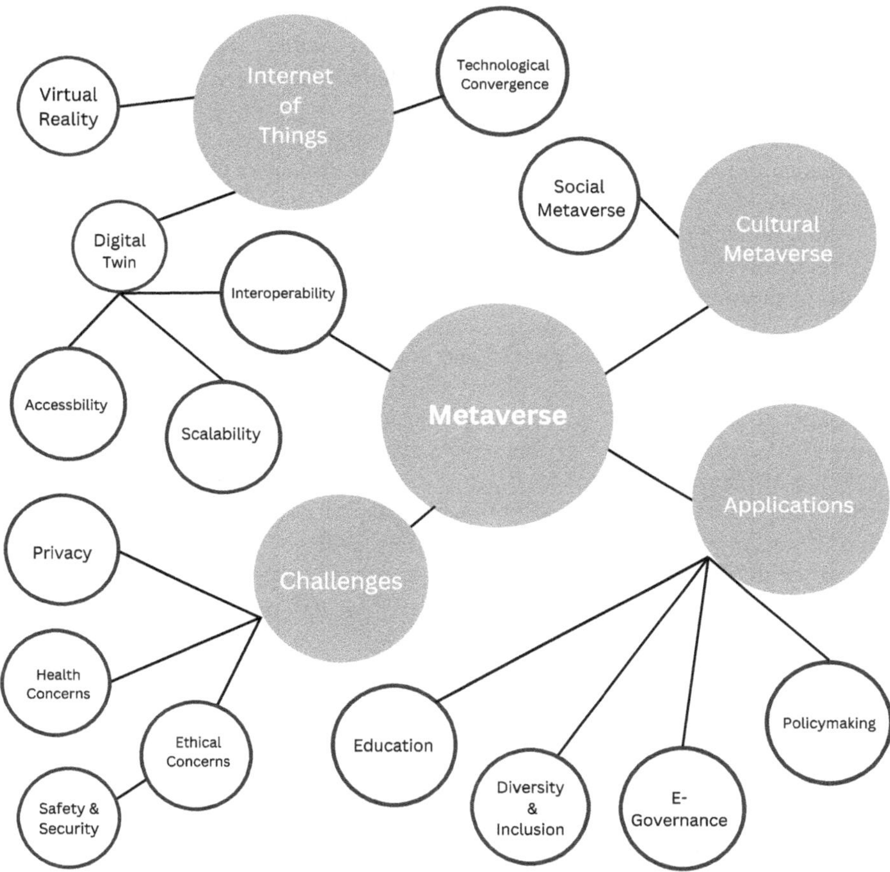

FIGURE 4.1 Application and contributions of the metaverse in developing the existing world of network and communication.

In a smart city, vast amounts of data are gathered from sensors, devices, and systems requiring a strong digital infrastructure. Businesses and city planners may acquire useful insights into traffic patterns, energy use, and environmental concerns by using the metaverse as a tool for visualizing and analyzing the data. This enables data-driven decision-making, efficient resource allocation, effective policymaking, and the identification of opportunities for improvement. It also provides a virtual sandbox for simulating and testing different scenarios within a smart city environment by creating virtual replicas of physical spaces. This allows users to run simulations and determine the possible effects of modifications or interventions.

AR can aid in visualizing future buildings, infrastructure projects, and urban designs, facilitating better decision-making and citizen engagement. The study into the cultural metaverse shows how interactive AR applications can lead to increased visitor engagement and revolutionize the art of digital storytelling. Blockchain technology can be used with the metaverse to provide smart contracts and safe transactions in smart cities. Smart contracts are self-executing contracts with the terms of the agreement directly written into code. Businesses and governments may automate and protect transactions, expedite administrative procedures, and guarantee openness and confidence in financial and governance systems by integrating blockchain-based technologies into the metaverse.

It is crucial to note that while the potential benefits of the metaverse in building smart cities and businesses are significant, the realization of its full potential relies on advancements in technology, infrastructure, and collaboration among various stakeholders. There are potential risks associated with this utopian portrayal of the metaverse in media. Ruolan Deng and others (2023) suggest that it may lead to a lack of critical engagement with the concept and its implications. They also point out that it may reinforce existing power structures and inequalities, particularly if the development of the metaverse is monopolized and dominated by a small group of powerful corporations.

Organizations are set to dramatically envision a world where the lines between physical and virtual reality blur. Immersive experiences, enabled by Virtual Reality (VR) and Augmented Reality (AR), will revolutionize the world of enterprise metaverse. Mass simulations and AI technologies will assist senior executives to develop the right course of action during challenging times, which will leverage data streams to forecast the future. This enterprise metaverse will connect and optimize experiences with the help of a powerful engine: Digital Twins.

4.4.2 Industry 4.0 and Digital Twin

The fourth industrial revolution and the underlying digital transformation, known as Industry 4.0, is set to drastically evolve industrial production. Interconnection, information transparency, and decentralized decision-making are three of the key design tenets of Industry 4.0. One of the most promising enabling technologies for achieving smart manufacturing and Industry 4.0 is the Digital Twin (DT)—the foundation of this new industrial revolution. It opens up a multitude of possibilities for the future of the digital and virtual corporate world. In the industrial sector, it has replaced the dated conventional strategy of 'first build, then tweak.' Digital Twins can offer the metaverse a more technologically advanced, virtual-based system design method that results in considerably more dynamic machinery or systems (Jacoby and Usländer, 2020).

The capacity to gather, handle, and organize the enormous quantity of data from linked equipment, devices, and sensors is referred to as information transparency. The idea of 'a formal digital representation of some asset, process, or system that captures attributes and behaviors of that entity suitable for communication, storage, interpretation, or processing within a particular context' addresses this requirement. The seamless integration of the digital and physical realms is a defining characteristic of DTs. Several DT applications have already been successfully used in a variety of industries, including production, prognostics, and health management.

DT and IoT both focus on resources including assets, devices, and actual or virtual entities. Within the context of IoT, resources are usually associated with Internet-connected devices that may communicate with consumers either directly or indirectly.

Both DT and IoT presume that machine-to-machine (M2M) communication between resources should occur without any human interaction.

Digital Twins can be incorporated into the metaverse to design products, integrate production, and provide actionable insight for informed decision-making. The global metaverse infrastructure can be improved using the Digital Twin. By examining its virtual or digital counterpart, it will be possible to ascertain and forecast the present and future states of any physical asset. Organizations and businesses may provide greater insight into product performance and enhance customer service by integrating Digital Twins in the metaverse.

4.5 SOCIALIZATION IN THE METAVERSE

The metaverse offers the possibility to represent the individual's physical body, but the user may also choose a completely different fictional appearance. If exposed to both virtual and physical reality since the early developmental years, one may perceive it as a singular reality (Henz, 2022). This can have a significant impact upon one's self-image and confidence. Based on Leon Festinger's 'Theory of Cognitive Dissonance,' a person perceives an inner pressure to have cognition, feelings, and behavior in line. If, due to society, individuals cannot live their beliefs in the physical world, the metaverse may serve as escapism.

Developed by Albert Bandura in 1986, Social Cognitive Theory emphasizes the role of observation, imitation, and reinforcement in learning and social behavior. In the metaverse, individuals have opportunities to learn new skills or modify behavior by observing and interacting with different avatars in virtual environments (Bandura, 1986). This nature of the metaverse allows for immediate feedback and reinforcement, shaping individuals' behavior and socialization experiences.

Social metaverse refers to several virtual worlds where interacting and socializing with other avatars serves as the primary goal. With Facebook's recent rebranding as "Meta" in 2021, there has been a renewed emphasis on the metaverse as a platform for commerce, education, socialization, and more. These virtual platforms, which are considered a "very immersive virtual environment," hold events ranging from meetings to music festivals and employ technology such as virtual reality goggles.

During the Connect 2021 conference, Meta's Codec Avatars software demonstrated how avatars could adapt to various lighting and situations, as well as exhibit a variety of different traits (Mitchell, 2021). More sophisticated qualities like detailed facial expressions and focused gaze, on the other hand, are still challenging to replicate in avatars. Virtual characters' resemblance to real-life characters may make consumers uncomfortable as they become more lifelike than before. The "uncanny valley" idea goes into detail about this. This idea claims that when people interact with objects that are lifelike, such as humanoid robots or animated characters, their obvious differences from real people may result in uneasiness. For many years, the uncanny valley has been studied in relation to robots and digital avatars, but there has never been an opportunity for a long-term investigation into how it may affect human behavior, because this level of technological realism has not yet been optimized on a large scale (Clark, 2021).

The metaverse's visual design is based on graphics methods, such as 3D scene building, non-player characters (NPCs), and player characters (Avatar), that unite the real and virtual worlds to create an interconnected cosmos. Using interactive technology, users may interact with visual elements, freely explore the metaverse, and enjoy an immersive experience. However, more guidelines and rules must be implemented in the actual world to increase user awareness of the virtual world.

Providing people with immersive experiences, appropriate and effective interaction, and easily accessible visual aids for quick awareness, understanding, and analysis is the key to exploring the metaverse. Users of the metaverse are exposed to a data-rich environment. Therefore, a thorough analysis of the material is necessary in such hypothetical situations. However, such innovative and exciting technologies are often complemented by cybersecurity risks like phishing, malware attacks, and hacking (Ramprakash Ramamoorthy, 2022). For example, The widespread use of cryptocurrencies and non-fungible tokens (NFTS) in the metaverse makes them potential targets for hackers and data theft. Additionally, it may be more challenging to protect intellectual property when individuals and businesses exist both in the real world and the metaverse.

4.6 THE METAVERSE IN INDIA

Tanishq's Titan becomes the first Indian company in this industry to enter the metaverse through the launch of its 'Rivaah'—a bridal jewelry collection called the 'Romance of Polki.' According to the Gem & Jewellery Export Promotion Council's website, Rivaah used the metaverse to launch its new collection with 'an immersive and customized 3D virtual experience.' At the 'experiential' launch, guests could rotate their 3D avatars to see products from different perspectives and in greater detail, as well as witness 14 of the collection's new jewelry designs as 3D photographs in a special zone.

The second example is Ajnalens, a Mumbai-based XR hardware and software company that collaborated with Tata Technologies and created a metaverse to teleport their trainees to job sites for industrial training. AjnaXR Station is a scalable VR training system designed in India that provides training across various modules by replicating the real world in the virtual world. This new pedagogy makes training more effective and engaging while lowering training time and cost. Apart from this, Tata Group's Tata Tea Premium made its debut in the metaverse by hosting a platform to celebrate Holi in the virtual space (PricewaterhouseCoopers, n.d). For this, customers were able to dress up in their favorite avatars and celebrate this festival with the other participants remotely. To add fervor to the party, a special performance by superstar, composer, and lyricist duo Sachet-Parampara was also included.

India has played a significant part in the development of the metaverse, much like the rest of the globe (Fallmann, 2022). The popularity and usage of the metaverse in India have grown significantly over the years. With the country's thriving tech sector and entrepreneurship culture, there are plans to create an all-encompassing virtual reality environment. Mahindra Tractors utilized virtual reality (VR) to offer customers immersive experiences, enabling them to virtually engage with tractors and assess their features before making purchasing choices. Similarly, Mahindra's motorsport division employed VR to provide Formula E fans with a comprehensive 360-degree view of races. Another instance involves Flipkart, which introduced an augmented reality (AR) feature called "Flipkart Camera." This feature enables users to virtually place and visualize furniture and appliances within their living spaces. These technological advancements have revolutionized the shopping experience, introducing new levels of immersion, interactivity, and personalization.

On India's 73rd Republic Day, nearly 20 million people across the world attended India's first metaverse concert. Daler Mehndi, a well-known Indian singer, was invited to commemorate India's digital debut on the metaverse. Digitalization has ensured its exponential growth in the global economy. Giant players like Infosys and TCS are largely investing in virtual worlds and have further highlighted the need for data privacy laws and regulatory bodies. According to a PwC report (2023), almost 70% of business executives in India plan to integrate the metaverse into their organizational activities. Consumers are receptive to embracing new technology, and businesses are substantially investing in the infrastructure needed to leverage the metaverse.

Metaverse and Digital Twin technology blur the line between the real and virtual world. The significant youth population of India has accelerated the growth of digital India since they are accustomed to digital interactions and entertainment. The rapidly expanding usage of video streaming and gaming platforms, the use of digital payments, and the vibrant entrepreneurial culture may all be seen as the metaverse's building bricks. It has been widely acknowledged that youth are more likely to adopt and rely faster on digital lifestyle. Similarly, the metaverse has drawn global eyeballs while posing an ethical dilemma around it.

One of the largest barriers to adopting metaverse in India is its existing Digital Divide. According to the Oxfam Report (2022), the Internet penetration in rural India remains only 31% in contrast to 67% in urban areas. Among mobile users in India, close to 40% do not have smartphones. Less than 32% of women in India own a mobile phone, compared to over 60% of men. Another barrier that developers must overcome while integrating the metaverse is to accommodate the differences. The linguistic diversity of India highlights the need for technology to be accessible for all, irrespective of all differences. A multi-language interface with additional features inclusive of any disability can contribute significantly in the adoption of the metaverse in India.

4.7 TECHNOLOGICAL LIMITATIONS

Immersive experiences allow users to escape the real world and transport themselves to a pleasurable virtual world. Addiction to VR can display symptoms to what is described as behavioral addiction, like excessive VR Gaming, withdrawal symptoms when offline, and loss of interest in other activities (Vainilavičius, 2022). It has further been associated with, among other mental health issues—digital isolation, stress, paranoid ideation, somatic symptoms, and psychosis. Additionally, users may experience tiredness and movement injuries in addition to simulated motion sickness as a result of visual imbalance. Long-term VR use can result in eye pain and cybersickness. Effective service in the metaverse requires mental balance and steadiness.

Given these limitations, it cannot be denied that virtual spaces have the potential to be used for effective psychiatric cure/solutions. The possibility of repeatedly examining the user experience in a safe setting can allow better diagnosis and meaningful interactions (Maples-Keller et al, 2017). For those who have trouble envisioning or visualizing, virtual reality removes a potential obstacle. Along with the detrimental impacts of technology, the use of virtual reality to treat PTSD, phobias, anxiety disorders, delusions, and hallucinations is critical to note (Maples-Keller, et al., 2017).

Even though the metaverse has immense potential, organizing and scaling it up presents several difficulties. Adoption to high-end technology like VR, for instance, is a major issue. The HTML protocol acted as a unifying framework for the growth of the Internet (Coles, 2022). A similar protocol is required to enable seamless travel between devices and metaverses. One of the most significant issues is the lack of compatibility and uniformity among the multiple metaverse systems. Alternatively, there are a few key companies with integration protocols or Application Programming Interfaces (API), similar to how Apple's iPhone and Google's Android operating systems dominate. Even though their devices are different, users may still communicate and exchange information. Distributing their work across both platforms is simple for content creators and app developers. These features must be made available to both users and content producers through the metaverse's platforms and devices. Businesses will have to explore this new platform for effective delivery architecture so there is a seamless product delivery and customer satisfaction similar to web-based transactions.

Both usability and utility are required for consumers and businesses to invest, participate, collaborate, and interact in the metaverse. Many businesses seek to tap into this sector's potential but are wary by its status as an emerging technology. The metaverse is still in its early phases of development, and finding the right talent pool familiar with the platform can be challenging. Furthermore, limited tools or resources are accessible in the metaverse to assist IT professionals in managing and optimizing their virtual infrastructure. In this case, businesses are hesitant to invest in the metaverse or virtual reality (VR) due to uncertainty surrounding potential profits and returns.

At present, the use of the metaverse is at a primitive stage, making it difficult to assess its broader implications on a larger scale. The ultimate judgment might not be reached right away in India because of multiple reasons. India, being a vast and diverse country, still faces challenges in providing robust digital infrastructure across all regions, limiting access to the metaverse for many individuals and businesses. Education plays a pivotal role in facilitating metaverse adoption. Without adequate knowledge and understanding, people may hesitate to explore or invest in these high-end technologies. Accessibility is another critical aspect. The metaverse should be accessible to individuals with disabilities or those who may face barriers due to socioeconomic factors. Ensuring inclusivity by developing metaverse platforms and applications accessible to a wide range of users is essential for widespread adoption. Addressing these barriers will require concerted efforts from the government, private sector, and educational institutions to invest in digital infrastructure, promote digital literacy, and ensure inclusivity in metaverse development.

Additionally, data privacy and security will be paramount in the metaverse, where users will be sharing vast amounts of personal information and engaging in various digital transactions. Implementing robust data protection measures, encryption protocols, and authentication mechanisms will be critical to maintain trust and secure users' information in the virtual realm.

4.8 DATA SAFETY AND SECURITY

It is predicted that users would spend a lot of time in the metaverse, and their activities will be tracked. The tracking of people's actions and movements will go hand in hand with the extensive collection of personal data. Determining accountability for data processing and data loss is crucial since network and communication is a key component of the metaverse concept. Other than that, there are serious challenges in payments and transactions that should be addressed. According to a report by Venture Beat (2022), it is projected that by 2026, approximately 25% of the global population will spend at least an hour each day in the metaverse, engaging in activities such as shopping, work, and socializing (Adebayo, 2022). However, this widespread adoption of the metaverse also brings forth concerns regarding social engineering attacks, including identity theft and false impersonation. The Center for Countering Digital Hate (CCDH) highlighted in 2021 that VRChat, a platform within the metaverse, was plagued by issues of harassment, racism, and sexually explicit content (Aijaz, 2022).

The metaverse presents additional challenges due to its reliance on virtual reality, augmented reality, machine learning, and AI technologies. These behavioral-learning tools collect substantial amounts of personal data, raising concerns about user privacy. The emergence of deep fakes, big data, and cyber-attacks poses risks to both brand reputation and consumers. As metaverse technologies gain traction, marketers need to stay informed about these developments. In a reality-bending world driven by data-driven advancements, it is crucial for corporations and major technology companies to assume responsibility for ensuring transparency and ethical utilization of data.

Based on research, consumers are more willing to share their data if they perceive that a company will use it to their advantage. Consequently, businesses should prioritize integrating privacy and data security measures into their products and services right from the start (Aijaz, 2022). This proactive approach may help mitigate potential risks associated with data breaches and alleviate concerns surrounding the acquisition and management of personal data by technology companies. In addition, lawmakers need to address significant concerns related to legislation and jurisdiction in this vast digital environment. As Meta and other companies continue to expand, regulatory and governance concerns will demand significant attention. The growth of the metaverse will give rise to new activities that have implications for compliance and legal frameworks, requiring careful consideration (Report State Laws Related to Digital Privacy, n.d). Ensuring adherence to laws and regulations will be crucial as the integration of the metaverse into our daily lives deepens.

4.9 DILEMMAS AND CONUNDRUM OF THE METAVERSE

The emergence of the metaverse has now made it possible to transfer everyday activities into a parallel digital space. It transcends the limitations of the physical world and moves beyond the spatial structures virtually. The metaverse makes use of technologies like virtual reality, augmented reality, and blockchain to support a virtual economy and promote social connections. Its components include 3D avatars, digital assets, and a variety of events. While the concept may seem as exciting as ever, there are major privacy concerns lying here.

Metaverse systems may track individuals in a far more intimate way than traditional social media. Companies may monitor physiological reactions and biometric data including facial expressions, vocal inflections, and vital signs in real time while participants are in their metaverse (Uberti, 2022). This enormous data allows businesses to obtain a better insight of their customers' behavior, which can then be leveraged to create highly targeted and personalized advertising campaigns. According to a report published by Venture Beat in 2022, a third of developers (33%) believe these technologies are the biggest hurdles the metaverse has to overcome.

Even though there is no comprehensive federal privacy law with clear rules on the collection, use, and sharing of personal data, five states in the US—California, Colorado, Connecticut, Utah, and Virginia—have implemented data privacy laws that include several provisions like the right for individuals to access and delete their personal information. As Google CEO Sundar Pichai made major announcements focused on artificial intelligence at the highly anticipated Google I/O event on May 10, 2023, it's important to go back to his words and note that, "Privacy cannot be a luxury good offered only to people who can afford to buy premium products and services. Privacy must be equally available to everyone in the world" (Pichai, 2019; The New York Times, 2019).

Existing online issues, such as snooping, bullying, data breaches, harassment, and hate speech, are anticipated to increase and get worse with newer digital technologies. These issues are amplified in nations like India, which lacks a dedicated data privacy law. In addition to concerns about data theft, the emergence of virtual interactions and the creation of ideas like digital avatars will make it more challenging to monitor cybercriminals and intercept illegal data. Indian law does not contemplate harassment through advances/groping in digital environments.

Self-regulation, ethical standards, and data protection are not enough to address the issues brought on by technology. To uphold the digital rights of users would require strong regulation and enforcement to protect people's privacy applicable to the metaverse. Legislators should exercise caution regarding the tendency of large technology corporations to assimilate their competitors prematurely, impeding the development of alternative platforms that prioritize individual rights and privacy over the prevailing surveillance-driven models.

4.10 FUTURE OF THE METAVERSE

The recent emergence and growing prominence of the metaverse have had a significant impact on businesses across various sectors. One notable impact of the metaverse on businesses is the potential for enhanced customer engagement and immersive brand experiences. By leveraging virtual reality and augmented reality technologies within the metaverse, companies can create compelling and interactive virtual environments for consumers. These experiences offer unique opportunities for product demonstrations, virtual storefronts, and personalized interactions, fostering deeper customer engagement and brand loyalty.

4.10.1 Metaverse for Business Ideas

In the fashion industry, companies like Gucci have expanded their presence on Roblox—a popular metaverse and gaming platform—with virtual art installations. The Gucci Garden within the Roblox platform consists of various themed rooms that not only pay tribute to the brand but also transcend the physical limitations (McDowell, 2021).

Assumptions around ChatGPT and DALL-E to be major game-changers for businesses are gaining global consensus. The 14 billion dollars' worth 'creator economy' now stands disrupted in the face of traditional roles like copywriting, graphic designing, and coding. With new generative language models, we may soon have machines taking over the human process of ideation and creation. These recent advances in AI have upskilled millions of people with the ability to create intelligent content.

The metaverse offers new possibilities for remote collaboration and work environments. As the COVID-19 pandemic accelerated the adoption of remote work, companies like Spatial developed virtual collaboration platforms that allow teams to meet, brainstorm, and collaborate in a shared 3D virtual space. This feature fosters a sense of presence and connectivity, despite physical distance.

4.10.2 Customer User Engagement

Furthermore, the metaverse provides businesses with novel avenues for advertising and marketing strategies. Through virtual spaces and communities within the metaverse, companies can deploy targeted advertisements and promotional campaigns tailored to specific user demographics and preferences. This personalized approach has the potential to yield higher conversion rates and increased customer satisfaction. As part of Coca Cola's 'Real Magic' campaign of 2021, the brand put up a 3D billboard at Guarulhos Airport in São Paulo where Johannes Vermeer's 'Girl with the pearl earring' artwork is revealed from behind a red curtain, leaning over to pick up a Coke bottle (Houston, 2023). The bottle is then shown cleverly going down a pipe, finding its way into a nearby vending machine, where the customer retrieves it.

Similarly, Frito-Lay's FIFA World Cup 2022 advertisement used the metaverse for a phygital experience. Customers participated in the Pass the Ball Challenge to win prizes, such as an NFT, by scanning a QR code on a snack bag. Users were prompted to take a selfie, which was then placed on a virtual football, after scanning the code. These examples highlight the diverse impact of the metaverse on businesses, ranging from customer engagement and remote collaboration to advertising opportunities and the emergence of new business models. It's worth noting that the metaverse is an evolving concept, and its impact on businesses continues to evolve as technologies and platforms develop further.

4.10.3 Legal Framework in the Metaverse

The legal aspects in the metaverse are an evolving area of interest. As the metaverse expands and becomes more integrated into our lives, several legal considerations arise. Protecting intellectual property like copyrights, trademarks, and patents related to virtual assets, virtual worlds, and virtual representations of real-world brands or objects is a significant concern. Clear guidelines and regulations are needed to address ownership, licensing, and enforcement of intellectual property rights within the metaverse.

Legal frameworks need to address issues related to the creation, ownership, transfer, taxation, and regulation of virtual assets and virtual currencies. This includes considerations for fraud prevention, money laundering, and ensuring fair and transparent virtual marketplaces. Just like in the physical world, instances of online harassment, cyberbullying, and hate speech can occur within the metaverse. Legal measures are required to combat such behaviors, protect users from harm, and establish consequences for perpetrators (Kazim Rizvi, 2022). Companies and platforms operating within the metaverse may also need to implement community guidelines and moderation policies to address these issues.

The metaverse transcends physical borders, which raises questions about jurisdiction and governance. Determining which legal jurisdiction applies to metaverse activities, resolving disputes, and enforcing regulations can be challenging. International cooperation and coordination are crucial for establishing consistent legal frameworks that apply to the metaverse. Data protection, self-regulation, and ethical principles are insufficient to limit the harms produced by technology. In this situation, governments must enact or revise data protection rules that restrict data collection. Governments should unambiguously classify this material as sensitive, strictly protected personal data under the law, even if it does not meet the high bar for classification.

4.11 CONCLUSION

According to research, people's "choices" to provide companies access to their personal information are frequently unconscious, subject to cognitive biases, and/or reversible. Authorities should require transparency and control over the gathered data as well as usage or disclosure regarding user behavior due to human limitations, dark patterns, legal gaps, and the complexity of existing data processing. People's rights must be safeguarded by accountable independent agencies, and data protection laws must be upheld. The control of the metaverse should not be monopolistic. Users should not feel compelled to use a specific platform to actively engage in society. To safeguard the variety of metaverse platforms and avoid monopolies over hardware and infrastructure, competition regulators must act.

The coexistence of technologies is essential for building a successful and safe metaverse. Each technology brings its unique capabilities and functionalities, contributing to enhanced immersion, advanced content creation, trust and security, seamless integration, and safety within the metaverse. Technologies such as AI and machine learning play a crucial role in generating and manipulating content within the metaverse. AI algorithms can facilitate real-time interactions, intelligent character behavior, and dynamic content creation. The coexistence of technologies enables seamless integration and interoperability within the metaverse. Different technologies need to work together to create a unified and interconnected virtual environment where users can move between different platforms, share content, and collaborate effortlessly. Interoperability allows for a consistent user experience and fosters the growth of the metaverse ecosystem.

The combination of technologies can also be used for ensuring safety and moderation within the metaverse. For instance, AI and machine learning algorithms can help detect and prevent harmful or inappropriate behavior, enabling proactive moderation and content filtering. These technologies can assist in creating a safe and inclusive environment for users, protecting them from harassment, hate speech, and other harmful activities. The landscape of digital governance and regulatory policy in India will certainly keep evolving as global best practices influence policymakers to make key decisions. The metaverse is the next big technological shift that is coupled with its own set of concerns; it would require planned regulation before a wide implementation. By leveraging the strengths of different technologies, developers and stakeholders can create a metaverse that offers engaging, secure, and immersive experiences for users.

REFERENCES

Adebayo, K. S. (2022, November 10). Why privacy and security are the biggest hurdles facing metaverse adoption. VentureBeat. https://venturebeat.com/virtual/why-privacy-and-security-are-the-biggest-hurdles-facing-metaverse-adoption/

Aijaz, S. (2022, April 12). How we can mitigate the potential threat to data privacy in the metaverse. VentureBeat. Retrieved June 9, 2022, from https://venturebeat.com/2022/04/12/how-we-can-mitigate-the-potential-threat-to-data-privacy-in-the-metaverse/

Bandura, A. (1986). *Social Cognitive Theory of Mass Communication* Taylor & Francis Online. https://www.tandfonline.com/doi/abs/10.1207/S1532785XMEP0303_03

Clark, M. (2021, October 28). Amid the fluff, Meta showed an impressive demo of its codec avatars. The Verge. Retrieved June 9, 2022, from https://www.theverge.com/2021/10/28/22751177/facebook-meta-codec-avatar-real-time-environment-rendering-neural-interface

Coles, T. (2022, June 1). Overcoming metaverse technology challenges: What to do before jumping in. ITPro Today: IT News, How-Tos, Trends, Case Studies, Career Tips, More. Retrieved June 9, 2022, from https://www.itprotoday.com/artificial-intelligence/overcoming-metaverse-technology-challenges-what-do-jumping

Dionisio, J. D., III, W. G., & Gilbert, R. (2013). 3D virtual worlds and the metaverse. *ACM Computing Surveys*, 45(3), 1–38. https://doi.org/10.1145/2480741.248075

Fallmann, D. (2022, May 4). Council post: Using Digital Twins and preparing for the metaverse. Forbes. Retrieved May 15, 2022, from https://www.forbes.com/sites/forbestechcouncil/2022/05/03/using-digital-twins-and-preparing-for-the-metaverse/?sh=36cce11029e2

Henz, P. (2022, October 6). The societal impact of the Metaverse - Discover Artificial Intelligence. SpringerLink. https://link.springer.com/article/10.1007/s44163-022-00032-6

Houston, A. (2023, March 24). Ad of the day: Coke makes an ad… about an ad. The Drum. https://www.thedrum.com/news/2023/03/24/ad-the-day-coke-makes-ad-about-ad

Jacoby, M., & Usländer, T. (2020, September 18). Digital Twin and internet of things-current standards landscape. MDPI. https://www.mdpi.com/2076-3417/10/18/6519

Kazim Rizvi, S. S. (2022, March 30). Legality of Metaverse in India & way forward for web3's sustainable evolution. TheQuint. Retrieved May 15, 2022, from https://www.thequint.com/voices/opinion/legality-of-metaverse-in-india-way-forward-for-web3s-sustainable-evolution#read-more

Kye, B., Han, N., Kim, E., Park, Y., & Jo, S. (2021). Educational applications of metaverse: possibilities and limitations. *Journal of Educational Evaluation for Health Professions*, 18, 32. https://doi.org/10.3352/jeehp.2021.18.32

Maples-Keller, J. L., Bunnell, B. E., Kim, S.-J., & Rothbaum, B. O. (2017). The use of virtual reality technology in the treatment of anxiety and other psychiatric disorders. *Harvard Review of Psychiatry*, 25(3), 103–113. https://doi.org/10.1097/hrp.0000000000000138

McDowell, M. (2021, May 17). Inside gucci and Roblox's new Virtual World. Vogue Business. https://www.voguebusiness.com/technology/inside-gucci-and-robloxs-new-virtual-world

Pichai, S. (2019, May 8). Google's Sundar Pichai: Privacy should not be a luxury good. The New York Times. https://www.nytimes.com/2019/05/07/opinion/google-sundar-pichai-privacy.html

Press, S. U. (n.d.). A theory of Cognitive Dissonance - Leon Festinger. Stanford University Press Home Page… https://www.sup.org/books/title/?id=3850

PricewaterhouseCoopers. (n.d.). 70% of businesses plan to integrate the metaverse into their organisational activities. PwC. https://www.pwc.in/press-releases/2023/70-percentage-of-businesses-plan-to-integrate-the-metaverse-into-their-organisational-activities.html

Ramprakash Ramamoorthy, Z. C. (2022, September 26). Why the metaverse is filled with security, privacy and Safety Issues. VentureBeat. https://venturebeat.com/security/why-the-metaverse-is-filled-with-security-privacy-and-safety-issues/

Report state laws related to Digital Privacy. National Conference of State Legislatures. (n.d.). https://www.ncsl.org/technology-and-communication/state-laws-related-to-digital-privacy#:~:text=Five%20states%E2%80%94California%2C%20Colorado%2C,of%20personal%20information%2C%20among%20others

Uberti, D. (2022, January 4). Come the metaverse, can privacy exist? The Wall Street Journal. Retrieved June 9, 2022, from https://www.wsj.com/articles/come-the-metaverse-can-privacy-exist-11641292206

Vainilavičius, J. (2022, October 3). Metaverse expansion puts virtual-reality addiction into focus. https://cybernews.com/editorial/metaverse-puts-virtual-reality-addiction-into-focus/

5 Intelligent Optical Networks

Challenges, Opportunities, and Applications

R. K. Jeyachitra
National Institute of Technology, Tiruchirappalli, India

S. Manochandar
CARE College of Engineering, Tiruchirappalli, India

M. Pradeep Doss
National Institute of Technology, Tiruchirappalli, India

5.1 INTRODUCTION

In the digital era, all day-to-day activities like education, business, medicine, entertainment, shopping, consumer, finance, etc., are connected to Information and Communication Technology (ICT). All these activities communicate directly or indirectly, but they must ensure full connection among them and hence share their data without any interruptions. Access and backbone networks perform these kinds of activities and make them completely connected. A huge amount of data is available and involved from these activities, and it should be shared among them whenever it is needed. This information is carried by the backbone network, and it links the access networks to provide stable connectivity among the users and user devices, clouds, etc. [1]. The optical network is one of the most powerful backbone networks in the telecommunication field since it provides high-speed, wide-bandwidth, long-distance transmission capability and security. AI can be used in optical networking in various applications such as optical network planning, connection establishment, network reconfiguration through virtual topologies, software-defined networking, optical burst switching, passive optical networks, and intra-data center networking. Various optimization algorithms can be used in the optical network to optimize the various process involved in the network, such as resource allocation, connection establishment, regenerator placement, network reconfiguration, fault detection, placement of Optical Network Unit (ONU), etc. The organization of this chapter includes an overview of ION, Challenges and Opportunities in ION, applications in ION, and various simulation tools.

DOI: 10.1201/9781003303114-5

5.2 OVERVIEW OF INTELLIGENT OPTICAL NETWORKS

There is an exponential growth in the data traffic in communication technologies and networks because of very rapid development in cloud computing, 5G/6G, Internet of everything, and network services such as cloud/edge networking, ultra-high video streaming, Augmented Reality/Virtual Reality (AR/VR), and Network Functions Virtualization (NFV) [2]. In the Cisco annual report [3], it was disclosed that the number of Internet users will grow from 3.9 billion in 2018 to 5.3 billion in 2023. The Nokia company has recently reported a significant growth in network traffic [4]; a future optical access network is the key enabler to address this capacity crunch. These rapid advancements and developments have created new challenges in traditional optical networks [5].

Thus, it is crucial to introduce automation and intelligence into communication networks [6] so as to handle a huge volume of situations and hence to guarantee the Quality of Service (QoS). Also, the next-generation fiber optic communication networks should be robust, spectral, and energy-efficient. Since the future optical network needs to be flexible, programmable, dynamic, and reconfigurable, the intermediate optical node's current state has to be learned, and this information will be used to predict its future. The ML technique helps in this process and adds intelligence to the optical network. An ION is a new and advanced network technology that replaces traditional optical network technology by introducing "intelligence" into the network so that it can provide the fast flexible establishment of connections, automatically detecting failures, and based on that, it can detect resources automatically. In addition to intelligent management of traffic in the network, it can provide other intelligent functions. Figure 5.1 depicts the characteristics of ION.

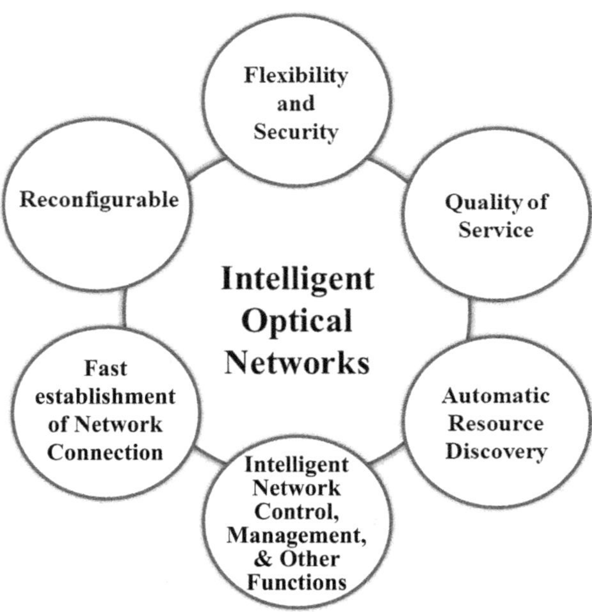

FIGURE 5.1 Characteristics of ION.

Intelligent networks can monitor intermediate nodes more efficiently. It must monitor multiple parameters simultaneously with great efficiency. It can be incorporated with ML techniques so that the system computation becomes easier and predicts more accurately. Intelligent monitoring in the optical network means monitoring the performance of physical layer parameters and signals, QoT estimation of light paths, and intelligent failure management of the optical networks. In addition, the optical network transforms into an ION because it has the characteristics of reconfigurability and flexibility. Also, the overall utilization of the network requires accurate optical signal quality monitoring for further improvements. The conceptual diagram of ION is represented in Figure 5.2.

The process of simulating human intelligence by machines acting more efficiently than humans is Artificial Intelligence. AI systems require training data to extract the patterns and correlations between the data, and based on that, makes predictions about future states or classifies the new data samples. Learning, reasoning, and self-correction are three cognitive skills in which AI can perform tasks better than humans. The main application of AI includes expertise in automated systems, natural language processing, and speech recognition and is not limited [7]. More consistent and accurate results are the essential features of AI.

Machine learning is a branch of AI that mainly uses data and algorithms in the same way humans learn, gradually improving accuracy. ML uses mathematical schemes and huge data processing to build procedures that find relationships between input and output data. It covers everything from "general intelligence," or the ability of a machine to think and act in the same way a human would, to specialized, task-oriented intelligence, which is where ML falls on the spectrum. The ML development process collects data and labels it before it is fed into a model and follows the Model Development, Training, Testing, and Refining, as in Figure 5.3.

Data collection and labeling is a time-consuming process. At the same time, there are several innovations in the ML space that leverage pre-trained models to offset some of the work and emerging tools to streamline data collection from real systems. After data collection, the next steps are model development, training, testing, and refinement. This phase is where a data scientist or engineer creates a program that ingests the mass of collected input data and transforms it into expected outputs using one or more approaches. Once the model is ready, it is deployed and available for real-time prediction against new data. In traditional ML, the model is deployed to a cloud service so that it can be called by a running application that provides the needed inputs and receives an output from the model.

5.2.1 Evolution of Optical Network Transmission Technology

Now we are in the digital technology world, including advancements in communication and networking. The history of optical networks starts with first-generation optical networks, including Synchronous Optical Networks (SONET) and Synchronous Digital Hierarchy (SDH). Then move on to the second-generation optical network that supports high data rates, namely Wavelength Division Multiplexing (WDM). To meet the demands of data traffic in the network, integrated routing and signaling for optical path and automation in the network discovered the intelligent networks

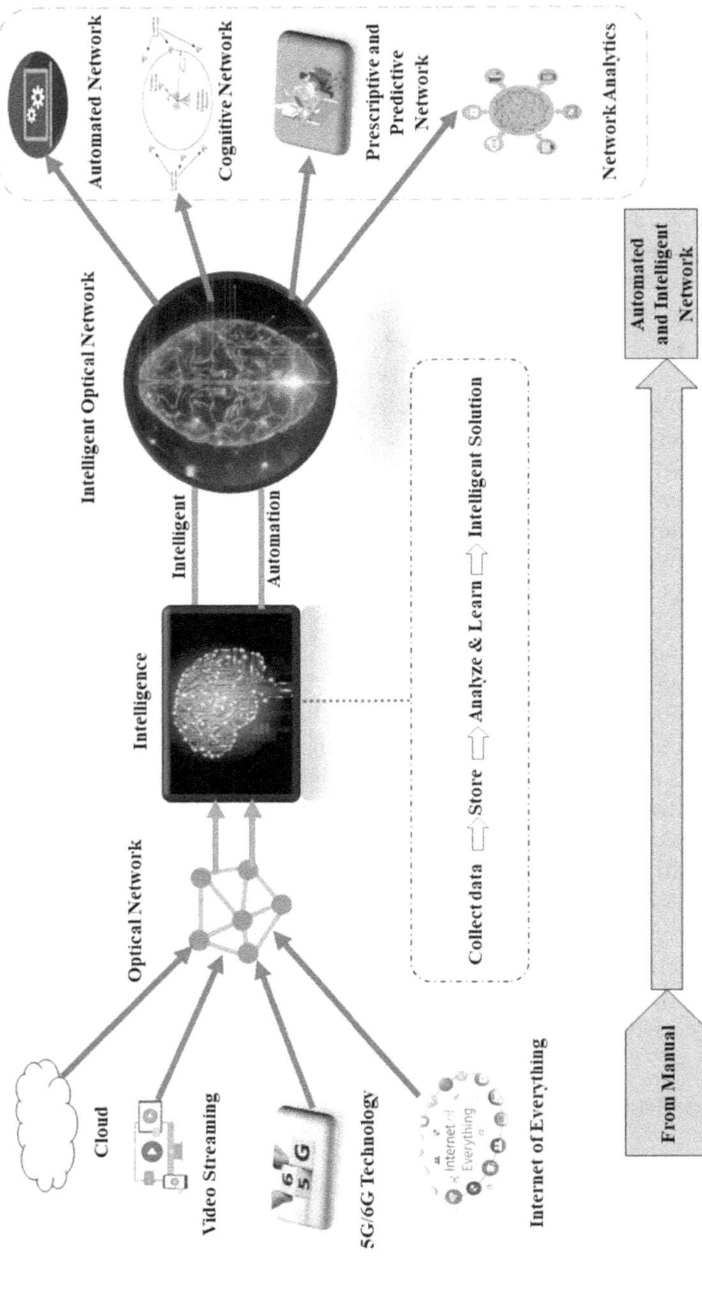

FIGURE 5.2 Conceptualized diagram of ION.

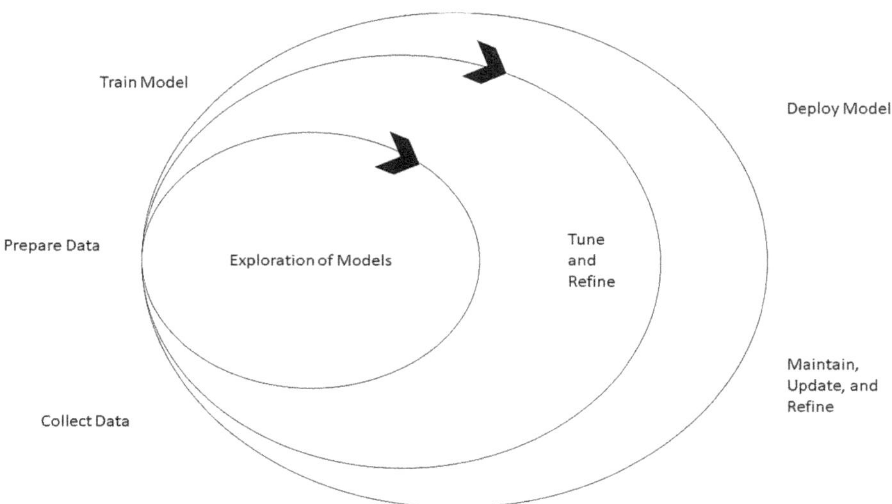

FIGURE 5.3 Different phases in machine learning.

as the third-generation optical networks. Then integration of intelligent, flexible, and programmable optical networks created the Elastic Optical Network (EON) and the Software Defined Optical Network (SDON) that supports 5G/6G communications [8]. The evolution of optical networks is shown in Figure 5.4.

According to the ZTE white paper "Beyond 400G Technical" in 2021, the B400G system control plane with various flexible grid technology and Software-Defined Network (SDN) technology will be evolving in ION for future networks. The evolution of autonomous networks is a promising milestone for the control and management of future optical networks. Innovation in multiple aspects is expected by using AI-assisted automated network operations.

5.2.2 Optical Network Functionality

A **point-to-point WDM scheme** used in long-range transmission can be implemented using an optical add-drop multiplexer. It has increased channel speeds, good signal integrity, reliability, and quick path restoration. Optical amplifiers, lasers, fibers, optical multiplexers, and demultiplexers are components in a WDM point-to-point system. In a point-to-point WDM network, optical signal to electrical signal and vice versa conversions are required [10].

In **Wavelength-Routed Networks (WRN)**, the routing at the network nodes depends on the wavelength of the incoming signal. The optical connection between a node pair is called a light path, and the optimization of a WRN includes a feasible route and wavelength for each light path. It helps in fast restoration and changes the underlying network without using any reconfiguration of the upper layers [10]. **Optical Burst Switching** comes in between the WDM and optical packet switching. In this network, the transmission takes place by optical bursts containing multiple

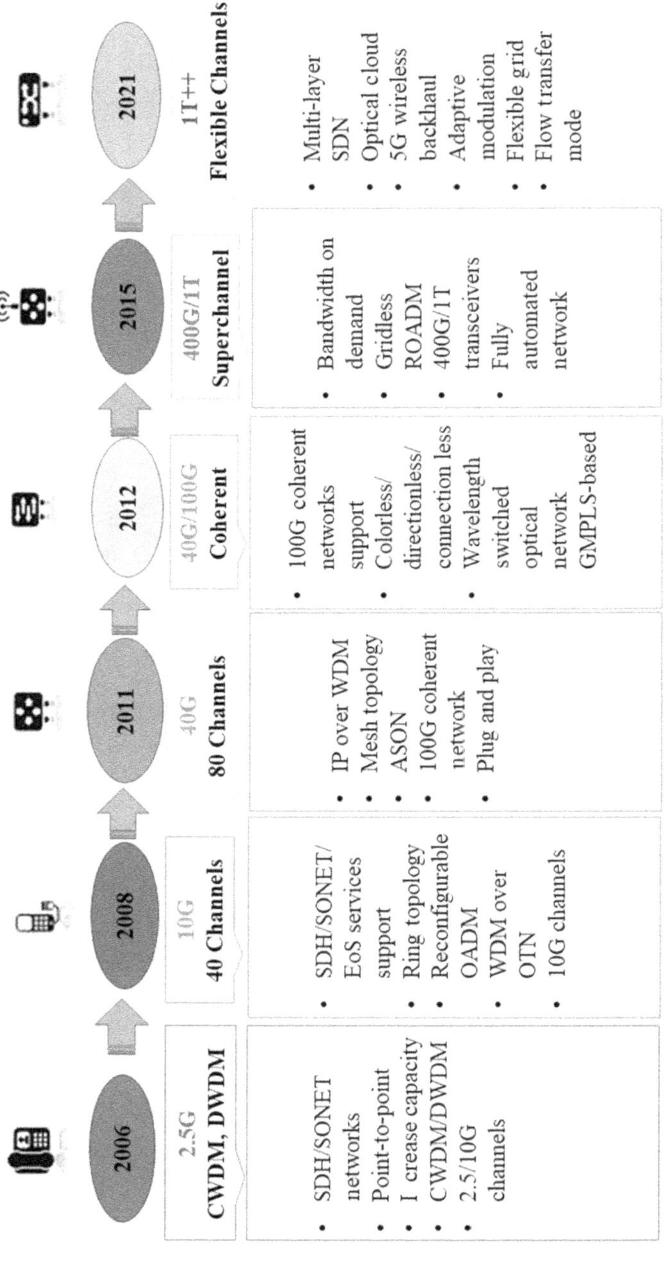

FIGURE 5.4 Evolution of optical networks [9].

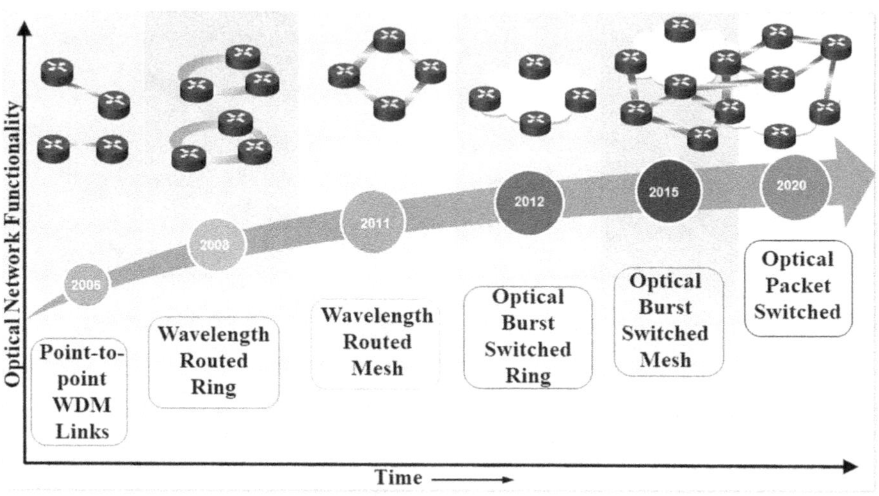

FIGURE 5.5 The optical networking functionality road map [10].

Internet Protocol (IP) going to the same reception [10]. **Optical Packet Switching** provides more flexibility than optical burst switching, so it is developed to be used in Local Area Networks (LAN). It has two forms: one is slotted, and another is unslotted. In slotted, every packet has the same or fixed length, and it operates synchronously. In unslotted, it is not necessary to have constant length. They can vary, and it operates asynchronously [10]. Figure 5.5 shows the road map of optical networking functionality.

5.2.3 Categories of Optical Networks

The **Active Optical Network (AON)** is a Point-to-Point (PTP) network structure, implying that each user has a fixed fiber optical line, and it is ended on an optical detector. In this system, electrical devices are generally deployed, such as switches or routers, managing the signal distribution and routing of data to the destination [11]. The following Figure 5.6 shows an AON system.

The **Passive Optical Network (PON)** is a Point-to-Multipoint (PMP) network structure. Optical fiber and passive components like optical splitters are used instead of active optical components such as optical amplifiers, optical routers, or shaping circuits. In a passive optical system, a single fiber connects a central optical line terminal to ONUs at the user end [11]. The following Figure 5.7 shows a PON system.

APON (ATM Passive Optical Networks) is the first version of PON based on Asynchronous Transfer Mode (ATM). The working of APON for transferring data is PMP for the downstream direction and multi-point to-point system for the upstream direction. **BPON (Broadband PON)** is the improved version of APON. It supports WDM, dynamic and higher data transmission capacity, and survivability. BPON is,

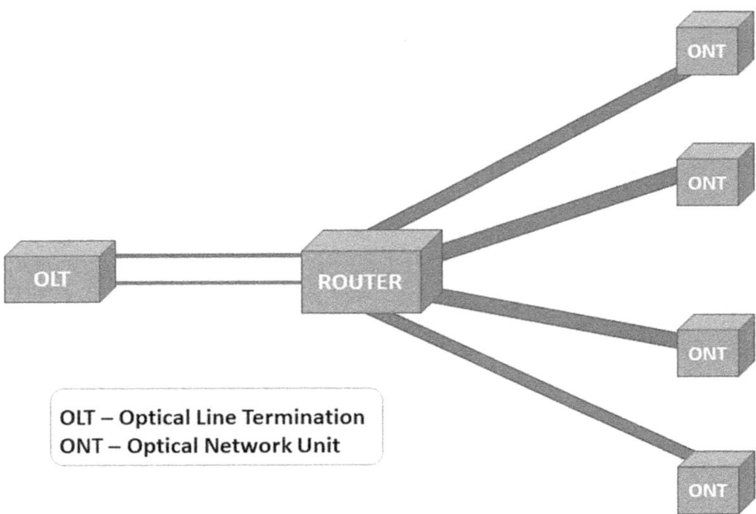

FIGURE 5.6 Active optical network.

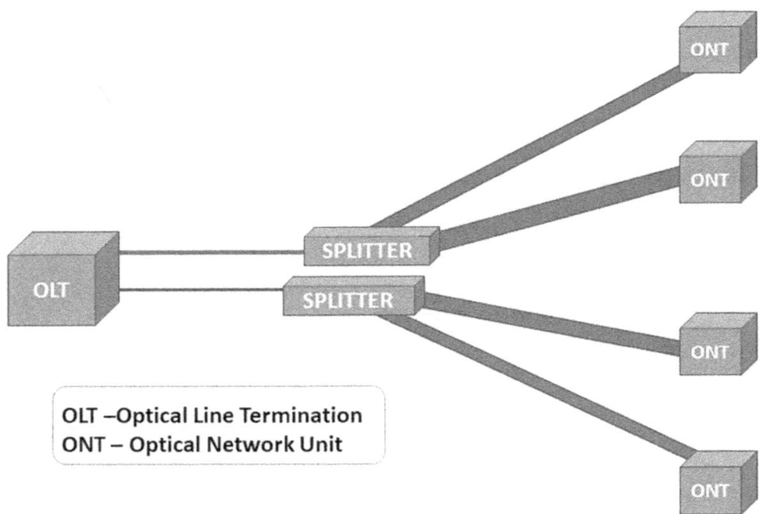

FIGURE 5.7 Passive optical network.

by and large, presented at 622 Mbps downstream and 155 Mbps upstream. Notwithstanding, its ATM design and transfer speed limit make it unsuitable for video transmission. BPON organizations will be switched over completely to EPON or GPON over the long haul [12].

EPON (Ethernet PON) is a popular version of PON, and it replaces the ATM cells from ethernet packets. Ethernet PON utilizes a two-layer network that utilizes IP to convey information, data, voice, and video. EPON withstands 1 Gbps symmetrical data transmission, which is why it is extremely well-known in present-day networks [12]. **GE-PON (for Gigabit Ethernet PON)** is yet to be developed; however, it needs different conventions through interpretation to help the local Generic Encapsulation Method (GEM) transport layer. This imitation upholds ATM, ethernet, and WDM conventions. It works on timing and brings down expenses by utilizing even 2.5 Gbps information streams. The intricacy is lower, and the cost is not exactly as GPON [12]. **GPON (Gigabit Ethernet PON)** is a 3-layer two network: ATM for voice, Ethernet for information, and Proprietary Encapsulation for voice. It provides 1.25 Gbps or 2.5 Gbps for downstream and upstream data transmissions versatile from 155 Mbps to 2.5 Gbps [12].

The **EON** is a kind of optical network where different information signals go through optical fiber channels by regulating them with optical signs (carrier signals). EON is an application of WDM that operates using very small spacings between transmittable frequencies, usually 12.5 GHz (or 0.2nm separation). It follows the sliceable bandwidth variable architecture, which consists of a bandwidth variable transponder and bandwidth variable cross-connects [13]. The EON model is shown in Figure 5.8.

The optical network coordinated with SDN, to be a specific SDON, can make available high transmission capacity with software assets for various occupants [14]. It isolates the controlling plane from the information plane and transforms the control way into a united and adaptable one. SDON is extending optical network knowledge—it addresses the way the control board of optical networks changes from changing knowledge to an exhaustive programmed one, where a variety of services and the executives' management are likewise [15]. The schematic diagram for SDON is represented in Figure 5.9. The recent development on SDON, namely the ORCHESTRA European project, uses software-defined optical performance monitors to improve the

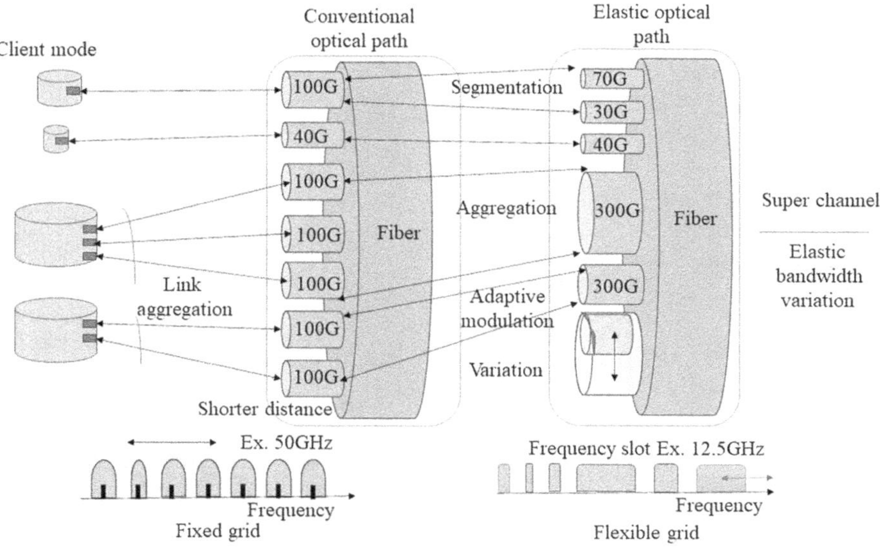

FIGURE 5.8 Network architecture of EON.

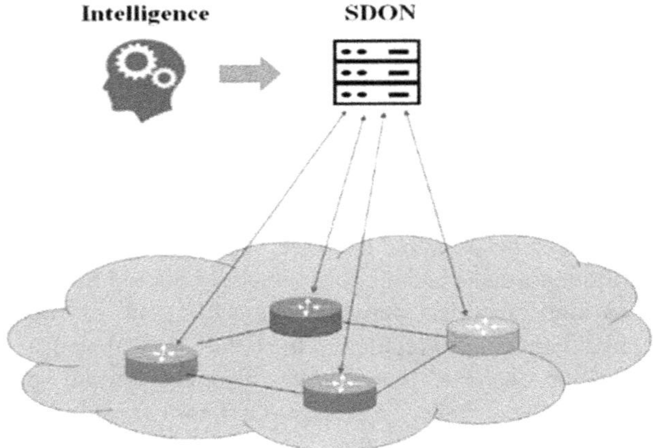

FIGURE 5.9 Schematic diagram of SDON.

observability of optical networks by using data analytics and understands the physical-layer conditions using feed cross-layer optimization algorithms. It develops automation mechanisms to achieve efficient and reliable optical networks [16].

5.2.4 Classification of Machine Learning

ML is widely classified into three subdivisions: Supervised, Unsupervised, and Reinforcement Learning, and it is shown in Figure 5.10.

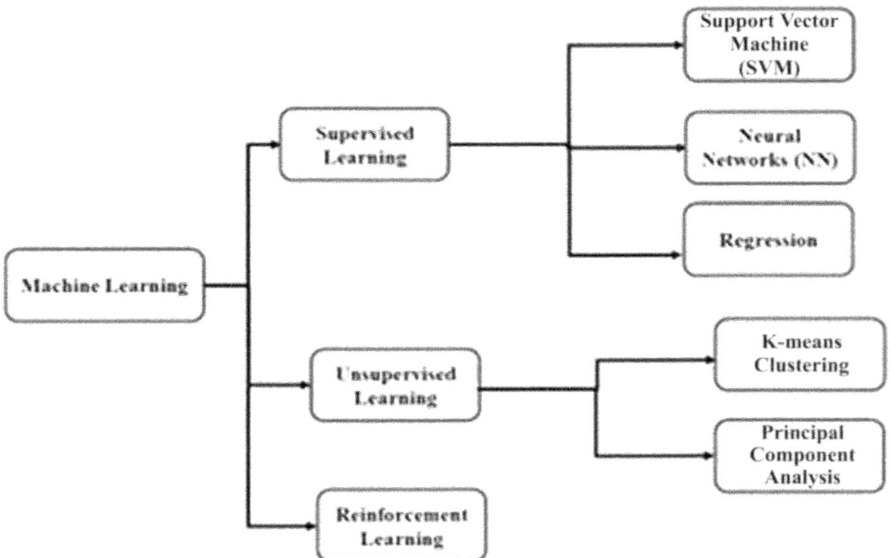

FIGURE 5.10 Classification of machine learning.

5.2.4.1 Supervised Learning

Supervised learning executes the expected output data when the input is given with a class label. In the classification problem, the objective of the model correctly predicts whether the class should be a binary or multi-class. In the regression problem, the model is used for forecasting the numerical value.

Support Vector Machine (SVM) is a type of supervised ML algorithm that focuses on solving classification problems. Support vectors are data points lying on the hyperplane. The classification is performed based on the support vectors [17]. The primal objective function to determine the parameters of SVM is given in equations 5.1a, 5.1b, and 5.1c, respectively.

$$\min \frac{1}{2}\|w\|^2 \tag{5.1a}$$

$$s.t. y_i(w^T x_i + b) \geq 1 \tag{5.1b}$$

$$L = \frac{1}{2}\|w\|^2 + \sum_{i=1}^{m} \alpha_i [1 - y_i \left(w^T x_i + b\right)] \tag{5.1c}$$

Where x_i is the input data, y_i is the class label, α_i is the Lagrangian multiplier, w and b are the unknown parameters to be estimated by solving the primal problem.

In primal problems, the number of constraints increases as the data points increase. To resolve this, the dual form to solve the SVM [18] is given below:

$$\max_{(\alpha_1,\ldots,\alpha_m)} \sum_{k=1}^{m} \alpha_i - \frac{1}{2} \sum_{i=1}^{m} \sum_{j=1}^{m} \alpha_i \alpha_j y_i y_j x_i^T x_j \tag{5.2a}$$

$$\sum_{i=1}^{m} \alpha_i \, y_i = 0 \quad \text{where } \alpha_i \geq 0, \forall_i = 1,2,\ldots,m \tag{5.2b}$$

Sequential Minimal Optimization (SMO) technique is used to solve Convex quadratic programming [19]. After identifying the support vectors and other parameter values, the decision function can be computed as

$$f(x) = \text{sign}\left\{w^T x + b\right\} = \text{sign}\left\{\sum_{i=1}^{m} \alpha_i y_i x_i^T x + b\right\} \tag{5.3}$$

The kernel function $K(x_i, x_j)$, where $\phi(x)$ is the new feature space is represented as

$$K(x_i, x_j) = \langle \phi(x_i), \phi(x_j) \rangle = \phi(x_i)^T \phi(x_j) \tag{5.4}$$

TABLE 5.1
Types of Kernel Function Used in SVM and Their Corresponding Expressions

Kernels	Functions
Polynomial	$K(x_i, x_j) = (x_i^T x_j + 1)^d$
Gaussian	$K(x_i, x_j) = \exp\left(-\frac{\lVert x_i - x_j \rVert^2}{2\sigma^2}\right)$
Laplacian	$K(x_i, x_j) = \exp\left(-\frac{\lVert x_i - x_j \rVert}{\sigma}\right)$
Rational Quadratic	$K(x_i, x_j) = 1 - \frac{\lVert x_i - x_j \rVert^2}{\lVert x_i - x_j \rVert^2 + c}$
Multiquadratic	$K(x_i, x_j) = \sqrt{\lVert x_i - x_j \rVert^2 + c}$
Wave	$K(x_i, x_j) = \frac{\theta}{\lVert x_i - x_j \rVert} \sin \frac{\lVert x_i - x_j \rVert}{\theta}$
Power	$K(x_i, x_j) = -\lVert x_i - x_j \rVert^d$
Log	$K(x_i, x_j) = -\log(\lVert x_i - x_j \rVert^d + 1)$
Bessel	$K(x_i, x_j) = \frac{J_{v+1}(\sigma \lVert x_i - x_j \rVert)}{\lVert x_i - x_j \rVert^{-n(v+1)}}$
Cauchy	$K(x_i, x_j) = \frac{1}{1 + \frac{\lVert x_i - x_j \rVert^d}{d}}$
Wavelet	$K(x_i, x_j) = \prod_{m=1}^{N} h\left(\frac{x_i - c}{a}\right)\left(\frac{x_j - c}{a}\right)$

Where d, σ, c, a, and θ are user-defined hyperparameters. Types of kernel function used in SVM and their corresponding expressions are listed in Table 5.1. The SVM classifier is written as

$$f(x) = w^T \varphi(x) + b = \sum_{i=1}^{m} \alpha_i y_i K(x_i, x_j) + \mathrm{b} \tag{5.5}$$

Neural Networks (NN) play a vital role in the field of data analytics. NN consists of a series of nodes, which behaves like a human brain of neurons. This provides the best solution for image processing, speech recognition, language processing, text mining, and web analytics. The most common terms used in NN are input nodes or input layer, where the data tuples are fed in this layer so that each node represents the variables used for the analysis. In this layer, no computation is required. The hidden layer is an intermediate layer connecting the input layers and output layers. It performs the computation and transfers the weights (W_{ij}) from the input layer to the following layer. The activation function generates a desired output in the output nodes or the output layer based on the weights and bias.

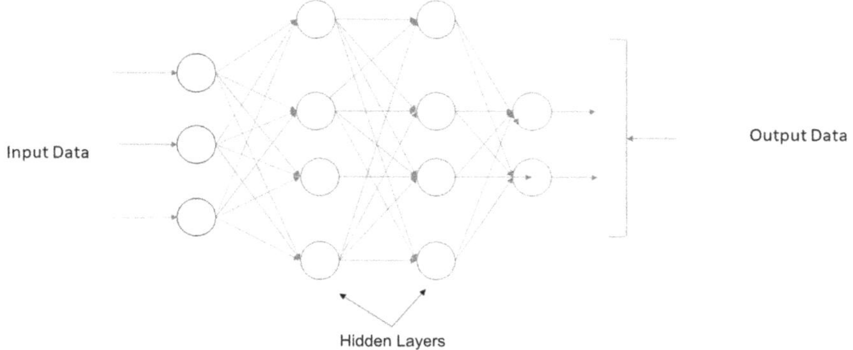

FIGURE 5.11 Architecture of neural network.

The network consists of connections; each connection transferring the output *i* neuron is connected to the input *j* neuron called weights. In the same vein, this sense *i* is the predecessor of *j* and *j* is the successor of *i*; each connection is assigned a weight W_{ij}. The activation function or transfer function is used to get the desired output from the given set of inputs. The common architecture of the NN is shown in Figure 5.11.

The various types of neural networks are Feedforward Neural Networks (FNN), Single layer perceptron's, multi-layer perceptron's, Convolutional Neural Networks (CNN), and Recurrent Neural Networks (RNN). Table 5.2 shows the commonly used activation functions.

TABLE 5.2
Commonly Used Activation Functions

Function	Equation	Derivatives
Identity	$f(x) = x$	$f'^{(x)} = x$
Binary Step	$f(x)=\begin{cases}0, & for\ x<0\\ 1, & for\ x\geq 0\end{cases}$	$f'^{(x)}=\begin{cases}0, & for\ x\neq 0\\ 1, & for\ x=0\end{cases}$
Logistic	$f(x)=\frac{1}{1+e^{-x}}$	$f'^{(x)} = f(x)(1 - f(x))$
TanH	$f(x)=\tanh(x)=\frac{2}{1+e^{-2x}}-1$	$f'^{(x)} = 1 - f(x)^2$
ArcTan	$f(x) = \tan^{-1}x$	$f'^{(x)}=\frac{1}{x^2+1}$
Rectified Linear Unit (ReLU)	$f(x)=\begin{cases}0, & for\ x<0\\ x, & for\ x\geq 0\end{cases}$	$f'^{(x)}=\begin{cases}0, & for\ x<0\\ 1, & for\ x\geq 0\end{cases}$
Parametric Rectified Linear Unit (PReLU)	$f(x)=\begin{cases}\alpha x, & for\ x<0\\ x, & for\ x\geq 0\end{cases}$	$f'^{(x)}=\begin{cases}\alpha, & for\ x<0\\ 1, & for\ x\geq 0\end{cases}$
Exponential Linear Unit (ELU)	$f(x)=\begin{cases}\alpha(e^{-x}-1), & for\ x<0\\ 1, & for\ x\geq 0\end{cases}$	$f'^{(x)}=\begin{cases}f(x)+\alpha, & for\ x<0\\ 1, & for\ x\geq 0\end{cases}$
SoftPlus	$f(x) = \log_e(1 + e^x)$	$f'^{(x)}=\frac{1}{1+e^{-x}}$

5.2.4.2 Unsupervised Learning

This technique builds the model with unlabeled input data. While compared to supervised learning, human intervention is less. It is commonly used for exploratory data analysis, which automatically recognizes data structure.

K-Means Clustering analysis tries to group the data based on similarity. Clustering algorithms are widely classified into partitioned, hierarchical, density-based, and grid-based methods. In the partitioning algorithm, k-means clustering is one of the most widely used methods [20]. The k-means algorithm is as follows:

ALGORITHM: K-MEANS ALGORITHM

1: Specify the number k of the cluster to assign.
2: Randomly initialize k centroids.
3: **repeat**
4: **expectation**: Assign each point to its closest centroid.
5: **maximization**: Complete the new centroid (mean) of each cluster until the centroid positions do not change.

Principal Components Analysis (PCA) is most used for dimension reduction or feature extraction techniques. PCA is one of the important processes in the pre-processing of data, and the main objective of the PCA is to convert the data from higher dimensional into lower dimensional simultaneously without losing more information [21]. PCA requires variance-covariance matrix or correlation matrix as an input. The principal components are extracted based on the eigen values and eigen vector of the input matrix. **Regression Analysis** is used for prediction or forecasting by establishing the relationship between the dependent and independent variables. The common applications for regression analysis are as follows: Stock price prediction, weather forecasting, sales forecast, and so on. **Support Vector Regression (SVR)** is the branch of SVM used for the prediction process. Generally, the Regression models minimize the error, i.e., the deviation between the actual values and predicted value, but SVR tries to construct the best-fitted line that lies within the threshold value. The distance between the hyperplane and the boundary line is the threshold value. The dual optimization problem [22] is represented as follows:

$$maximize \begin{cases} -\frac{1}{2}\sum_{i,j=1}^{l}\left(\alpha_i-\alpha_i^*\right)\left(\alpha_j-\alpha_j^*\right)\left(x_i,x_j\right) \\ -\varepsilon\sum_{i=1}^{l}\left(\alpha_i+\alpha_i^*\right)+\sum_{i=1}^{l}y_i\left(\alpha_i-\alpha_i^*\right) \end{cases} \tag{5.6a}$$

$$\sum_{i=1}^{l}\left(\alpha_i-\alpha_i^*\right)=0 \text{ and } \alpha_i,\alpha_i^* \in \left[0,C\right] \tag{5.6b}$$

Where ε is the threshold value. The expression for calculating w and predicted values are as follows:

$$w = \sum_{i=1}^{l} \left(\alpha_i - \alpha_i^*\right) x_i \tag{5.7a}$$

$$f(x) = \sum_{i=1}^{l} \left(\alpha_i - \alpha_i^*\right)\left(x_i, x\right) + b \tag{5.7b}$$

For non-linear problems, the objective function and constraints are as follows:

$$maximize \begin{cases} -\frac{1}{2}\sum_{i,j=1}^{l} \left(\alpha_i - \alpha_i^*\right)\left(\alpha_j - \alpha_j^*\right) k\left(x_i, x_j\right) \\ -\varepsilon \sum_{i=1}^{l} \left(\alpha_i + \alpha_i^*\right) + \sum_{i=1}^{l} y_i \left(\alpha_i - \alpha_i^*\right) \end{cases} \tag{5.8a}$$

$$\sum_{i=1}^{l} \left(\alpha_i - \alpha_i^*\right) = 0 \text{ and } \alpha_i, \alpha_i^* \in \left[0, C\right] \tag{5.8b}$$

$$w = \sum_{i=1}^{l} \left(\alpha_i - \alpha_i^*\right) \varphi\left(x_i\right) \tag{5.8c}$$

$$f(x) = \sum_{i=1}^{l} \left(\alpha_i - \alpha_i^*\right) k\left(x_i, x_j\right) + b \tag{5.8d}$$

where $k(x_i, x_j)$ is the kernel functions and $f(x)$ is used to determine the predicted values.

In **Time Series Analysis**, data are collected at different time periods on the same variable. In time series exponential smoothing, Auto-Regressive (AR), Moving Average (MA), and Auto-Regressive Moving Average (ARMA) are represented in equations 5.9, 5.10, 5.11, respectively. Auto-Regressive Integrated Moving Average (ARIMA) models are the most widely used models for time series forecasting. In general, exponential smoothing models are used to explore the trend and seasonality characteristics in the time series data, while the autocorrelation in the data is described by the ARIMA model.

$$Y_t = \beta_1 * y_{t-1} + \beta_2 * y_{t-2} + \beta_3 * y_{t-3} + \ldots + \beta_p * y_{t-p} \tag{5.9}$$

$$Y_t = \alpha_1 * \varepsilon_{t-1} + \alpha_2 * \varepsilon_{t-2} + \alpha_3 * \varepsilon_{t-3} + \ldots + \alpha_q * \varepsilon_{t-q} \tag{5.10}$$

$$Y_t = \beta_1 * y_{t-1} + \alpha_1 * \varepsilon_{t-1} + \beta_2 * y_{t-2} + \alpha_2 * \varepsilon_{t-2} + \beta_3 * y_{t-3} + \alpha_3 * \varepsilon_{t-3} + \ldots + \beta_p * y_{t-p} + \alpha_q * \varepsilon_{t-q} \tag{5.11}$$

Where α_i, $i = 1, 2, ...p$, and β_j, $j = 1, 2,.....q$ are coefficients.

5.2.4.3 Reinforcement Learning

This model is mainly used for decision-making process-based reward and punishing systems and depicted in Figure 5.12. Its learning process highly relies upon interaction with the environment. For repeated action, the intelligent agents provide a reward, similarly for the refrain from actions that are punished. In other terms, it is a trial-and-error method to obtain the best outcome through experience.

Where T is the set of iteration times, T includes a sequence of discrete time steps and at each time step t, the agent completes an iteration with the environment. S is the finite set that includes the possible states of the environment. A is the finite set, $A(s_t)$ is the set of actions available in state s_t. $P_t(s_{t+1}/s_t, a_t)$ denotes the state transition probability that environment transfers from s_t to s_{t+1} under action $a_t \in A(s_t)$. pi is the policy that maps from S and action A, and $\pi(s, a)$ denotes the probability of acting under state s.

The hidden Markov Model (HMM) is one of the statistical models that is highly used in ML. It depends on a probabilistic model that tries to explore the probabilistic characteristic of any random process. The basics of HMM is a Markov chain process where the observing or learning process is carried out from its hidden states [23]. Some of the applications of HMM are speech synthesis, speech recognition, part-of-speech tagging, handwriting recognition, machine translation, activity recognition, forecasting, sequence classification, and text analytics.

Genetic Algorithm (GA) is one of the most evolutionary algorithms, which helps to solve both constrained and unconstrained optimization problems. GA's main advantages are good global searching ability, learning from the closest optimum solution. GA is one of the most powerful tools for optimization and ML [24]. Five phases in a GA are initial population, fitness function, selection of best chromosome, crossover, and mutation. The flow chart for the GA is shown in Figure 5.13.

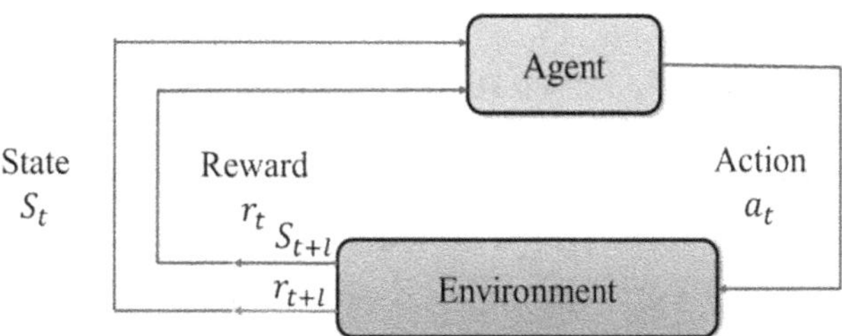

FIGURE 5.12 Conceptual diagram of Reinforcement Learning.

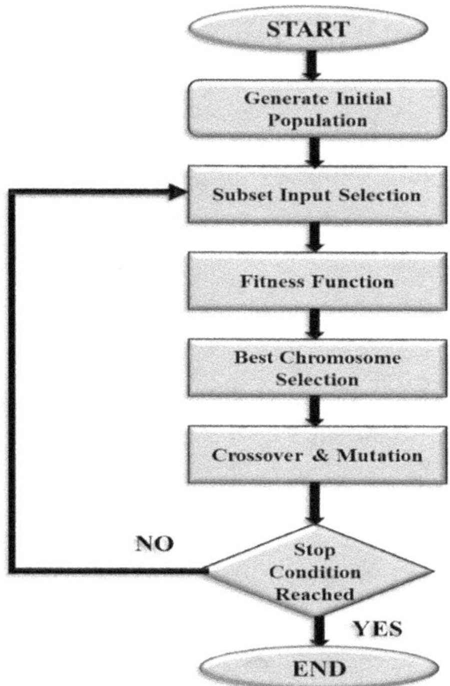

FIGURE 5.13 Genetic algorithm flowchart.

Deep Learning is a subfield of ML, which is mainly inspired by the structure and function of human brain activities with the base of Artificial Neural Networks (ANN). The classification of DL is shown in Figure 5.14. DL systems require powerful hardware because they require huge amounts of data and simultaneously solve more complex mathematical calculations for obtaining a predictive model.

5.2.5 Techniques for Generation of Data and Modeling Images

Generative Adversarial Networks (GANs) use unsupervised learning to train two different models parallelly. The network is forced to perform efficient representation of training data, making it effective for generating data similar to the training data. **Variational Autoencoders (VAEs)** have a specific encoding/decoding scheme, so that the generated image and the original image are compared. Thus, they have the ability to code for specific types of images producing the generated images.

Recurrent Neural Networks (RNN) is the family of NN mainly used for processing sequence data such as sentences, text data, and audio. RNN is different from other feed-forward networks because they send information over time steps. **Convolutional Neural Networks (CNN)** learn higher-order

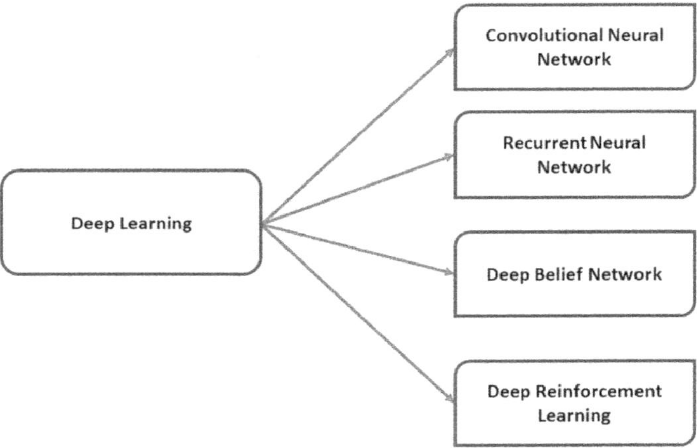

FIGURE 5.14 Classification of deep learning.

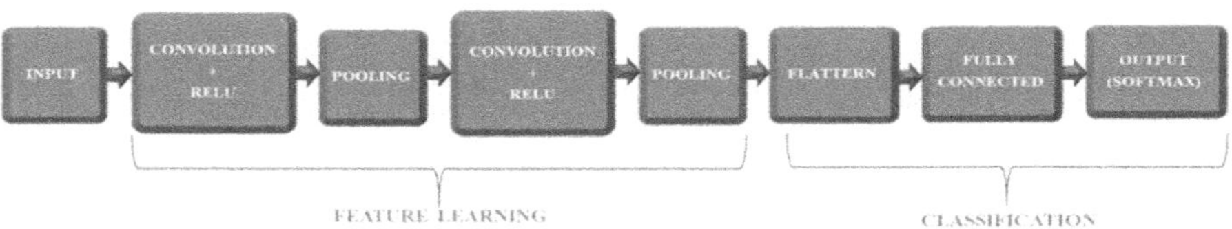

FIGURE 5.15 CNN flow diagram.

features in the data through convolutions. CNN are well suited for object recognition and accurate image classification and are shown in Figure 5.15. **Deep Belief Networks (DBNs)** consist of layers of Restricted Boltzmann Machines (RBMs). A feed-forward network is used for the training and fine-tuning phases.

5.2.6 Metrics-Classification Metrics, Regression Metrics, and Rand and Jaccard Indices

A confusion matrix is a tabular visualization of the actual class label versus the predicted class label. The total number of instances in the actual class is represented in each row of the confusion matrix, and the total number of instances in the predicted class is represented in each column. The confusion matrix for binary classification is presented in Table 5.3.

Root Mean Squared Error (RMSE) corresponds to the square root of the Mean Square Error (MSE). It addresses a few downsides of MSE and retains the MSE characteristics. It also holds the penalization of smaller errors done by MSE using square root [25].

TABLE 5.3
Confusion Matrix for Binary Classification

Actual \ Predicted	Positive	Negative
Positive	True Positive (TP)	False Negative (FN)
Negative	False Positive (FP)	True Negative (TN)

TABLE 5.4
Evaluation Metrics for Classification

Metrics	Equation	Objective
Classification Accuracy	$Accuracy = \frac{TP + TN}{TP + FP + TN + FN}$	To evaluate the performance of the supervised learning algorithm
Precision	$Precision = \frac{TP}{TP + FP}$	To measure the goodness of the ML model in terms of predicting
Recall/True Positive Rate	$Recall = \frac{TP}{TP + FN}$	To measure the ML model's ability in terms of detecting positive samples
F1 Score	$F1 = \frac{2 * Precision * Recall}{Precision + Recall}$	The harmonic mean of the precision and recall
Area Under the Curve (AUC-ROC)	$FPR = \frac{FP}{FP + TN}$	It makes use of True Positive Rates (TPR) and False Positive Rates (FPR)

TABLE 5.5
Evaluation Metrics for Regression

Metrics	Equation	Objective
Mean Absolute Error (MAE)	$MAE = \frac{\sum_{i=1}^{n} \lvert y_i - \hat{y}_i \rvert}{n}$	To explore how the predicted values are deviated from the actual values [25]
Mean Squared Error (MSE)	$MSE = \frac{\sum_{i=1}^{n} (y_i - \hat{y}_i)^2}{n}$	Finds the average of the squared difference between the target value and the value predicted

Various metrics used to evaluate the performance of classification and regression techniques are tabulated in Tables 5.4 and 5.5. In addition, some of the common performance metrics to evaluate the clustering algorithm are discussed below.

Cluster Cohesion represents the summation of similarity between each pair of data points within the same cluster and is represented in the equation below:

$$\text{Cohesion}(\mathrm{C_k}) = \sum_{\substack{e_p \epsilon \mathrm{C_k} \\ e_q \epsilon \mathrm{C_k}}} \text{similarly}(e_p, e_q) \tag{5.19}$$

Where $k = 1, 2, \ldots, r$ clusters separation between two clusters can be computed by summating the distance between each pair of records falling within the two clusters, and both the records are from different clusters.

$$Separation\left(C_j, C_k\right) = \sum_{\substack{e_p \epsilon C_j \\ e_q \epsilon C_k \\ j \neq k}} \text{distance}\left(e_p, e_q\right) \tag{5.20}$$

A set of clusters having high cohesion within the clusters and high separation between the clusters is considered to be good.

Rand and Jaccard Indices are more often used to compare clustering algorithms. Rand and Jaccard Indices are similar to sensitivity or specificity calculation using a confusion matrix. For a set of data elements $E = \{e_1, e_2, \ldots e_n\}$, we can define two different clustering algorithms X and Y, which divide them into r clusters each having $\{X_1, X_2, \ldots X_r\}$ clusters and $\{Y_1, Y_2 \ldots Y_r\}$ clusters. Combine all X clusters or all Y clusters and you will get your complete E set again.

$$Rand = \frac{A + B}{A + B + C + D} \tag{5.21}$$

$$Jaccard = \frac{A}{A + C + D} \tag{5.22}$$

Where 'A' denotes the number of pairs of data elements in E that are in the same set in the clustering algorithm and the same set in the clustering algorithm; 'B' denotes the number of pairs of data elements in set E that are in different sets in X clustering algorithm and different sets in Y clustering algorithm; 'C' denotes the number of pairs of data elements in E that are in the same set in X clustering algorithm and different sets in the Y clustering algorithm; 'D' denotes the number of pairs of data elements in E that are in different sets in X clustering algorithm and the same set in Y clustering algorithm. The maximum *Rand* and *Jaccard Indices* value shows the best method. Thus, the overview of optical networks and several machine learning algorithms are discussed.

5.3 CHALLENGES AND OPPORTUNITIES IN INTELLIGENT OPTICAL NETWORKS

Several challenges are to be addressed to build optical networks for the latest demand and requirements. The challenges due to various parameters like the blocking rate, speed due to fault location, resource utilization, accuracy, and security in ION are brought into the light. The research challenges of ION were discussed in detail in the above section. These challenges lead to the opportunity in the field of ION, which is highlighted in this section.

5.3.1 Challenges in Intelligent Optical Networks

Optical amplifications have challenges, such as an increase in capacity and dynamic network operations. The solutions to be adopted are higher energy efficiency, increased dynamic adaptability, and reduced cost. Wavelength channels must be provided in software-defined networks—WDM for Flexible rerouting [26]. The challenges in **System designs** are increments in the bit rate. The main conformation for increment in capacity is the high symbol rate with lowering of cost, largest constellation multiplicity, bearing the errors available in the application, and subcarrier multiplicity for the desired service rate. The super-channel—along with optical switching having multiple carriers—faces challenges related to nonlinearities in the transceiver, power consumption, linearity of electro-optic components, fiber bandwidth, and agility in the channel.

Free-space optics (FSO) consume wider bandwidth of higher frequencies and allows reconfiguration of nodes directly, where distance is not considered, but is susceptible to change in the environment, turbulence in the atmospheric, dust, smog, and vibrations causing laser beam attenuation [27]. FiWi (Wireless optical networking) needs specialized multilayer networking procedures, and advancements are considered with the latest technologies like femtocells network [28], multitier and heterogeneous cellular structures [29, 30], energy savings [31, 32], and machine-to-machine communication [33, 34] where several challenges remain.

Heterogeneous access networks require a high bit rate for the networks in the next generation to support 5G mobile front-haul communication networks. Although **PON networks** offer major advantages including good spectral efficiency, sensitivity, tolerance to chromatic dispersion, and design of modulation formats to bring the output power comparable to discrete lasers, they need several stages of amplification for multiwavelength sources, resulting in an expanded expense [35]. In optical networks, while devices are integrated with higher bandwidth, limitation in optical fiber communication includes carrier restricted spaces [35]. Routing and spectrum assignment procedures in spectral contiguity increase Bandwidth Blocking Probability (BBP) [36, 37]. The challenges that must be addressed in EON for the current scenario are hardware development, network control and management, spectrum management, and disaster management [38]. Depleting spectral resources and increasing power consumption are two critical issues in EON.

Automatically Switched Optical Networks are efficient and offer low cost-per-bit transportation. However, the service providers need to offer value-added services with innovation. The converged backbone transformation solution addresses this problem, which features the close integration of optical core transport networks [39] and IP. **SDNs** face several challenges from virtual machine allocation, controller placement, scalability, performance, security, interoperability, and reliability in service request flow balancing [40]. The network administrator can remove the GMPLS from the core network. SDNEON is deployed in the public network infrastructure.

The control plane, which is based on the SDN framework, will rely on a controller centralized to the network, such that spectrum defragmentation and path computation is responsible for light path establishment and management, mostly recovery, benefiting GMPLS/PCE [41]. **The Autonomous Network** faces challenges in the openness in the dataset, explainability of model, and lack of system reliability, which otherwise would make the optical network more efficient and flexible according to the service requirements and the status of the network [33]. Virtualization, telemetry frameworks, AI, and open architectures enable paradigm shifts toward self-driving networks. The accuracy of locating faults and eliminating interferences, when there is a spatial correlation between the alarm information and interaction between the faults, is challenging [42].

Machine learning is deployed in an optical network, and the difficulty in collecting the real datasets forces to use synthetic datasets for training. Considering the maintenance and operation while deploying ML, it is hard to interpret the trained models although training is transparent, and while troubleshooting problems, the interventions of public un-interpretability will occur. Also, ML endures tracing failures because of different operational actions. The deployment in the dynamic network for the ML model assessing, requirements of multiple datasets with training, brings in an increase in complexity and computational cost. The generalization metric is of high significance since the one-trained model adopts a specific environment, and the new network problem brings changes in the environment. In a 5G embedded optical network when there is a need to connect with other devices in real-time and computational complexity requirements, resource allocation and power cost are challenging. The availability of field data remains a serious problem in the demonstration of ML applications in light path QoT estimation and QoT forecast in real-world networks [43].

Open Dataset Access: Only a few datasets are available for optical networks. Synthetic datasets are difficult and lack credibility. Data imbalance problems occur in the real network dataset collection. Datasets in Internet Service Provider (ISP) cannot be made public as there are privacy issues in real telecommunication data. **Model Interpretability and Traceability**: The ML algorithms work in the black box mode. Despite an open and transparent training process, the models which are trained are uninterpretable and network operators find it difficult to troubleshoot the network. Tracing the output features toward the input is a difficult task for multiple input parameters (Figure 5.16).

Model Generalization Ability: For a dynamic network environment in an ION, the computational cost increases when there is a change in the network environment and model retraining. **Algorithm Computational Complexity**: Time sensitivity is important in the real network environment; different algorithms are used to identify the sub-optimal solutions and compare them in a particular threshold time. **System security and reliability**: No performance guarantee using the ML models leads to security issues, and performance degradation leads to reliability issues [44]. DL-based network traffic

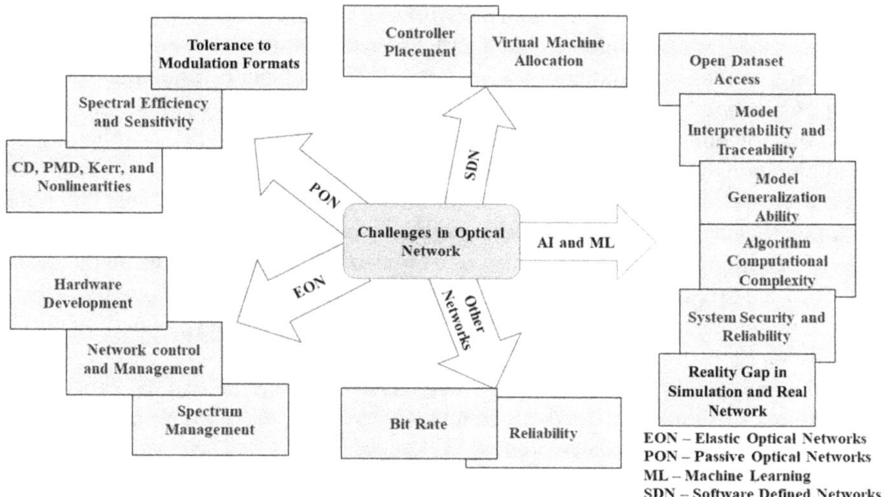

FIGURE 5.16 Challenges in Optical Network and ION in different aspects.

classifier has a critical issue of not being robust, due to the existing methodologies that require continuous research in this area [45]. Security and privacy are the primary concern in building the **Reality gap between the simulation and real networks**; also, the performance of the emulation and the real network is very difficult to evaluate [46].

5.3.2 Opportunities in Intelligent Optical Networks

AI technologies provide new opportunities to solve the challenges in optical networks. ML and DL methods can obtain higher accuracy and lower complexity than conventional models. The current opportunities in evolving intelligent optical networks make them autonomous. Emerging technologies such as blockchain can be integrated with optical networks, making them more reliable. As literature shows that AI supports optical communication and networking in many ways, there is scope for AI to support Space Division Multiplexing (SDM) technology using multimode/multicore fibers apart from the automation of network management operations, which can address the Internet capacity crunch [47]. Optical network automated operation and maintenance against the service objective is addressed by an Intent-Defined Optical Network (IDON) architecture working towards artificial intelligence, which has introduced a self-adapted generation and optimization (SAGO) policy. In the optical transport network, this policy was proposed to avoid manual intervention [48].

The application of AI in optical networks is still in its infancy. There is more work needed in the application of AI in QoT estimation and performance monitoring, optical network attack and intrusion detection, and identifying and localizing optical network attacks in large-scale network scenarios [49]. Automated network management operations of heterogeneous optical networks are another area where AI can be explored. Emerging technologies like tactile internet, Industry 4.0, etc., require high bandwidth, low latency, and network security. This issue can be attempted by combining 5G technology with high-speed optical fiber technology. This requires the joint allocation of computing devices and networking resources [49]. Another important application where researchers are now working is power-efficient AI-enabled optical network-on-chip (ONoC) as a substitute for the electronic NoC with computation time efficiency. As far as network automation is concerned, it can be achieved in terms of fast lightpath provisioning, margin optimization, and proactive maintenance [50]. They are possible if the real network field data are available—and with the help of ML techniques, it is possible to estimate and forecast QoT [43]. Now it is necessary to work on getting real-time data.

ML techniques developed in other fields, such as image and signal processing, bioinformatics, biomedical fields, recommender system, text mining, and so on to address other field issues can be applied to ION problems [5]. The basic requirement of an intelligent future optical network is a multitasking algorithm to jointly perform OPM/MFI [5]. Also, the existing Modulation Format Identification (MFI) classification algorithms show low classification accuracy when high-order modulation formats are used; this indicated that researchers should develop new MFI methodologies to handle new high-order modulation formats like M-ary Quadrature Amplitude Modulation (QAM). Telecom operators such as Ericsson, AT&T, and Nokia accepted the slicing network as an emerging technology. A lot of work was carried out for this, but there is a need for the development of automatically providing the type of dynamic controllers for optimized network resources where the situation changes over time. Also, there is a scope for analyzing the developed frameworks concerning their energy consumption and communication [51].

The performance of an ION depends on the accuracy, minimum error related to its analysis and usage of the system memory, and how the system automatically models and provides a cost-effective solution. But from the studies, it is observed that one of the major difficulties involved in the ML approach of analysis towards ION is the generation and availability of synthetic data sets since obtaining experimental-based data related to optical networks needs very huge financial investment. Hence, it is time to find an effective and possible solution for the availability of data sets publicly [52].

Data collection for ML-based QoT estimation in optical networks is a time-consuming process and is very difficult as well. One way to explore this is to build a virtual digital twin network of the real network for the application of ML in QoT estimation [53]. QoT prediction for multicast using transfer learning, refinement of input parameters, and adaptability for QoT prediction model are other possible areas for improvement in ML-based QoT estimation [54]. There is an opportunity for applying epsilon greedy Q-learning in transport optical networks to speed up the convergence rate of traffic prediction in large optical networks [55]. One of the hot topics in the industry towards building an intelligent management system for optical communication and network is telemetry in long-haul coherent communication systems [56]. To avoid overfitting problems in ML in the field of optical communication in heterogeneous conditions, it is difficult to simulate data for various scenarios.

This requires self-adaptive learning techniques to resolve this problem, and more ML techniques specifically adapted for optical communication should be developed [56]. A benchmark data set is required to test the performance of ML in optical communication, which the industry should provide.

5.4 APPLICATIONS IN INTELLIGENT OPTICAL NETWORKS

Applications of optical networks are vast and enormous and are shown in Figure 5.17. Intelligent optical networks have the ability to predict traffic. Thus, they facilitate the proper working of a network in an efficient manner as far as network control and resource optimization related to the service provider and resource allotments are concerned. Resource allotment is used for allocating multiple resources in optical networks, such as wavelength routing, frequency bands, and core fibers. These help in proper check of physical layer connections and estimation of QoT signals, thereby facilitating failure management in an optical network.

ML is used in optical systems in the fields of optical transmission and networks. It is applicable for the multi-functions of service providers and resource allotment, which supports continuously monitoring of the network's performance in an intelligent manner. With human-in-the-loop network control and management, optical networks are becoming smart for next-generation programmable networking and automation in optical networks. Areas of application in AI and ML collectively are transmission of optical signals, performance monitoring, QoT, and planning operation in an optical network—and also includes network traffic prediction, path computation, and resource allocation.

5.4.1 Artificial Intelligence in Optical Networks

AI is a fascinating technology in the fields where the decision-making process is autonomous, e.g., medical, automobiles, home appliances, communication, and industrial applications, which demand continuous increase in the control and management of the systems and services [57]. The working principle of AI is based on the representation of data followed by learning, using the biological neural network's structure; this is helpful to reproduce the analysis and learning process automatically,

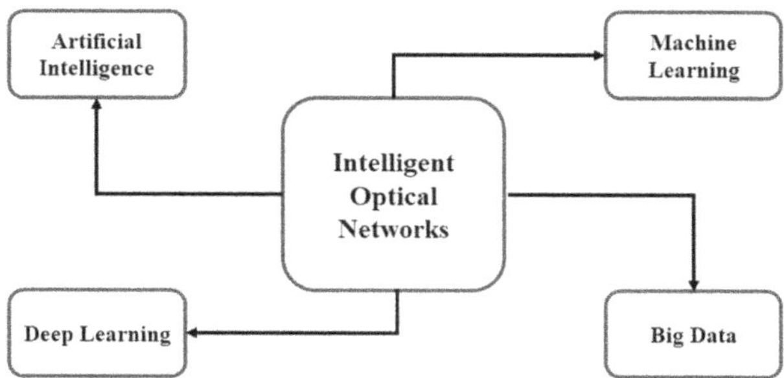

FIGURE 5.17 Applications of an intelligent optical network.

which makes the system more intelligent. The problems the conventional optical network experiences are difficulty in network planning, network operation, network maintenance, and inefficient capacity expansion due to manual operations. It is time-consuming and unsatisfactory in the network optimization process. It is also difficult to identify faults in an optical network, positioning, and fixing that requires field experts to address the issues.

AI is applied in the optical networks making the network automated, and thus less dependent on manual intervention; it is shown in Figure 5.18. AI techniques are introduced in optical networks for network control and management, including intelligent decision-making and automated operations. AI also addresses issues such as connection establishment, self-configuration, self-optimization, estimation, and prediction with the help of the present state of network and historical data [47]. This helps to improve network configuration and device operation in a network, such as optical performance monitoring, recognition of modulation format, mitigation of fiber nonlinearities, and QoT estimation, leading to higher network performances.

Network planning and resource management: To meet the current demand of increasing the data rate to end users, network planning and resource allocation is of high interest. Optical network planning is performed by the optimum design of the physical layer network and ensuring survivability with minimum cost. AI techniques such as GA [58] and Ant Colony Optimization (ANT) [59] are reported to achieve optical network planning tasks with guaranteed network survivability. An efficient resource allocation process using AI in a flex-grid optical network based on a representation learning-based strategy combined with the admission strategy optimizes the slice scaling problem and improves resource utilization [42]. GA is used for the optimum network resource allocation. This can be achieved by proposing a novel network architecture for a flex-grid optical core and a joint routing of dynamic WDM in a ring network [60, 61]. Particle swarm optimization technique and ANT algorithms are used to solve the resource allocation problem in terms of restrictions in QoS and transmitted power in optical WDM/OCDM networks [62, 63]. Other AI algorithms, such as k-means clustering and Markov decision processes, are used in optical network resource allocation problems [64, 65].

Intelligent fault detection and localization in optical networks: The backbone optical network is an inevitable transmission medium to satisfy Internet data traffic by providing colossal bandwidth. When deployed in the real network scenario, it involves optical amplifiers, transceivers, optical switches, ROADM, and an extended length of fiber links, which are prone to failures that lead to disruptions of QoS in the network [66]. These network failures require efficient fault diagnosis with intelligence at an affordable cost. The implementation of AI technology in the fault diagnosis application of optical networks is trending. AI technology automatically identifies which optical node device is causing the problem, and it locates the fiber breakdown through training, where the machine gains intelligence. Hence, it reduces the troubleshooting period of fault diagnosis in

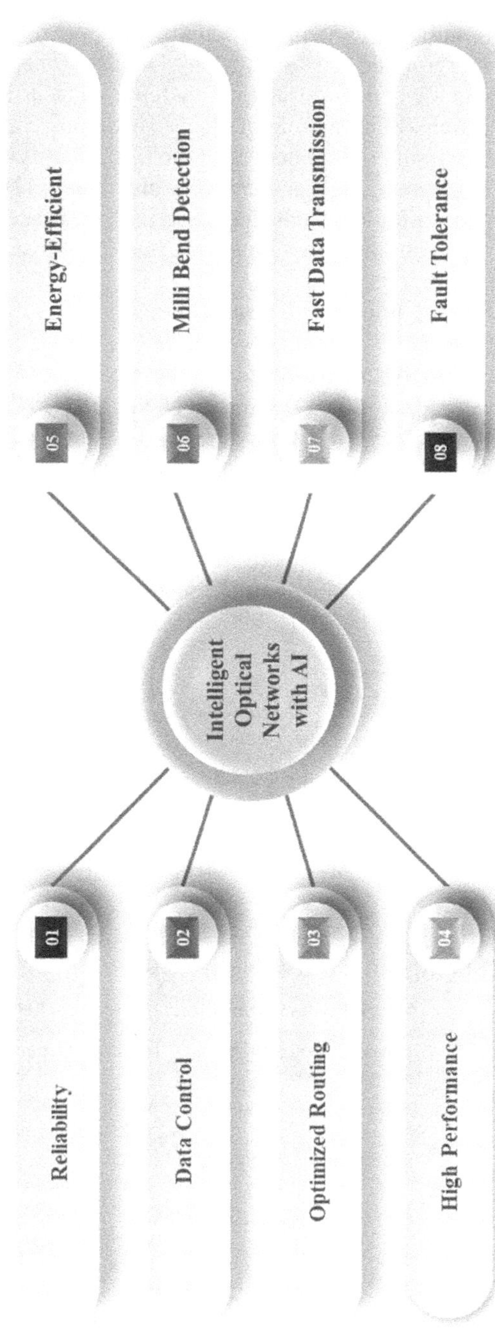

FIGURE 5.18 Intelligent optical network with AI.

the optical network. It is also used to predict network faults. In a 5G optical network, frequent network failures and alarm signals will disappear due to the increase in the number of equipment and make it more difficult to locate network faults in large-scale scenarios. Many DL-based methods for fault localization using gradient down strategy were reported [67]. This approach required global search ability and no gradient constraints with the help of AI technology [68]. In the virtual network topology, identifying failures to improve the QoS is done by Bayesian network clustering [42]. SDN-based networks failure detection and fault diagnosis in optical access networks were performed with cognition-based methods [69] and Bayesian inference networks [70], respectively.

Traffic classification and flow prediction: The optimum utilization of network resources by network operators must have knowledge about network traffic and forecasting. AI methods address the challenges in traffic classification and flow prediction. AI, Bidirectional NN, and LSTM are used to improve data center traffic [42]. A Genetic Expression Programming (GEP) algorithm was formulated for QoS traffic prediction with limited packet transmission [71]. The balance load in ONU and PON Network architecture using GA was proposed, and it reduced congestion in ONU [72]. HMM and NN were used to predict network traffic in WDM networks and the flow classifier at the edge of the SDN network, respectively [73, 74]. Application of AI in optical networks is listed in Table 5.6.

5.4.2 Machine Learning in Optical Networks

In physical layer, applications of ML in optical networks are OPM, QoT estimation, short reach equalization and mitigation of fiber nonlinear, MFI, amplifier control, and digital twins for optical networks. ML approach in the network layer applications for failure management is discussed in Section 5.4.2.2.

TABLE 5.6
Application of AI Technology in Optical Networks

Ref.	Application	Algorithm/Technology	Remarks
[75]	Intelligence in optical networking	Advanced ML algorithms for optical networks	Traffic prediction, virtual topology design, failure management, flow classification, and path computation
[76]	Edge nodes of optical label switching networks	Improved AI traffic aggregation algorithm	Predicts the size of the transport block, reduces the amount of service data, and increases efficiency
[77]	Accuracy of fiber bend estimation	ANN	Detecting a fiber bend prior to a fiber break of a small diameter in the transmission line
[78]	OSNR and data rate in network layer	Adaptive neural fuzzy inference system	Efficient data rate improvement

5.4.2.1 ML Approaches to Physical Layer Applications

Optical Performance Monitoring (OPM) is an important process in the optical network management of next-generation networks. It measures and monitors the real-time parameters related to the optical network at the intermediate nodes and gives feedback to the network controller for the effective assessment of network failure and to identify faults in the network and hence improve the network performance. OPM is one of the important enabling techniques for future optical network control and management. Simultaneous monitoring and accurate measurement of multiple parameters in an intermediate optical node in an optical link make the network intelligent [79]. The OPM parameters to be monitored are Optical Signal-to-Noise Ratio (OSNR), optical power, fiber attenuation loss, linear fiber distortions such as Chromatic Dispersion (CD), Polarization Mode Dispersion (PMD), nonlinear fiber distortions like Self-phase Modulation (SPM), Four-Wave Mixing (FWM), Cross-Phase Modulation (XPM), Raman scattering, Brillouin scattering, laser line width, modulation format recognition, optical amplifier noise receiver noise, Laser Relative Intensity Noise (RIN), and extinction ratio, etc. A review of various ML approaches for OPM in the optical network is given in Table 5.7.

In **QoT Estimation**, light path connection in an optical network is a representation of the data transmission in a light signal form transmitting from source to user destination. In the optical network, there will be a massive connection between the source and destination nodes. QoT is an indication of the availability of connections between them. Hence, Qot estimation will guarantee the successful connection and proper utilization of resource allocation [89]. Table 5.7 represents the application of AI, ML, and DL in the physical layer of optical network QoT estimation.

Short reach equalization and fiber nonlinear noise mitigation are the types of optical communication schemes that cover a distance of less than 100 kms. The applications of the short-reach communication access network are intra datacenters and inter datacenters. Tackling of fiber nonlinear distortions in optical networks needs attention for quantifying the number of distortions added or for the actual mitigation of these disablements for better QoT. The application of ML techniques in short-reach optical communication and nonlinear distortions are discussed in Table 5.8.

5.4.2.2 ML Approaches to Network Layer Applications

Smart intelligent operations and overall maintenance of optical networks require network failure management. This involves failure prediction, failure identification, failure detection, failure localization, failure estimation, and recovery [105]. A review of various ML approaches in the network layer is given in Table 5.9.

5.4.3 Big Data Analytics in Optical Networks

Currently, big data analytics is a trending technology that plays a major role in transforming huge amounts of data into useful information and also provides correlations among the data, as shown in Figure 5.19. By using predictive models in ML alongside big data, it is useful for industries to identify their future trends and massive growth in their businesses. Applications such as telecommunication, cloud

TABLE 5.7
ML Approaches for OPM and QoT in the Physical Layer

Ref.	ML Algorithm	Network Scenario	Methodology	Remarks
		ML - Optical Performance Monitoring (OPM)		
[80]	Multi-Tasking Learning (MTL) -CNN	Elastic and cognitive optical networks	MTL-CNN technique with the optical spectrum for OSR and OSNR monitoring	High accuracy in optical signal recognition low error in OSNR monitoring
[81]	SVM, random forest algorithm and CNN	Mode Division Multiplexed Optical Networks	Optical networks based on few modes fiber. In-phase Quadrature Histogram (1D & 2D features).	Accurately estimated OSNR, CD, and mode coupling coefficient values
[82]	ML	Optical networks	Multi-parameter monitoring using a pair of low bandwidth coherent receivers	Cost-effective solution for joint monitoring of multiple parameters in an ION
[83]	ML regression algorithm	Disaggregated optical networks	Signal and filter-related monitoring	The total reduction of the estimated SNR penalty estimation of filter
[84]	CNN	Optical transport network	Estimation of OSNR	Autonomous self-driving optical networks
		QoT Estimation		
[51]	NN	Sliceable optical network	Centralized and distributed QoT models	Improved network performance
[85]	ML, DNN	Disaggregated optical networks	QoT computation for light paths with dynamic traffic loading	Precise estimation of total GSNR
[86]	ML Regression Model	Low-margin optical network planning	ML model on precomputed per-channel Self Channel Interference (SCI)	Spectral optimization for network planning and high accuracy
[87]	Polynomial Regression, Decision Tree (DT), and NN	General optical network	Live environment estimation of QoT	Optimization of variable bandwidth and baud rate and better leverage in their capacity
[88]	ML	Optical network	Estimation model based on gaussian processes	Minimize estimation uncertainty, reductions in training set size

computing, medical, social network services, and financial services generate large amounts of data. Optical networks with large capacity, high speed, and great flexibility support big data services with great potential [113]. Table 5.10 discusses big data analytics in optical networks and how it improves optical network performances.

Block chain technology is another scope for improving the security in optical networks. Taking advantage of data consistency and the distributed nature of blockchain

TABLE 5.8
Other ML Approaches in the Physical Layer

Ref.	ML Algorithm	Network Scenario	Methodology	Remarks
		ML – Short Reach Equalization and Nonlinear Mitigation		
[90]	Fully fledged feed forward NN	Short reach optical transmission	Hybrid method of spectral slicing and digital reservoir computing	Good short reach extension, lower complexity, and faster training
[91]	Deep NN	Coherent 92GBd dual polarization 64QAM	Comparison of soft DNN equalizers and Volterra non-linear equalizers	Improved performance and lower complexity
[92]	FNN and RNN	Pulse amplitude modulation (PAM4) - 50-Gb/s and 20-km	NN-based equalizers aided by TL	Future optical switched data center
[93]	Gaussian noise (GN) model	Open optical networks	Particle swarm optimization technique and the steepest descent algorithm for optimization of launch power	QoT estimation, data throughput
[94]	DNN	General optical network	80GBd DP-64QAM soft demapper	Handle mixed signal memory and colored noise effects
[95]	MLP	OPC-aided fiber-optic communication	Optical phase conjugation in 64QAM over 400 km SSMF	Reduce residual nonlinear impairments
		ML - Modulation format recognition		
[96]	Fast density-peak-based pattern recognition	EON	Different energy level features	Simple and flexible with lower OSNR requirement
[97]	CNN	Heterogeneous optical networks	Analyze the optical signals	Improved performance
[98]	CNN	Super-channel optical networks	Histogram	Modulation formats identification
[99]	2D Stokes plane	EON	Graph-based 2D stokes plane analysis	Robust and exhibits good resilience
[100]	CNN	Optical network	Modulation schemes RZ-OOK, NRZ-OOK, 4-PAM, RZ-DPSK, 4-QAM, and 16-QAM	Estimation of OSNR, BER, and Q factor Accuracy is improved
		ML - Amplifier Control and Digital Twins for optical networks		
[101]	DL algorithms	Cognitive optical network	Digital Twin in optical communication for fault management	Improved control efficiency and stable operation; promotes intelligence
[102]	ANN	SDN	Field data for the no-failure network state	Recognize double failures
[103]	Multiobjective optimization (KNN)	All optical networks	Surrogate model for adaptive control of optical amplifier problem	KNN for regression-based ML technique
[104]	Evolutionary algorithm	Software-defined WDM	Vendor-agnostic framework for QoT estimation to control EDFA	Maximum average GSNR

TABLE 5.9
ML Approaches in the Network Layer

Ref.	ML Algorithm	Network Scenario	Methodology	Remarks
		Network Traffic Prediction, Classification, and Path Computation		
[106]	Single-Layer Perceptron (SLP-NN)	Underground optical network	Linear regression implemented in the Python sci-kit learning library	Predicted accurately with high efficiency
[107]	SLR	Flexible optical networks	XT-aware dynamic simulations	Lowers blocking and improves data rate in the network
[108]	GA	Ultra-wideband optical networks	Channel launch power stimulated Raman scattering	Significant throughput with improved performance
[105]	DL	Optical transport network	Bi-directional LSTM	The average accuracy and F1 score are predicted
[109]	LSTM - NN	Passive optical network	LSTM-NN prediction model, along with the Swish function	Efficient bandwidth allocation, prediction, and high accuracy
[110]	LSTM - RNN	SDN	LSTM encoder-decoder model with Adam optimizer	Improvement in packet latency and loss
[111]	Disaggregation-Functional Block	Flexible grid	ILP-based universal path computation system	Reduced path computation time
[112]	LSTM - NN	Data center networks	Real server-generated traffic streams	High accuracy is larger for bursty, real-time server traffic flows
[45]	DL-based ANT	Optical networks	Packet classification and flow content classification	Robustness of the network traffic classifiers

technology, it can combine with other distributed networks. Since the edge optical network is open and intelligent, it needs to be more secured by integrating the blockchain decentralized technology. It improves reliability and security of optical networks. Blockchain technology of the optical network solves the trust problem like slice leasing, privacy protection and security of data for user nodes, efficient service provisioning, better throughput, and efficient failure recovery, which brings intelligence into the network. AI can be used in optical networking in various applications such as optical network planning, connection establishment, reconfiguration of network through virtual topologies, SDN, optical burst switching, passive optical networks, and intra data center networking. This paves the way to ION and large data classification in an optical network by resorting to big data and security improvisation using blockchain technology to fulfill all the needs of data communication.

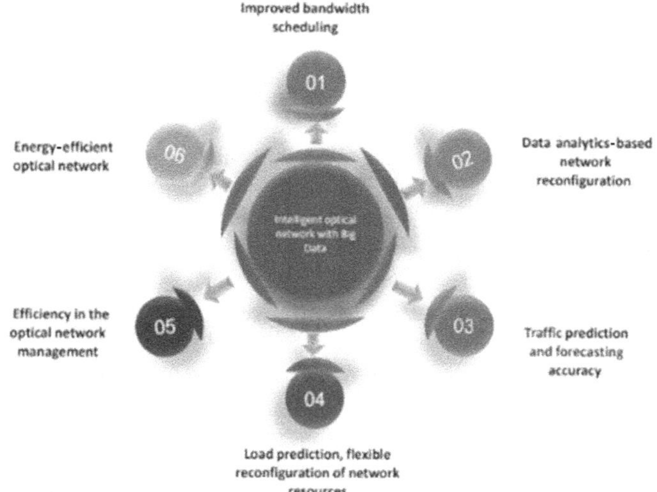

FIGURE 5.19 Big data are transforming optical network intelligence.

TABLE 5.10
Big Data Analytics in Optical Networks

Ref.	Application	Algorithm/Technology	Remarks
[114]	Big Data Services, SDON	Bayesian classification-based service awareness mechanism	High matching degree between services and SDON improved bandwidth scheduling
[115]	Big data services	Bayesian traffic prediction algorithm	Better traffic prediction accuracy and improved efficiency
[116]	Big Data, Virtual Network Topology (VNT) Reconfiguration	Heuristic algorithm	Traffic prediction and network reconfiguration
[117]	5G fronthaul network - Virtual base stations	Dynamic clustering and load balancing combined with LSTM and K-means algorithm	Accurate traffic forecasting, prediction of load, flexible reconfiguration of network resources, and improves user experiences
[118]	Optical network management, big data computation, HDFS	Hadoop distributed file system for data processing	Improved efficiency optical network management
[119]	Optical mesh network, green optical network	3 Heuristic algorithms - First-fit, incremental grooming, and predictive and incremental grooming	Effective reduction in energy and cost. Efficiently support Big Data transfer.
[120]	IP/WDM networks	Mixed integer linear programming model	Energy-efficient processing and reduction in network power consumption

5.5 VARIOUS SIMULATION TOOLS FOR INTELLIGENT OPTICAL NETWORKS

Simulation tools play a significant role in the design and analysis of the model that is proposed. New ideas can be developed and tested with the help of simulation software. A detailed performance analysis is possible with the available tools. This leads to the development of innovative solutions to real-world problems. Hence, the simulation process is an inevitable one in the model development. Simulators are available with various component blocks, testing blocks with different features and characteristics. The physical test beds for any model are expensive and time-consuming, even for small-scale network development. Also, these experimental arrangements must be more flexible and are limited in their programmability [121].

Many simulation tools are available for research and practitioners. These simulation tools may be open-source versions or commercial versions [122]. In this section, both types of tools will be discussed. It is suggested that, depending on the specification, their requirements, and their application, simulation tools can be selected. The simulation tools provided in this section are based on the reported literature. Some of the simulation tools for optical network study are given: OptSim, VPI Photonics, OptiSystem, GNPy, etc. The simulation tools may be open-source software, which can be downloaded from their official website freely. Then commercial software must be purchased with a license for high data rates [122]. In this section, some of the optical communication- and network-related software are presented. The description, benefits, and applications are given. The details are collected from the software's official website, existing literature, and its user manual.

There are two ways to address the physical reality problems. One approach is the hardware experimentation method. Along with the available data and based on the design requirement, one may design and develop the experimental setup directly. Another approach is the simulation-based method. In this approach, with the available data and design requirements, first, the model will be developed in the simulation platform. Again, in this approach, model construction can be directed with the simulation tool or it can be equivalently modeled mathematically. The simulation-based approach is flexible for testing new ideas. It is expandable when the analysis moves from small-scale to large-scale testing. But in the case of experimentation setup, it is less flexible than the simulation-based approach and is an expensive and time-consuming process. Figure 5.20 shows the classification of physical reality approaches, and Table 5.11 lists the simulation software and tools for network design.

The following section discusses various simulation software in optical communication and networks:

OptSim is helpful for the design and simulation of optical systems at the signal propagation level [123]. The simulation tool is easy to use with GUI and can also be integrated to lab measurement equipment. It provides good accuracy and usability. Applications are FTTx/PON-BPON, WDM-PON, G(E)PON, RSOA-based bi-directional PON, OCDMA/OTDM, and coherent PON.

OptiSystem is a system-level simulation tool for optical network study and can be interpreted. This software will facilitate network-level applications. It is

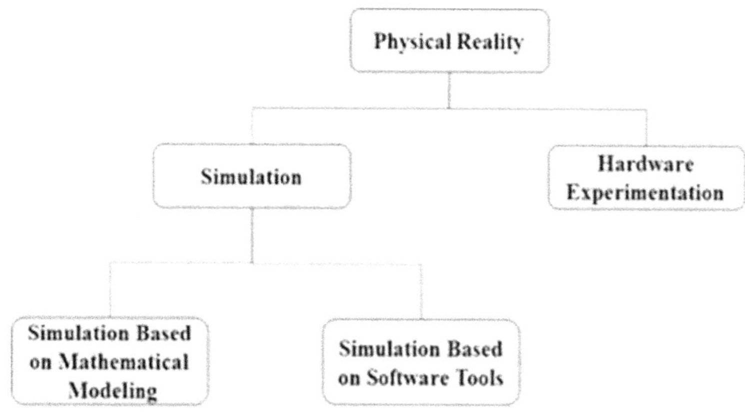

FIGURE 5.20 Classification of physical reality approaches.

a simulation package for optical communication systems, which includes the design, testing, and optimization of optical links in the physical layer of optical networks [124]. Applications are SONET/SDH rings, OTDM, CWDM, DWDM, PON, and OCDMA.

OptiSPICE is a simulation program with Integrated Circuit Emphasis. The simulation and construction of optoelectronic layouts such as optical interconnects, laser drivers, electronic equalizers, and transimpedance amplifiers, etc., can be performed using this software [125]. It is studied for the design of transmitter circuit with modulation where co-simulation of channel and receiver circuits are implemented in OptiSystem.

VPI Photonics was initially termed as Photonic Design Automation (PDA) and is used to develop complex photonic networks and design methodologies. Optical systems enhance the speed of the design of photonic systems and subsystems for all types of transmission systems [126]. For existing fiber plants, the technology upgrade and component substitution strategies need to be developed [126]. It is used for NRZ modulation versus RZ, chirped-RZ, duobinary, vestigial side band, DPSK, DQPSK, multilevel quadrature amplitude, and phase shift keying and other optimum countermeasures against dynamics in reconfigurable DWDM networks due to power transients and optical access network architectures.

Fiber Optic Network Management Software is helpful for fiber optic network operators in planning, designing, and building the network. It helps us plan and design optical networks to build new plans and estimate the costs and design connectivity of the fiber optic network. It draws redlines, creates work orders, and coordinates with crews on the ground [127]. This software provides better fiber data records and helps manage fiber network operations.

Ensemble Simulator (ADVA™) is a network simulation tool for training, planning, development, testing, and verification in a virtual networking

TABLE 5.11
List of Simulation Software and Tools for Network Design

Simulation Tool	Layer	Layer Description	Advancement/Network Components
Network Simulator	Application	CBR, VBR, FTP, Web, Telnet, API	M2M communication, IOT, 5G, and 5G beyond networks. It is easy to Interface with MATLAB, SNS3, and Mininet.
	Transport	TCP, UDP	
	Network	DSR, DSDV, AODV, IPv4, IPv6, OLSR	
OMNET++	Application	Telnet	It interfaces with third-party hardware like CloudSim and iCanCloud
	Transport	TCP, SCTP, UDP	
	Network	OSPF, BGP, Ipv4, Ipv6	
OPNET	Application	CBR, VBR, FTP	It supports LTE, WiMAX, and sensor networks
	Transport	TCP, UDP	
	Network	ICMP, ICMPv4, OSPF, Ipv4, Ipv6	
QUALNET	Application	CBR, VBR, FTP	QualNet's Human-in-the-Loop (HITL) interface upholds dynamic collaborations to adjust the tasks of a running situation
	Transport	TCP, UDP, RVSP, RTP, DiffServ, MPLS	
	Network	RIPv2, BGP, Flooding, Fisheye, DSR, DSDV, AODV, WRP, LAR, OSPFv2	
MININET	Application and Control	SDN-based	It supports integration with AMAZON EC2
LTESIM	Physical	LTE-Sim has wide support for all LTE-based entities	Bears the cost of multiple times preferable web association over the past age networks in all wireless network gadgets
NETSIM	Application	FTP, Database, Voice, Video	The outer point of interaction permits NetSim to pass the momentary organization information to other tools like MATLAB and SUMO
	Transport	TCP, UDP	
	Network	RIP, OSPF, AODV, DSR	
COOJA SIM	Physical	IEEE 802.15.4 (LR-WPANs), uIPv6, and uIPv4	Cooja Simulator is an organization test system for Wireless Sensor Networks incorporated with the AWS IoT stage to impart
CONTIKI OS	Internet	Allows Internet Protocol Suite (TCP/IP with IPv6)	It is mainly focused on low-power IOT devices

environment [128]. It provides a virtual environment for engineers to safely test and validate any new service with a digital network twin. The ensemble simulator gives the answer for intelligent agents thoroughly testing and verifying any new configuration before they are applied to the production network.

MATLAB software is a programming tool to design and analyze systems and models that bring transformation to the world. The main feature of this software is a matrix-based language that allows the computation of mathematical

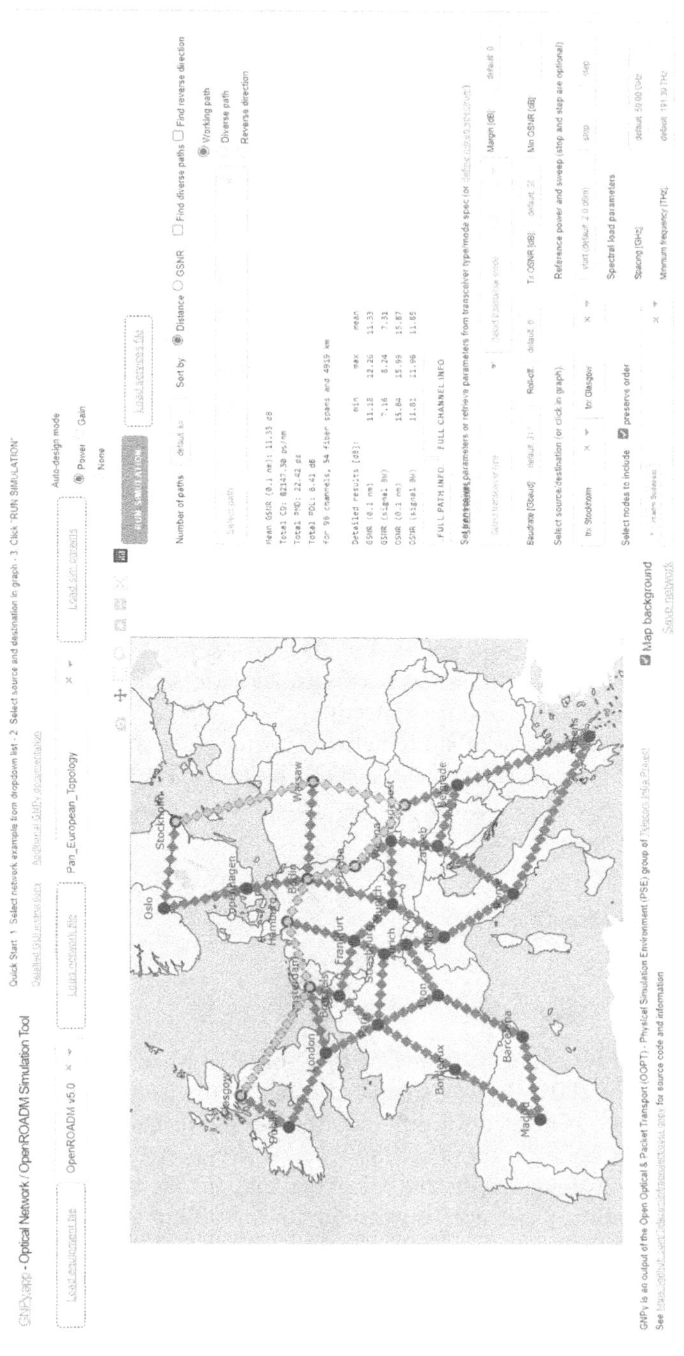

FIGURE 5.21 Calculation of QoT using GNPy application for European network topology.

expressions. This tool generates network nodes from the given matrix and provides multiple toolboxes support for networks and communication. Optical Fiber Toolbox (OFT) package functions are available to calculate guided modes in optical fibers by considering tapered microfibers (nanofibers). It provides solutions for both strong and weak guidance fibers. Material dispersion is well considered. There are several other Toolbox/Blockset related to optical communication systems (SOFTDM Ver 1.5) Update Matlab 2010a [129]. Networks setup can be created using this toolbox.

GNPy is an open-source network planning tool for open optical networks. Gaussian Noise Model is used in the optical route planning library. This software is for abstracting the physical layer of an optical network and is developed by the physical simulation environment (PSE) team in a consortium of the Telecom Infra Project (TIP). This tool helps predict the QoT in a network and to address physical layer upgrades. It generates a synthetic set of data for lightpath QoT computation in optical networks [130]. The simulated schematic of OpenROADM 4.0 in GNPy application for calculating OSNR in different scenarios is shown in Figure 5.21.

Qtool is a physical layer model (PLM) used to estimate the QoT of either new or reconfigured connections. The analyses that can be carried out using Qtool are BER, SNR, etc. [131].

T-API is abbreviated as Transport Application Programming Interface, which is standard API [132]. It is mainly used to monitor and integrate optical transport networks with higher-level applications such as interconnection CORD, supports virtual transport networks, and also supports high bandwidth network slicing enabling connectivity.

OPenConfig is defined as a group of network operators. It is a model-driven network management API mainly designed by users. The network operators aim is to move their networks toward good programmable and dynamic environments by using SDN [133].

5.5.1 Simulation Study

5.5.2 Result and Inferences

A 16-channel WDM system in the frequency range of 193.1 MHz to 194.6 MHz with a channel spacing of 100 KHz was simulated using VPItransmissionmaker v11.0. Optical spectrums are generated for different transmitting distances. The values of transmission power, noise power, and OSNR have been derived from the optical spectrum and given as input data to the ML-based classification techniques for the estimation of transmitting distance between 50 km to 200 km and a dispersion of 9^{-6} or 16^{-6}. The classification learner app supported by MATLAB R2022A has been used for implementing the classification techniques. From the results generated, it is clear that techniques such as support vector machine (SVM) and neural network (NN) provide 100% accuracy for estimating the transmission distance in the WDM system.

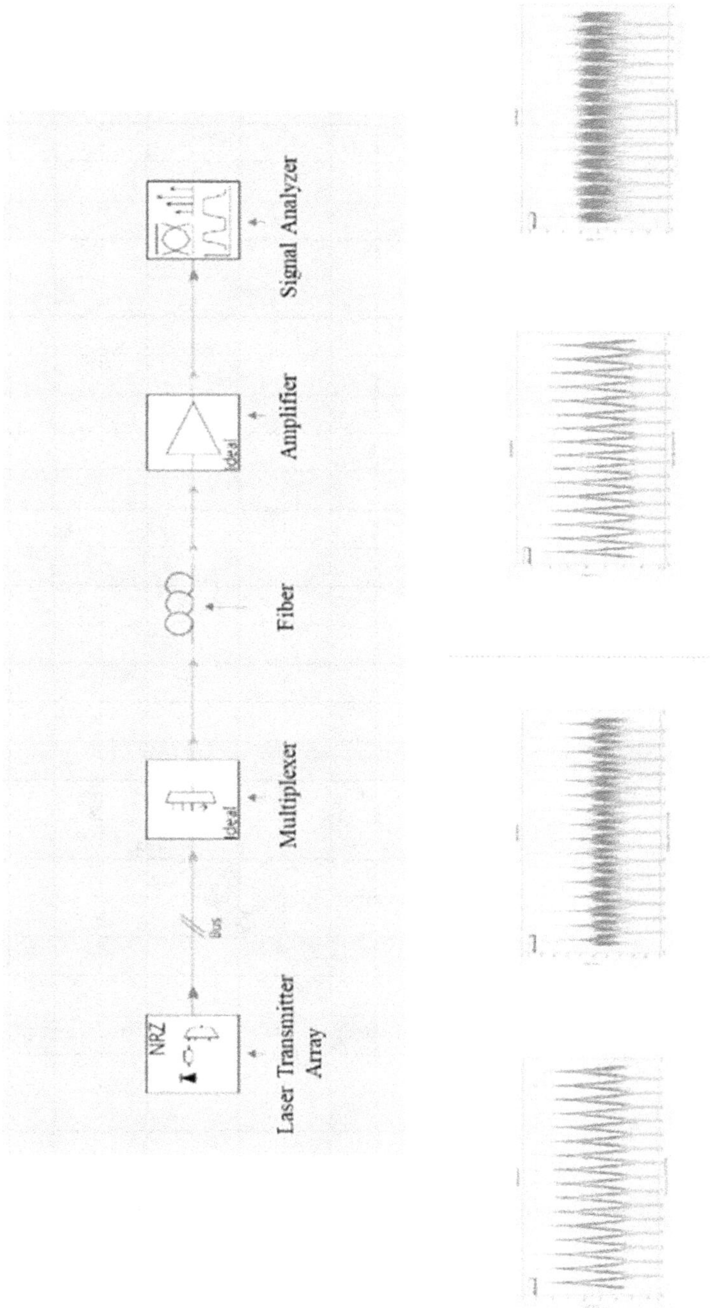

FIGURE 5.22 16-channel transmission system using VPI software.

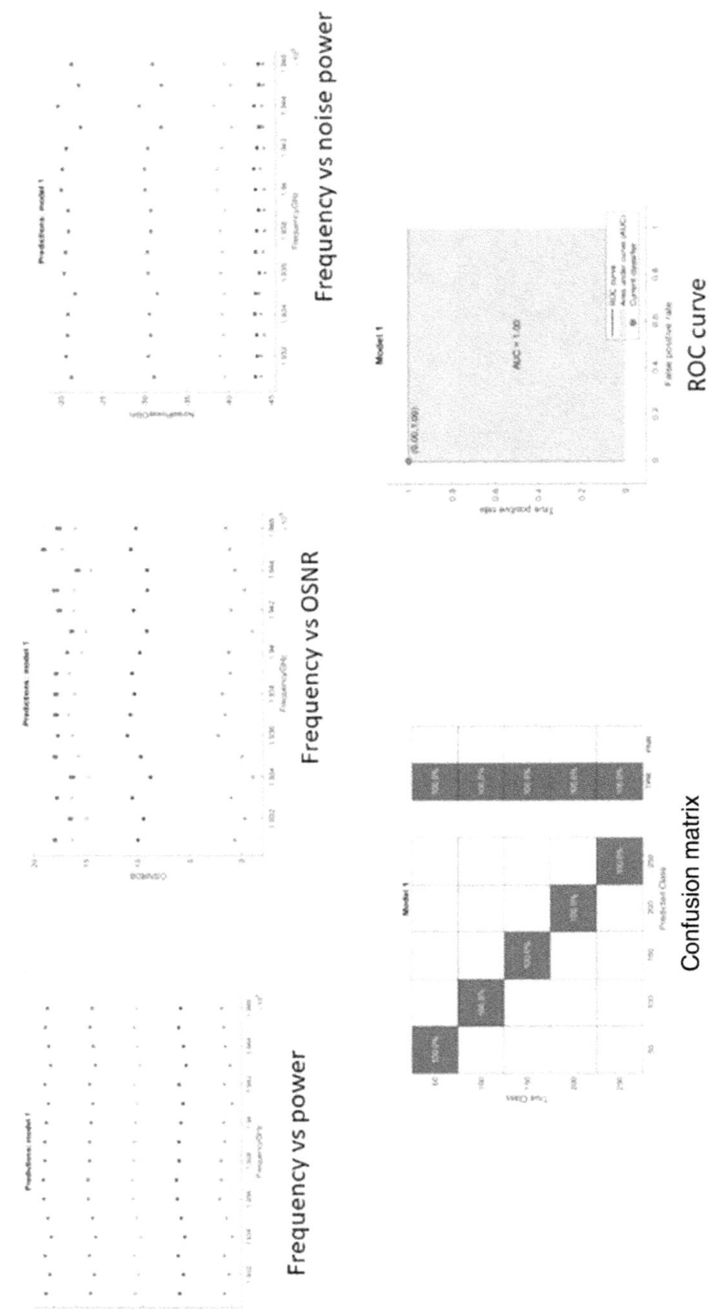

FIGURE 5.23 Calculation of optical parameters and ML metrics.

5.6 CONCLUSION

It is necessary to build an intelligent optical network for the futuristic network environment. AI plays a vital role in constructing and supervising a next-generation network effectively. The existing technology and intelligent optical networks are classified, and their corresponding challenges are discussed. The opportunities in ION are explored, and that creates new avenues. Applications of ML in the physical and network layer of the optical networks give a broad insight into intelligent optical networks. A comprehensive review of optical communication and network simulation tools under various network scenarios is studied, enabling new research solutions. The simulation study drives a better understanding of intelligent optical networks. Thus, this chapter gives wisdom about the future of the optical networks industry.

ACKNOWLEDGEMENT

The authors gratefully acknowledge the financial support received from the Science and Engineering Research Board (SERB), Government of India, Grant No: EEQ/2019/000647 for the purchase of the VPItransmissionmaker software.

REFERENCES

1. Dibakar Das et al., "A non-intrusive failure prediction mechanism for deployed optical networks", *International conference on communication systems and Networks*, 2021.
2. D. Simeonidou, R. Nejabati and S. Azodolmolky, "Enabling the future optical Internet with OpenFlow: A paradigm shift in providing intelligent optical network services," *2011 13th International Conference on Transparent Optical Networks*, Stockholm, Sweden, 2011.
3. CISCO, Cisco Annual Internet Report (2018–2023); CISCO: San Jose, CA, 2020.
4. Nokia, Deepfield Network Intelligence Report Networks in 2020; Nokia: Espoo, Finland, 2020.
5. W. S. Saif et al., "Machine learning techniques for optical performance monitoring and modulation format identification: A survey," *IEEE Communications Surveys & Tutorials*, vol. 22, no. 4, pp. 2839–2882, 2020.
6. Szostak et al., "Machine learning classification and regression approaches for optical network traffic prediction" *Electronics*, vol. 10, 2021.
7. M. S. Leeson, R. J. Green, M. D. Higgins and E. L. Hines, "Intelligent Systems Engineering: Optical network applications," *2011 13th International Conference on Transparent Optical Networks*, Stockholm, Sweden, 2011.
8. X. Liu, H. Lun, M. Fu, Y. Fan, L. Yi, W. Hu, Q. Zhuge, AI-based modeling and monitoring techniques for future intelligent elastic optical networks. *Applied Sciences*, vol. 10, p. 363, 2020.
9. I. Tomkos et al., "The evolution of optical networking", *Proceedings of the IEEE*, Vol. 100, No. 5, May 2012.
10. Hyytia Esa, "Resource Allocation and Performance Analysis Problems in Optical Networks", 2004.
11. www.fiberopticshare.com/ftth-access-networks-aon-vs-pon.html
12. www.carritech.com/news/different-types-pon-passive-optical-networks/
13. www.wiki.pathfinderdigital.com/wiki/elastic-optical-networks-eons

14. Tao Liua, He Wang, Yuqing Zhang, "A traffic anomaly detection scheme for non-directional denial of service attacks in software-defined optical network", *Computers and Security*, 2022.
15. www.intechopen.com/chapters/54939
16. K. Christodoulopoulos et al., "Toward efficient, reliable, and autonomous optical networks: the ORCHESTRA solution," *The Journal of Optical Communications and Networking*, 2019.
17. S. Manochandar, M. Punniyamoorthy, "Scaling feature selection method for enhancing the classification performance of Support Vector Machines in text mining", *Computers & Industrial Engineering*, vol. 124, 2018.
18. Mehryar Mohri, Afshin Rostamizadeh, and Ameet Talwalkar, *Foundations of Machine Learning*, MIT Press, 2nd Edition, 2018.
19. Christopher M. Bishop, *Pattern Recognition and Machine Learning*, Springer, 2006.
20. S. Manochandar, M. Punniyamoorthy, & R.K. Jeyachitra, "Development of new seed with modified validity measures for k-means clustering", *Computers & Industrial Engineering*, vol. 141, p. 106290, 2020.
21. Gu, R., Yang, Z., & Ji, Y, "Machine learning for intelligent optical networks: A comprehensive survey", *Journal of Network and Computer Applications*, vol. 157, p. 102576, 2020.
22. Smola, A.J., & Schölkopf, B, "A tutorial on support vector regression", *Statistics and Computing*, vol. 14, pp. 199–222, 2004.
23. Siepel, A., & Haussler, D. (2005). Phylogenetic Hidden Markov Models. In: *Statistical Methods in Molecular Evolution. Statistics for Biology and Health.* Springer, New York, NY.
24. Ding, S., Su, C. & Yu, J. An optimizing BP neural network algorithm based on genetic algorithm. *Artificial Intelligence Review* vol. 36, pp. 153–162, 2011.
25. Zhou, L., Zhao, P., Wu, D. et al. "Time series model for forecasting the number of new admission inpatients", *BMC Medical Informatics and Decision Making* vol. 18, p. 39, 2018.
26. Le, Le Thi, Hoang Nguyen, Jie Dou, and Jian Zhou. "A Comparative Study of PSO-ANN, GA-ANN, ICA-ANN, and ABC-ANN in Estimating the Heating Load of Buildings' Energy Efficiency for Smart City Planning" *Applied Sciences*, vol. 9, no. 13, p. 2630, 2019.
27. M. Hasan, E. Hossain and D. Niyato, "Random access for machine-to-machine communication in LTE-advanced networks: issues and approaches," *IEEE Communications Magazine*, vol. 51, no. 6, pp. 86–93, June 2013.
28. A. Laya, L. Alonso and J. Alonso-Zarate, "Is the random access channel of LTE and LTE-A suitable for M2M communications? A survey of alternatives," *IEEE Communications Surveys & Tutorials*, vol. 16, no. 1, pp. 4–16, First Quarter 2014.
29. Anastasopoulos, M.P., Tzanakaki, A., Rofoee, B.R. et al. Optical wireless network convergence in support of energy-efficient mobile cloud services. *Photonic Network Communications* vol. 29, pp. 269–281, 2015.
30. Z. Hasan, H. Boostanimehr and V. K. Bhargava, "Green cellular networks: A survey, some research issues and challenges," in *IEEE Communications Surveys & Tutorials*, vol. 13, no. 4, pp. 524–540, Fourth Quarter 2011.
31. Tania Panayiotou, Giannis Savva, Ioannis Tomkos, and Georgios Ellinas, "Decentralizing machine-learning-based QoT estimation for sliceable optical networks," *Journal of Optical Communications and Networking*, vol. 12, pp. 146–162, 2020.
32. Ihtesham Khan, Muhammad Bilal, and Vittorio Curri, "Assessment of cross-train machine learning techniques for QoT-estimation in agnostic optical networks," *OSA Continuum*, vol. 3, pp. 2690–2706, 2020.

33. Ji, Y., Gu, R., Yang, Z. et al. "Artificial intelligence-driven autonomous optical networks: 3S architecture and key technologies," *Science China Information Sciences*, vol. 63, p. 160301, 2020.
34. Elias Giacoumidis, Yi Lin, Michaela Blott, Liam P. Barry, "Real-time machine learning based fiber-induced nonlinearity compensation in energy-efficient coherent optical networks," *APL Photonics*, vol. 5, no. 4, p. 041301, 2020.
35. Thanh Tuan Tran, Minje Song, Minhyup Song, Dongsun Seo, "Highly flat optical frequency comb generation based on pulse carving and sinusoidal phase modulation," *Optical Engineering*, vol. 58, no. 7, 076103 (11 July 2019).
36. Iyer, Sridhar and Singh, Shree Prakash. "Effect of channel spacing on the design of mixed line rate optical wavelength division multiplexed networks," *Journal of Optical Communications*, vol. 40, no. 1, 2019, pp. 75–82.
37. Oleg G. Morozov et al., "External amplitude-phase modulation of laser radiation for generation of microwave frequency carriers and optical poly-harmonic signals: an overview," *Proc. SPIE 9807, Optical Technologies for Telecommunications 2015*, 980711 (26 March 2016).
38. Bijoy Chatterjee, Eiji Oki, *Elastic Optical Networks: Fundamentals, Design, Control, and Management* (pp. 169–175), CRC Press, 2020.
39. D. Dahan and U. Mahlab, "OSNR system margin monitoring technique for coherent transparent optical networks," *2014 16th International Conference on Transparent Optical Networks (ICTON)*, Graz, Austria, pp. 1–4, 2014.
40. Jammal M, Singh T, Shami A, Asal R and Li Y, Software defined networking: State of the art and research challenges, *Computer Networks*, vol. 72 p 74–98, 2014.
41. J. R. de Almeida Amazonas, G. Santos-Boada, S. Ricciardi and J. Solé-Pareta, "Technical challenges and deployment perspectives of SDN based elastic optical networks," *2016 18th International Conference on Transparent Optical Networks (ICTON)*, Trento, Italy, 2016.
42. H. Yang et al., "Intelligent Optical Network with AI and Blockchain," *2019 18th International Conference on Optical Communications and Networks (ICOCN)*, Huangshan, China, 2019, pp. 1–3.
43. S. Allogba, S. Aladin and C. Tremblay, "Machine-Learning-Based Lightpath QoT Estimation and Forecasting," *Journal of Lightwave Technology*, vol. 40, no. 10, pp. 3115–3127, 2022.
44. R. M. Morais, "On the suitability, requisites, and challenges of machine learning [Invited]," *Journal of Optical Communications and Networking*, vol. 13, no. 1, pp. A1–A12, 2021.
45. A. M. Sadeghzadeh, S. Shiravi and R. Jalili, "Adversarial network traffic: Towards evaluating the robustness of deep-learning-based network traffic classification," *IEEE Transactions on Network and Service Management*, vol. 18, no. 2, pp. 1962–1976, 2021.
46. Rentao Gu, Zeyuan Yang, Yuefeng Ji, "Machine learning for intelligent optical networks: A comprehensive survey," *Journal of Network and Computer Applications*, vol. 157, 2020, 102576.
47. J. Mata et al., "Application of Artificial Intelligence Techniques in Optical Networks," *2018 IEEE Photonics Society Summer Topical Meeting Series (SUM)*, Waikoloa, HI, 2018, pp. 35–36.
48. Yang, H., Zhan, K., Yao, Q. et al. "Intent defined optical network with artificial intelligence-based automated operation and maintenance", *Science China Information Sciences*, vol. 63, 160304 (2020).
49. Javier Mata et al., "Artificial intelligence (AI) methods in optical networks: A comprehensive survey", *Optical Switching and Networking*, vol. 28, 2018, pp. 43–57.

50. Zhensheng Zhang, James Fu, Dan Guo and Leah Zhang, "Lightpath routing for intelligent optical networks," *IEEE Network*, vol. 15, no. 4, pp. 28–35, 2001.
51. T. Panayiotou, G. Savva, I. Tomkos and G. Ellinas, "Decentralizing machine-learning-based QoT estimation for sliceable optical networks," *Journal of Optical Communications and Networking*, vol. 12, no. 7, pp. 146–162, 2020.
52. Yvan Pointurier, "Machine learning techniques for quality of transmission estimation in optical networks," *The Journal of Optical Communications and Networking*, vol. 13, pp. B60–B71, 2021.
53. J. Hao et al., "Intelligent Scheduling Method of Optical Transmission Network Based on Digital Twin," *2023 IEEE 6th Information Technology, Networking, Electronic and Automation Control Conference (ITNEC)*, Chongqing, China, 2023.
54. Lu Zhang, Xin Li, Ying Tang, Jingjie Xin, Shanguo Huang, "A survey on QoT prediction using machine learning in optical networks", *Optical Fiber Technology*, vol. 68, p. 102804, 2022.
55. Aaron Chen, Jeffrey Law, and Michal Aibin. "A survey on traffic prediction techniques using artificial intelligence for communication networks" *Telecom* vol. 2, no. 4, pp. 518–535, 2021.
56. Huazhi Lun, Xiaomin Liu et al., "Machine-learning-based telemetry for monitoring long-haul optical transmission impairments: methodologies and challenges [Invited]," *Journal of Optical Communications and Networking*, vol. 13, pp. E94–E108, 2021.
57. D. Benjamin, R. Trudel, S. Shew and E. Kus, "Optical services over the intelligent optical network," *IEEE Communications Magazine*, vol. 39, no. 9, pp. 73–78, 2001.
58. M. Morgan, "An ant colony approach to regenerator placement with fault tolerance in optical networks," *2015 7th International Workshop on Reliable Networks Design and Modeling (RNDM)*, Munich, Germany, 2015.
59. Rui Manuel Morais, Claunir Pavan, Armando Nolasco Pinto, and Cristina Requejo, "Genetic Algorithm for the Topological Design of Survivable Optical Transport Networks," *Journal of Optical Communications and Networking*, vol. 3, pp. 17–26, 2011.
60. L. Velasco, P. Wright, A. Lord, and G. Junyent, "Saving CAPEX by extending flexgrid-based core optical networks toward the edges [Invited]," *Journal of Optical Communications and Networking*, vol. 5, pp. A171–A183, 2013.
61. Ignacio de Miguel, Reinaldo Vallejos, Alejandra Beghelli, and Ramón J. Durán, "Genetic Algorithm for Joint Routing and Dimensioning of Dynamic WDM Networks," *Journal of Optical Communications and Networking*, vol. 1, pp. 608–621, 2009.
62. Fábio Renan Durand and Taufik Abrão, "Energy-efficient power allocation for WDM/OCDM networks with particle swarm optimization," *Journal of Optical Communications and Networking*, vol. 5, pp. 512–523, 2013.
63. M. de Paula Marques, F. R. Durand and T. Abrão, "WDM/OCDM energy-efficient networks based on heuristic ant colony optimization," *IEEE Systems Journal*, vol. 10, no. 4, pp. 1482–1493, 2016.
64. S. G. Petridou, P. G. Sarigiannidis, G. I. Papadimitriou and A. S. Pomportsis, "On the use of clustering algorithms for message scheduling in WDM star networks," *Journal of Lightwave Technology*, vol. 26, no. 17, pp. 2999–3010, 2008.
65. Ronald Romero Reyes and Thomas Bauschert, "Adaptive and state-dependent online resource allocation in dynamic optical networks," *Journal of Optical Communications and Networking*, vol. 9, pp. B64–B77, 2017.
66. S. Ponmalar and S. Sundaravadivelu, "Design of high speed optical switches for intelligent optical networks," *2008 International Conference on Computing, Communication and Networking*, Karur, India, 2008.

67. Ruiz et al., "Service-triggered failure identification/localization through monitoring of multiple parameters." In *ECOC 2016; 42nd European Conference on Optical Communication*, pp. 1–3. VDE, 2016.
68. Harinder Singh et al., "Artificial intelligence based quality of transmission predictive model for cognitive optical networks", *Optik*, vol. 257, 2022.
69. Xu Zhang et al., "Failure recovery solutions using cognitive mechanisms for Software Defined Optical Networks," *2016 15th International Conference on Optical Communications and Networks (ICOCN)*, Hangzhou, 2016.
70. S. R. Tembo, S. Vaton, J. L. Courant and S. Gosselin, "A tutorial on the EM algorithm for Bayesian networks: Application to self-diagnosis of GPON-FTTH networks," *2016 International Wireless Communications and Mobile Computing Conference (IWCMC)*, Paphos, Cyprus, 2016.
71. I. S. Hwang, J. Y. Lee and A. T. Liem, "Genetic expression programming: A new approach for QoS traffic prediction in EPONs," *2012 Fourth International Conference on Ubiquitous and Future Networks (ICUFN)*, Phuket, Thailand, 2012.
72. C. T. Huang, "The Study of Balance Traffic Load with Genetic Algorithm for PON," *2008 International Conference on Intelligent Information Hiding and Multimedia Signal Processing*, Harbin, China, 2008.
73. K. Chitra and M. R. Senkumar, "Hidden Markov model based lightpath establishment technique for improving QoS in optical WDM networks," *Second International Conference on Current Trends In Engineering and Technology*, 2014.
74. M. Glick and H. Rastegarfar, "Scheduling and control in hybrid data centers," *2017 IEEE Photonics Society Summer Topical Meeting Series (SUM)*, San Juan, PR, 2017.
75. F. Musumeci et al., "An overview on application of machine learning techniques in optical networks," *IEEE Communications Surveys & Tutorials*, vol. 21, no. 2, pp. 1383–1408, Second quarter 2019.
76. M. Kaidan et al., "Intelligent Data Flow Aggregation in Edge Nodes of Optical Label Switching Networks," *2019 3rd International Conference on Advanced Information and Communications Technologies (AICT)*, Lviv, Ukraine, 2019.
77. T. Tanaka, W. Kawakami, S. Kuwabara, S. Kobayashi and A. Hirano, "Intelligent Monitoring of Optical Fiber Bend Using Artificial Neural Networks Trained With Constellation Data," *IEEE Networking Letters*, 2019.
78. S. Vishwakarma and R. K. Jeyachitra, "Analysis of OSNR and Data Rate Selection Using ML Techniques for Optical Networks," *2020 5th International Conference on Devices, Circuits and Systems (ICDCS)*, Coimbatore, India, 2020.
79. Sindhumitha Kulandaivel, R.K. Jeyachitra, "Combined image Hough transform based simultaneous multi-parameter optical performance monitoring for intelligent optical networks," *Optical Fiber Technology*, 2023.
80. C. Yu, H. Wang, C. Ke, Z. Liang, S. Cui and D. Liu, "Multi-task learning convolutional neural network and optical spectrums enabled optical performance monitoring," *IEEE Photonics Journal*, vol. 14, no. 2, pp. 1–8, 2022, Art no. 7217808.
81. W. S. Saif, A. M. Ragheb, T. A. Alshawi, & S. A. Alshebeili (2021). Optical performance monitoring in mode division multiplexed optical networks. *Journal of Lightwave Technology*, vol. 39, no. 2, 491–504.
82. D. Wang et al., "Optical performance monitoring of multiple parameters in future optical networks," *Journal of Lightwave Technology*, vol. 39, no. 12, pp. 3792–3800, 2021.
83. F. Locatelli et al., "Spectral processing techniques for efficient monitoring in optical networks," *Journal of Optical Communications and Networking*, vol. 13, no. 7, pp. 158–168, 2021.

84. Takahito Tanimura et al., "Convolutional neural network-based optical performance monitoring for optical transport networks," *Journal of Optical Communications and Networking*, vol. 11, pp. A52–A59, 2019.
85. A. D'Amico et al., "Enhancing Lightpath QoT Computation With Machine Learning in Partially Disaggregated Optical Networks," in *IEEE Open Journal of the Communications Society*, vol. 2, pp. 564–574, 2021.
86. J. Müller et al., "A QoT Estimation Method using EGN-assisted Machine Learning for Network Planning Applications," *2021 European Conference on Optical Communication (ECOC)*, Bordeaux, France, 2021.
87. R. Pousa, P. Georgieva, J. Pina, P. Cruz and P. André, "Machine Learning Approach for Online Monitoring of Quality of Transmission Performance Indicators in Optical Fiber Networks," *2021 European Conference on Optical Communication (ECOC)*, Bordeaux, France, 2021.
88. D. Azzimonti, C. Rottondi and M. Tornatore, "Reducing probes for quality of transmission estimation in optical networks with active learning," *Journal of Optical Communications and Networking*, vol. 12, no. 1, pp. A38–A48, 2020.
89. Dawei Wang, Qi Sui, and Zhaohui Li. "Toward universal optical performance monitoring for intelligent optical fiber communication networks," *IEEE Communications Magazine*, vol. 58, no. 9, pp. 54–59, 2020.
90. F. Da Ros, S. M. Ranzini, H. Bülow and D. Zibar, "Reservoir-Computing Based Equalization With Optical Pre-Processing for Short-Reach Optical Transmission," *IEEE Journal of Selected Topics in Quantum Electronics*, vol. 26, no. 5, pp. 1–12, 2020.
91. M. Schädler, G. Böcherer and S. Pachnicke, "Soft-Demapping for Short Reach Optical Communication: A Comparison of Deep Neural Networks and Volterra Series," *Journal of Lightwave Technology*, vol. 39, no. 10, pp. 3095–3105, 2021.
92. Zhaopeng Xu, Shuangyu Dong, Jonathan H. Manton, and William Shieh, "Low-complexity multi-task learning aided neural networks for equalization in short-reach optical interconnects," *Journal of Lightwave Technology*, 2022.
93. Daniel Semrau, Eric Sillekens, Polina Bayvel, and Robert I. Killey, "Modeling and mitigation of fiber nonlinearity in wideband optical signal transmission [Invited]," *Journal of Optical Communications and Networking*, vol. 12, pp. C68–C76, 2020.
94. M. Schaedler, F. Pittalà, S. Calabrò, G. Böcherer, C. Bluemm and S. Pachnicke, "Recurrent Neural Network Soft Demapping for Mitigation of Fiber Nonlinearities and ISI," *2021 Optical Fiber Communications Conference and Exhibition (OFC)*, San Francisco, CA, 2021.
95. M. Kamalian-Kopae, A. A. Ali, K. Nurlybayeva, A. Ellis and S. Turitsyn, "Neural Network-Enhanced Optical Phase Conjugation for Nonlinearity Mitigation," *Optical Fiber Communications Conference and Exhibition (OFC)*, 2022.
96. Lin Jiang et al. "Blind density-peak-based modulation format identification for elastic optical networks," *Journal of Lightwave Technology*, vol. 36, pp. 2850–2858, 2018.
97. X. Fan et al., "Joint optical performance monitoring and modulation format/bit-rate identification by CNN-based multi-task learning," *IEEE Photonics Journal*, vol. 10, no. 5, pp. 1–12, 2018.
98. W. S. Saif, A. M. Ragheb, B. Nebendahl, T. Alshawi, M. Marey and S. A. Alshebeili, "Performance Investigation of Modulation Format Identification in Super-Channel Optical Networks," *IEEE Photonics Journal*, 2022.
99. L. Yang et al., "Modulation format identification using graph-based 2D stokes plane analysis for elastic optical network," *IEEE Photonics Journal*, vol. 13, no. 1, pp. 1–15, 2021.

100. V. S. Ghayal, R. K. Jeyachitra Efficient eye diagram analyzer for optical modulation format recognition using deep learning technique. In: Sengodan, T., Murugappan, M., Misra, S. (eds) *Advances in Electrical and Computer Technologies*. Lecture Notes in Electrical Engineering, vol. 672. Springer 2020.
101. D. Wang et al., "The role of digital twin in optical communication: Fault management, hardware configuration, and transmission simulation," *IEEE Communications Magazine*, vol. 59, no. 1, pp. 133–139, 2021.
102. K. S. Mayer et al., "Demonstration of ML-assisted soft-failure localization based on network digital twins," *Journal of Lightwave Technology*, vol. 40, no. 14, pp. 4514–4520, 2022.
103. C. J. A. Bastos-Filho et al., "Investigating the Creation of a Surrogate Model for Adaptive Control of Amplifier Operating Point Using Machine Learning," *2020 22nd International Conference on Transparent Optical Networks (ICTON)*, 2020.
104. G. Borraccini et al., "QoT-E Driven Optimized Amplifier Control in Disaggregated Optical Networks," *2021 Optical Fiber Communications Conference and Exhibition (OFC)*, 2021, pp. 1–3.
105. Chunyu Zhang, et al., "Potential failure cause identification for optical networks using deep learning with an attention mechanism," *Journal of Optical Communications and Networking*, vol. 14, pp. A122–A133, 2022.
106. Adekoya Adebayo Felix Nyarko-Boateng, Weyori Benjamin Asubam, Predicting the actual location of faults in underground optical networks using linear regression, Engineering Reports, 2020.
107. M. Klinkowski et al., "Machine learning assisted optimization of dynamic crosstalk-aware spectrally-spatially flexible optical networks," *Journal of Lightwave Technology*, 2020.
108. S. Nallaperuma et al., "Parameter optimisation for ultra-wideband optical networks in the presence of stimulated Raman scattering effect," *2021 International Conference on Optical Network Design and Modeling (ONDM)*, 2021.
109. X. Zhu, O. Xu and G. Li, "Prediction accuracy improvement of passive optical network traffic by a LSTM model with a new activation function," *2020 7th International Conference on Information, Cybernetics, and Computational Social Systems (ICCSS)*, Guangzhou, China, 2020.
110. S. K. Singh, R. Proietti, C.-Y. Liu and S. J. B. Yoo, "Multi-Cluster Reconfiguration with Traffic Prediction in Hyper-Flex-LION Architecture," *2022 Optical Fiber Communications Conference and Exhibition (OFC)*, 2022.
111. Kiyo Ishii and Shu Namiki, "Scalability of integer linear programming path computation for functional block-based disaggregation supporting a flexible grid mechanism [Invited]," *Journal of Optical Communications and Networking*, vol. 14, pp. A134–A142, 2022.
112. Mihail Balanici and Stephan Pachnicke, "Classification and forecasting of real-time server traffic flows employing long short-term memory for hybrid E/O data center networks," *Journal of Optical Communications and Networking*, vol. 13, pp. 85–93, 2021.
113. I. Tomkos, C. Kachris, P. S. Khodashenas and J. K. Soldatos, "Optical networking solutions and technologies in the big data era," *2015 17th International Conference on Transparent Optical Networks (ICTON)*, Budapest, Hungary, 2015.
114. D. Wang, Y. Song and C. Zhao, "Bayesian classification based service-awareness in software defined optical network for big data services," *2017 IEEE 2nd International Conference on Big Data Analysis (ICBDA)*, Beijing, China, 2017.
115. C. Zhao, H. Chang and Q. Liu, "Bayesian algorithm based traffic prediction of big data services in OpenFlow controlled optical networks," *2017 IEEE 2nd International Conference on Big Data Analysis (ICBDA)*, Beijing, China, 2017.

116. L. Gifre, L. M. Contreras, V. López and L. Velasco, "Big data analytics in support of virtual network topology adaptability," *2016 Optical Fiber Communications Conference and Exhibition (OFC)*, Fukuoka, CA, 2016.
117. H. Lu, M. Zhang, M. Wang, C. Song, D. Wang and L. Guan, "Big-Data-Driven Dynamic Clustering and Load Balancing of Virtual Base Stations for 5G Fronthaul Network," *2019 24th OptoElectronics and Communications Conference (OECC) and 2019 International Conference on Photonics in Switching and Computing (PSC)*, Fukuoka, Japan, 2019.
118. U. B. Chander, "Analysis on Implementation of Optical Network Management Solution with Hadoop Architecture," *2018 Tenth International Conference on Advanced Computing (ICoAC)*, Chennai, India, 2018.
119. K.-I. Kitayama et al., "Green, smart optical packet switching network with flow control for data centers," *2013 IEEE Photonics Society Summer Topical Meeting Series*, Waikoloa, HI, 2013.
120. A. M. Al-Salim et al., "Energy Efficient Tapered Data Networks for Big Data processing in IP/WDM networks," *2015 17th International Conference on Transparent Optical Networks (ICTON)*, 2015.
121. Hossam Mahmoud Ahmad Fahmy, *Concepts, Applications, Experimentation and analysis of wireless sensor networks*, 2nd Edition, Springer 2021.
122. Irin Dorathy, M. Chandrasekaran, "Simulation tools for mobile ad hoc networks: a survey", *Journal of Applied Research and Technology*, vol. 16, no. 5, pp. 437–445, 2018.
123. www.synopsys.com/photonic-solutions/optsim.html
124. www.optiwave.com/optisystem-overview
125. www.optiwave.com/optispice-overview/
126. www.vpiphotonics.com/Tools/OpticalSystems
127. www.ospinsight.com
128. www.adva.com/en/products/automated-network-management/ensemble-simulator
129. David Steckler (2023). Optical Communications Systems (SOFTDM Ver 1.5) Update Matlab 2010a (https://www.mathworks.com/matlabcentral/fileexchange/11359-optical-communications-systems-softdm-ver-1-5-update-matlab-2010a), MATLAB Central File Exchange. Retrieved May 19, 2023.
130. I. Khan, M. Bilal, M. Umar Masood, A. D'Amico and V. Curri, "Lightpath QoT computation in optical networks assisted by transfer learning," *Journal of Optical Communications and Networking*, vol. 13, no. 4, pp. B72–B82, 2021.
131. K. Christodoulopoulos et al., "Toward efficient, reliable, and autonomous optical networks: the ORCHESTRA solution [Invited]," *Journal of Optical Communications and Networking*, vol. 11, no. 9, pp. C10–C24, 2019.
132. www.wiki.opennetworking.org/display/OTCC/TAPI
133. www.openconfig.net/

6 Machine Learning for Non-Orthogonal Multiple Access

Abhinav Singh Parihar and Shubham Bisen
IIT Indore, Indore, India

Vimal Bhatia and Ondrej Krejcar
University of Hradec Kralove, Kralove, Czech Republic

6.1 INTRODUCTION: BACKGROUND AND DRIVING FORCES

The radio access architecture for beyond fifth-generation (B5G) networks has grown to rely heavily on non-orthogonal multiple access (NOMA) principles [1, 2]. Despite the fact that numerous multiple access methods have been developed for B5G networks, power-domain NOMA (PD-NOMA) [3, 4]; sparse code multiple access (SCMA) [5–8]; pattern division multiple access (PDMA) [9]; low density spreading (LDS) [10]; lattice partition multiple access [11]; and interleave division multiple access (IDMA) [12], all these methods are depending on the same underlying principle, where several users are served by a single orthogonal resource block (e.g., time/frequency/spreading code). Unlike NOMA, which allows for multiple users to share the same block of orthogonal resources, orthogonal multiple access (OMA) systems like time division multiple access (TDMA), frequency division multiple access (FDMA), and orthogonal frequency division multiple access (OFDMA) assign orthogonal resource block to users.

This fundamental difference in resource allocation exemplifies the spectral inefficiency of OMA, as demonstrated in the situation considering one user must be served despite having poor channel conditions, e.g., it is possible that the user either has high-priority data or hasn't received service in a while. In this scenario, employing OMA ensures that a user, regardless of unfavorable channel conditions, monopolizes the entire available bandwidth resource. This results in a negative impact on the system's throughput and spectrum efficiency. However, implementing NOMA in such a situation allows users with superior channel conditions to effectively share the same bandwidth resources with weaker users, while simultaneously ensuring that users experiencing poor channel conditions are still adequately served. Consequently, NOMA exhibits significantly higher system throughput compared to OMA [13, 14] if user fairness has to be guaranteed.

Furthermore, academic studies have demonstrated that NOMA not only achieves improved spectral efficiency but also effectively accommodates extensive connectivity,

DOI: 10.1201/9781003303114-6

which is crucial to guarantee that the Internet of things (IoT) features can be supported by the upcoming 5G network [15, 16]. While the utilization of NOMA in networks is a recent development, analogous subjects have been studied for a considerable period. Notably, components central to NOMA, including superposition coding and successive interference cancellation (SIC), have already been devised over two decades ago [17, 18]. However, the principle of NOMA, which involves the elimination of orthogonality, was not employed in previous generations. In this regard, the underlying theories of NOMA and code division multiple access (CDMA) are substantially distinct. In CDMA, users are separated based on the distinctions in their spreading codes. In contrast, NOMA leverages the utilization of the same code for multiple users. Consequently, CDMA requires a significant margin between the chip rate and the desired data rate.

Conventionally, due to NOMA's compatibility with various communication technologies, it has the capability to be seamlessly incorporated into both current and forthcoming wireless systems. For example, extensive research has demonstrated the coexistence of NOMA with traditional OMA techniques comprising TDMA and OFDMA. Because of this, the concept of NOMA has been put forth as a proposal within the framework of the third-generation partnership project long-term evolution advanced (LTE-A) standard [19], referring as a Multi-User Superposition Transmission. Specifically, the principle of NOMA enables the simultaneous service of two users on a single OFDMA subcarrier within the LTE resource blocks without necessitating any modifications. Additionally, NOMA has recently been incorporated into the upcoming digital TV standard (ATSC 3.0), known as Layered Division Multiplexing (LDM) [20]. Specifically, the NOMA principle and the superposition of several data streams improve the spectral efficiency of TV broadcasting. The mentioned instances vividly demonstrate the broad applicability of NOMA, extending beyond B5G networks to encompass various existing and forthcoming wireless systems.

6.2 PD NOMA

A promising approach for efficiently implementing multiple access is to leverage the power domain by applying the NOMA principle to individual orthogonal resource blocks, which represent a single carrier. In this section, we will begin by discussing PD NOMA, also known as non-orthogonal multiple access, which arises as a result of this phenomenon. This is followed by a description of a PD NOMA variant known as cognitive radio-inspired NOMA (CR-NOMA), which may precisely fulfill the users' different quality of service (QoS) criteria. Multiple access is accomplished in PD NOMA through allocating varying power levels to different users, facilitating the inclusion of multiple users within the same resource block [21, 22]. Multiuser superposition transmission (MUST) illustrates the PD NOMA scheme designed for downlink transmission with two users that are simultaneously served by a base station (BS) on the same OFDMA subcarrier. Consider a two-user NOMA system, where the source (BS) is serving two users, UE_1 and UE_2. Let the channels of the two users be denoted as h_i, where $i \in [1, 2]$, assuming that $h_1 \leq h_2$. The BS assigns corresponding power coefficients to the messages from the users and superimposes them, represented as a_i, $i \in [1, 2]$. The fundamental concept in PD NOMA is to assign a higher power allocation to users experiencing weaker channel, i.e., $a_1 \geq a_2$ and $a_1 + a_2 = 1$ if$|h_1| \leq |h_2|$. By treating UE_2's message as interference, UE_1 achieves a rate of $log_2\left(1+|h_1|^2\frac{a_1^2}{|h_2|^2 a_2^2+\frac{1}{\rho}}\right)$ bits/channel used (bpcu) by successfully decoding its own message directly, where ρ denotes the transmit signal-to-noise ratio (SNR). On the contrary, UE_2 executes SIC, meaning that it detects UE_1's signal first and then filters it out of its observation prior to detecting its own signal. This approach attains an achievable rate of $log_2\left(1+\frac{|h_2|^2 a_1^2}{\rho}\right)$.

- **Example 1**

  Let's consider a scenario with a high SNR, denoted as $\rho \to \infty$. To illustrate this, let's assume that $\rho\ |h_2|^2 \to 0$, indicates a significant fade in UE_1's channel. The total rate attained by NOMA is approximated as follows:

$$log_2\left(1+\frac{a_1^2|h_1|^2}{a_2^2|h_1|^2+\frac{1}{\rho}}\right)+log_2\left(1+a_1^2\rho|h_2|^2\right)\approx log_2\left(1+\frac{a_1^2}{a_2^2}\right)+log_2\left(\rho a_1^2|h_2|^2\right)$$
$$=log_2\left(\rho|h_2|^2\right).$$

  In contrast, the total achievable rate of OMA can be expressed in the following manner:

$$\frac{1}{2}log_2\left(1+\rho|h_1|^2\right)+\frac{1}{2}log_2\left(1+\rho|h_2|^2\right)\approx\frac{1}{2}log_2\left(\rho|h_2|^2\right).$$

  The superiority of NOMA over OMA in terms of performance becomes evident when observing above equations.

- **Example 2**

  Considering UE_1 is an IoT device with a minimal data rate requirement, while UE_2 with a substantial data rate demand. Each user is assigned to a subcarrier when OFDMA is employed, which is a kind of OMA. The example demonstrates a case where OMA exhibits a low spectral efficiency; this is because UE_1 receives higher bandwidth than it actually requires, while allocated bandwidth for the broadband user is insufficient. On the contrary, NOMA promotes spectrum sharing, allowing the broadband user to access the subcarrier utilized by the IoT device. This facilitates extensive connectivity and effectively meets the varied QoS needs of users, making NOMA an effective solution.

One could suggest that the disadvantage mentioned above can be avoided by using OMA with efficient resource allocation. Even though both NOMA and OMA utilize efficient resource allocation, substantial performance enhancements have been observed with NOMA. Furthermore, it is important to acknowledge that adaptive resource allocation in OMA introduces dynamic changes to the attributes of orthogonal resource blocks. However, this approach may necessitate impractically short time slots, making it infeasible in practical implementations.

6.3 CR NOMA

When users experience poor channel conditions, standard PD NOMA adjusts power allocation to ensure fairness among users. However, PD NOMA fails to ensure the QoS guarantees of users effectively. Cognitive radio (CR) NOMA, another form of PD NOMA, effectively guarantees all or some of the QoS requirements of the users. CR-NOMA operates by treating NOMA as a particular instance of cognitive radio. Its objective is to fulfill the specified QoS demands of users through optimized power allocation.

To illustrate CR-NOMA, we can examine a scenario where a base station (BS) employs the NOMA principle to transmit data to two downlink users, following the approach described in [13]. In this example, the primary user is considered to have inferior channel conditions within a cognitive radio network. As a result, this user has specific data rate requirements that need to be fulfilled. The power allocation policy is constrained by this data rate requirement, which can be explained as follows:

$$log_2\left(1+\frac{|h_1|^2 a_1^2}{|h_1|^2 a_2^2+\frac{1}{\rho}}\right) \geq R_1,$$

given that R_1 represents the desired data rate for UE_1, the power allocation policy inspired by cognitive radio can be formulated by considering the aforementioned constraint as follows:

$$a_1^2 = \min\left\{1, \frac{\left(|h_1|^2+\frac{1}{\rho}\right)}{|h_1|^2} 2^{R_1}\right\}.$$

This implies that for a large R_1, all the power is assigned exclusively to UE_1. The purpose of the aforementioned CR-NOMA power allocation policy is to guarantee that UE_1 consistently receives sufficient power to fulfill its QoS requirements. Subsequently, the remaining power will be used to serve UE_2. On the basis of Example 2 from the previous section, it is easy to explain the advantages of CR-NOMA. Specifically, the utilization of OMA implies that the low-rate IoT device exclusively utilizes a single resource block. By utilizing CR-NOMA, an extra user can be accommodated to this resource block in addition to the IoT device with its specified QoS requirement, which improves the total system throughput.

6.4 MULTI-CARRIER NOMA

A hybrid form of NOMA, known as multi-carrier NOMA, can be considered, where users are segregated into multiple groups. Under this configuration, users of the same group are simultaneously served within a shared orthogonal resource, employing the NOMA principle. On the other hand, distinct groups are allocated separate orthogonal resources. The primary goal of implementing hybrid NOMA is to mitigate system complexity. Implementing NOMA in a single orthogonal resource block by grouping all users in the network presents challenges. This arises due to the fact that the user with the strongest channel is required to detect every message from other users prior to detecting its own, leading to heightened complexity and decoding delays. Taking multi-carrier NOMA as an illustration, users are segregated into multiple groups, which may not be mutually exclusive. Each group of users is allocated a dedicated subcarrier, employing the NOMA concept to mitigate intra-group interference. Distinct user clusters are allocated to separate subcarriers, effectively mitigating interference between groups. As a result, to accommodate additional users beyond the capacity of the current subcarriers and provide ubiquitous connectivity, the system must operate under an overloaded condition—a hybrid NOMA scheme can achieve this. It should be noted that overloading occurs with hybrid NOMA at a lower complexity because there is a finite number of users in each subcarrier. Multi- carrier NOMA stands out as an appealing hybrid NOMA option when compared to other hybrid NOMA types, mostly owing to the likelihood that OFDMA would be used in the upcoming B5G network.

Understanding the effect of user pairing on system performance is a crucial stage in the hybrid NOMA implementation process. User pairing is studied in [13, 23, 24], wherein two users are served by utilizing only one orthogonal resource. It has been observed in research that the greatest performance improvement over OMA is attained by merging two users from a group with highly contrasting channel conditions. It should be emphasized that the two users respond to NOMA in quite different ways. For example, in the case of a strong channel user, NOMA is preferred as it attains a higher individual data rate than OMA, resulting in superior performance. However, the weak channel user may experience a lower NOMA rate compared to OMA, although the decline in performance may be alleviated by implementing the CR-inspired strategy covered in the preceding section.

To verify the performance advantages emphasized by the theoretical studies concerning user grouping, it is imperative to have practical design solutions for multi-carrier NOMA. These design solutions play a crucial role in translating the theoretical advantages into actual implementation. As the complex challenges of subcarrier and power allocation are intricately interconnected, it is difficult to develop suitable resource allocation algorithms, as illustrated in [25, 26]. By employing monotonic optimization techniques, the non-convex optimization problem presented in [25] is effectively addressed, leading to an optimal solution for joint subcarrier allocation, user grouping, and power allocation. The significance of this optimal scenario lies in its ability to provide an upper bound on the multi-carrier NOMA performance. A successive convex optimization-based low-complexity suboptimal solution is also given in [25]. The resource allocation method employed yields performance improvements that are similar to those achieved by the optimal approach.

6.5 COOPERATIVE NOMA

The first kind of cooperative NOMA takes into account the interaction between users, where one user serves as a relay to forward information to another. The following factors inspire this cooperative NOMA. NOMA systems have redundant information that can be used for cooperative transmission. Such a case is the two-user downlink scenario, wherein the stronger user needs to decode the information of the weaker user prior to detecting its message. As a result, the strong user can help the weak user by serving as a standard relay. Consider Example 1 explained earlier, where the achievable rate of the weak user is $log_2\left(1+\frac{|h_1|^2 a_1^2}{|h_1|^2 a_2^2+\frac{1}{\rho}}\right)$, which is adversely hampered by the strong user's co-channel interference. Hence, data rate for the weak user can be increased by cooperative transmission.

In [27], a cooperative NOMA protocol has been proposed that relies on collaboration between users. For instance, let's consider a scenario involving two users where cooperative transmission is executed in two phases. The BS broadcasts the superimposed signal of the users during the first phase. The stronger user takes on the role of a relay, utilizing short-range communication technologies like Bluetooth or Wi-Fi to transmit the message of the weaker user during the second phase. Furthermore, it is demonstrated in [27] that cooperative NOMA performs better than cooperative OMA even when short-distance communications are not utilized. The reason for incorporating short-range communications is that, without them, cooperative NOMA can be accomplished within two time slots. In comparison, cooperative OMA necessitates three time slots, with the base station utilizing two of them to transmit the signals to the users individually, and an extra slot dedicated to the stronger user to serve the weaker user. Furthermore, to enhance the spectral efficiency, full duplexing relaying can be employed, as illustrated in [28–30]. In these works, full duplexing is used so that the strong user can simultaneously decode signals from the BS and transmit relay data. By doing this, the drawback of half-duplex relaying is avoided, which calls for a specific time slot for relay transmission. Importantly, the utilization of full duplexing is not limited to cooperative NOMA scenarios. It has been demonstrated to be effective in enhancing the throughput of joint designs for communications in non-cooperative NOMA scenarios, as evidenced in [31].

6.5.1 Employing Dedicated Relays

In an alternative approach to cooperative NOMA, dedicated relays are employed to support the users. This variant of cooperative NOMA was specifically devised to efficiently serve users located near the cell boundary. The benefits of that result are demonstrated by the example that follows. Assume that two users who are near to the cell edge can be assisted by a special relay. When employing cooperative OMA, four time slots are necessary for transmission. In this approach, BS transmit data to the relay in two time slots, followed by two additional time slots to transmit the data to the NOMA users from the relay. On the other hand, cooperative NOMA achieves the same transmission with only two time slots: one for broadcasting NOMA signals from the BS to the designated relay and another for transmitting NOMA signals from the relay to the NOMA users. The fact that cooperative NOMA requires only two time slots instead of four allows us to instantly infer its superior spectral efficiency. The research presented in [32] investigates the advantages of incorporating buffer-assisted relaying in NOMA systems with a dedicated relay. It is crucial to recognize that the concept of employing dedicated relays in cooperative NOMA is flexible and has been applied in diverse scenarios, encompassing varying numbers of transmitters and receivers [33]. Another rationale for utilizing dedicated relays is the prevalence of idle users in wireless networks, particularly in deployments such as sports stadiums or convention centers. To assist active users and improve system coverage, these idle users might be used as specialized relays. When there are several accessible specialized relays, relay selection, which was initially addressed for NOMA in [34], becomes a crucial issue. The main finding indicates that the traditional max-min criterion is only suitable for cooperative networks and is not the best choice. As a result, a two-stage relay selection technique has been studied, prioritizing users based on their QoS requirements rather than channel conditions. Specifically, in the primary stage, relays that ensure the users' performance under stringent QoS criteria are selected and organized into a subset. During the secondary stage, the relay that achieves the highest rate is chosen from the eligible relay subset to serve the other user, who can be opportunistically served. This method of relay selection not only surpasses the performance of the max-min scheme but also reduces the likelihood of experiencing a system-wide outage, as shown in [34].

6.6 MILLIMETER WAVE NOMA

Millimeter wave (mmWave) transmission is widely acknowledged as a significant underlying technology for B5G networks [35]. The limited availability of spectrum resources below 6 GHz is a driving factor behind the development of both mmWave communications and NOMA in wireless communication. In contrast to NOMA, which improves spectrum efficiency, mmWave makes use of the less-utilized mmWave frequency spectrum. In mid-2016, a notable breakthrough in mmWave communications took place with approval from the US Federal Communications Commission. This approval involved the allocation of more than 10 GHz in the mmWave bands, within the framework of B5G [36]. Even with the ample spectrum resources accessible in the mmWave bands, the adoption of NOMA remains vital in mmWave networks for two key reasons. Firstly, NOMA plays a crucial role in facilitating extensive connectivity within mmWave networks, particularly in high-density environments like sports complexes with a large user base. By adopting NOMA, it becomes possible to efficiently accommodate a substantial number of users with varying QoS needs simultaneously, a feat that cannot be achieved with OMA. Second, the advantage from employing the mmWave bands will be swiftly eclipsed by the exponential rise in the need for novel data services such as augmented reality, mixed reality, and games. For instance, a data rate of 1000 Gbits/s is needed to transmit a high-quality data according to [37]. NOMA can greatly enhance the spectrum efficiency of mmWave communications and meet increasing requirements [38].

It is important to note that some characteristics of mmWave propagation also aid its combination with NOMA [39]. Specifically, users in mmWave networks may have highly correlated channels because of the directionality of mmWave transmission. For illustration, consider a mmWave network where each user has a single antenna and the base station is outfitted with M antennas. In a network characterized by a dense user deployment, it might be possible that many users will utilize the same normalized LOS direction; this indicates a high degree of correlation between their channels [40]. Within OMA schemes, the presence of channel correlation leads to a reduction in the multiplexing gain and subsequently affects the overall spectral efficiency. However, the adoption of NOMA can result in the dirty paper coding (DPC) performance when users' channel vectors are significantly correlated, as demonstrated by the notion of quasi-degradation in [41]. In other words, NOMA greatly enhances system throughput and is well-suited with the properties of mmWave transmission.

6.7 DETECTION IN NOMA: FROM SIC TO DEEP LEARNING

The primary concept behind NOMA is to serve multiple users over the same resource block utilizing the power diversity at the transmitter node. In the downlink NOMA system, the superposition coding is performed at the base station by allocating a different amount of power to users [1]. Due to multiple users sharing the same resource block, the desired user suffers from multi-user interference. Thus, multi-user detection technique at the receiver is required to mitigate the multi-user interference. SIC is the most prevalent algorithm for canceling the interference in the NOMA. In SIC, the decoding is performed iteratively at a particular user, from tthe strongest user signal to the weakest in succession. Once the strongest user signal is detected, it is canceled from the received signal to enhance the instantaneous signal to interference plus noise ratio (SINR) for the remaining symbols. Similarly, in the uplink NOMA system, the base station simultaneously receives the signal comprising all the transmitted users. Subsequently, the NOMA users' signal is decoded using SIC at the base station.

The performance of the NOMA system in terms of error rate considering SIC receiver is analyzed in [42–50]. In [42], the authors considered an uplink NOMA system, and the bit error rate (BER) expression is derived using the quadrature phase shift key (QPSK) scheme. In [43], closed form BER expression is derived for uplink NOMA system with two users using the binary phase shift key and QPSK. In reference [45], the authors derived the exact BER for downlink NOMA systems over Nakagami-m channels using QPSK. In [46], the authors analyzed the downlink NOMA system with symbol-level SIC and obtained a closed-form expression for the symbol error rate (SER) using Quadrature Amplitude Modulation (QAM). The work presented in [47] derived the pairwise error probability expression for downlink NOMA considering symbol-level SIC [49] focused on a downlink coordinated relay NOMA system and derived a closed-form expression for the average SER (ASER) using QAM over a Rayleigh faded channel. In [50], the authors investigated a downlink NOMA system with an arbitrary number of users and modulation order using the QAM scheme over a Rayleigh faded channel.

Since SIC is a layered algorithm, it suffers from error propagation at each layer due to the wrong decisions in the earlier detection layer. In practical systems, error propagation is inevitable, and thus SIC is generally imperfect [51, 52]. Therefore, the SIC will not be able to detect the remaining user's signal perfectly if any prior user fails to decode. As a result, imperfect SIC significantly degrades the performance of NOMA users. The analysis of the NOMA system with imperfect SIC is investigated in [53]. In [53], the authors consider uplink and downlink NOMA systems, and BER performance is analyzed considering perfect and imperfect SIC.

While SIC decoding can be implemented with low complexity, it has its own disadvantage. For example, the works in [42, 43, 53] studied the error rate performance, which suffers from error floor and thus reduces the reliability of the NOMA transmission. In the recent work a different decoding scheme has been proposed [54–57] that removes the error floor under stringent conditions on the number of users and power allocation. Joint multi-user detection techniques have recently been attracted as detectors for NOMA systems [56]. In [56], the authors considered an uplink NOMA system with a joint maximum likelihood detector, and the upper bound of BER is derived considering a two-user NOMA system. The joint maximum likelihood detector significantly improved the performance with the increased computational complexity at the receiver.

In recent times, there has been a growing interest in leveraging deep learning (DL) techniques in the field of wireless communication. Numerous studies have explored the use of DL to enhance channel estimation and signal detection processes. Applying DL in NOMA systems shows promise in overcoming the limitations of the SIC method. In [58], DL is investigated for multiuser detection in a MIMO-NOMA system. The study showcases the effectiveness of DL in replacing the SIC receiver, resulting in lower computational complexity. Furthermore, DL-based algorithms have been applied to NOMA, and the investigation of multiuser detection can be found in [59, 60].

6.8 PRACTICAL IMPLEMENTATION OF NOMA

6.8.1 Modulation and Coding for NOMA

For NOMA, efficient channel modulation and coding methods are essential to achieving the theoretically expected possible rates in reality. In the NOMA systems described in references [61,62], pulse amplitude modulation is employed with gray labeling and turbo codes. Channel codes other than turbo codes are also used with NOMA; for example, check [11]. [63] studies the effect of finite-alphabet inputs on NOMA-assisted Z-channels. The incorporation of advanced coding and modulation techniques into NOMA has not only enhanced its performance but has also spurred the development of new variations of the technology. Examples include network-coded multiple access (NCMA) [64] and LPMA (Low Power Multiple Access) [11].

6.8.2 Imperfect CSI

One of the main challenges to achieving performance improvement in NOMA is imperfect CSI [65, 66]. In general, three types of imperfect CSI are defined: limited channel feedback, partial CSI, and channel estimation errors.

Inadequately designed channel estimation algorithms and noisy data are the main contributors to channel estimate errors and lead to performance degradation of NOMA network, as these errors cause uncertainty in the user's order. In [67], the effects of channel estimate errors on PD NOMA and SCMA are investigated, and [68] studies the pilot transmission design for NOMA schemes. These investigations have shown that, compared to OMA, NOMA is actually more resistant to errors in channel estimate. In the scenario where only statistical CSI is accessible at the transmitter, the resource allocation problem for multi-carrier NOMA is investigated in [69]. This study establishes the decoding order for SIC and leverages the heterogeneity of QoS requirements and statistical CSI. The proposed suboptimal power allocation and user scheduling strategy for NOMA significantly outperforms OMA and achieves optimal performance. The effect of small-scale fading is greater than the large-scale path loss as presented in works [70, 71] after utilizing partial CSI.

Limited feedback has the advantage of reducing system overhead, compared to the scenario when each receiver relays the transmitter with all channel data [72–74]. As an illustration, look at the one-bit feedback NOMA method proposed in [72]. A threshold value is broadcast by the BS to the users in the network. Users compare their received signal strength with the threshold, indicating whether it is above or below. They then provide feedback to the BS by sending a binary value of 1 or 0, based on the result. For NOMA networks, where the nodes have multiple antennas, this one-bit feedback is especially important, e.g., [75, 76]. The user simply has to send one bit back to the BS, not the entire vector or matrix. For one-bit feedback methods, the threshold selection is obviously critical, and in [72], optimal thresholds are designed for systems with various power limitations.

6.9 NOMA WITH MACHINE LEARNING

New designs are required to address more difficult issues in the 6G network since 6G network scenarios are dynamic and more complicated. Many current research studies lack effectiveness in optimizing problems due to their limited focus on controllable variables. However, the utilization of machine learning in the design of NOMA presents a promising alternative. Unlike traditional algorithms, ML-based approaches offer a more comprehensive framework capable of addressing a wide range of problems, including those with numerous controllable variables and temporal correlations [93]. The role of ML in NOMA-based networks is further classified as:

- **Reinforcement Learning (RL) for NOMA-based Networks**
  RL empowers NOMA-based networks to swiftly adapt to dynamic environments and continuously enhance their decision-making capabilities [94, 95]. Application of RL in NOMA- based networks include:
  Dynamic user selection based on CSI and QoS requirement.
  Resource allocation.
- **DL for NOMA-based Networks**
  DL algorithms utilize neural networks to automatically extract features from input data and make informed decisions. DL-based NOMA networks are adopted to overcome limitations with lower complexity, offering

solutions that effectively address these challenges. Application of DL in NOMA includes [96, 97]:

DL-aided channel state estimation and signal detection for NOMA network.
Power optimization in NOMA network using DL.
DL-aided data rate and bit error rate optimization in NOMA network.

6.9.1 Different Aspect of ML in NOMA Networks

In this subsection, we will discuss different aspects of the DL-based NOMA system. In NOMA, multiple users share the same resource based on their channel condition and QoS requirement. Sharing the same resources among multiple users causes intra-user interference, which results in performance degradation of NOMA users. The intra-user interference can be minimized by employing optimized power allocation. Further, as mentioned in the previous section, users not properly paired lead to the degradation of channel capacity, which nullifies the advantage of using NOMA over OMA. Different optimization techniques, ML and DL algorithms, are proposed in the literature for optimal power allocation and user pairing. In [98], the author introduced a novel approach to user pairing by matching users with the highest channel gains to those with the next highest gains. Further, the author has explored various game theory algorithms and machine learning techniques to address the challenge of user pairing in multiple-user scenarios. In [99], the authors presented a closed-form approach for optimal power allocation in conjunction with a predefined sub-channel assignment. Building upon this, they employed a conventional deep reinforcement learning (DRL) algorithm known as Deep Q-Network (DQN) to explore an optimal user pairing scheme. The application DL in NOMA has led to significant advancements in power allocation, with deep neural network (DNN) architectures leading the way. In the context of power allocation optimization, [100] proposes a deep reinforcement learning (DRL) approach, specifically by utilizing an artificial neural network (ANN) to handle channel assignments.

The SIC is one of the important parts of decoding NOMA users' symbols. Since the SIC is iterative deciding, as the number of users increases, the receiver complexity increases. Further, in the practical system, the cancellation of the previously decoded signal is not perfect, leading to the imperfect SIC. In [101], a DNN is employed to approximate the SIC receiver. In the context of MIMO-NOMA systems, the joint optimization of precoding and SIC decoding is performed to minimize the overall mean square error, aiming to enhance the accuracy of decoding the intended signal for each user.

6.10 FUTURE CHALLENGES

6.10.1 NOMA with Heterogeneous Networks

An ultra-dense network (UDN) is a wireless network with more access points (APs) than users actively using the network. Another effective method of using spectrum is to deploy a large number of BSs for wide spatial reuse [77, 78]. Future network generations are increasingly adopting heterogeneous networks (HetNets), which involve the coexistence of various types of base stations (BSs) to cater to different requirements. HetNets consist of both macro-BSs (MBS) and small BSs (SBS) deployed within the same area, enabling them to address a wide range of diverse needs. The MBSs can offer extensive coverage; however, the SBSs increase network capacity and data rates because of their limited coverage areas. The literature, such as in [22, 24, 79, 80], has extensively researched NOMA combined with such HetNets. Additionally, heterogeneous UDNs mandate that MBS and SBS deployments adhere to UDN network requirements, i.e., it is necessary to have more BSs than users. This indicates that the MBS and SBS will have a very dense deployment in a heterogeneous UDN [81]. However, despite the fact network densification increases capacity, the total interference brought on by interfering BSs restricts the throughput performance that can be attained. The interference from nearby users increases as BSs are deployed more densely [82]. On the plus side, dense installations guarantee that each BS serves fewer users. Additionally, the distance between users and BSs decreases, which minimizes the effect of path loss. Extensive research is underway to develop the architecture of heterogeneous UDNs, aiming to manage network operations effectively and mitigate the growing interference challenges associated with dense networks. For user access management and control, the APs are linked to a server. The number of active connections can be increased even more by implementing NOMA protocols, which multiplex the signals of various users over a single RB.

NOMA can be implemented by multiplexing multiple APs over the same RB, especially when the number of users and the number of APs are roughly comparable. This approach allows for efficient utilization of resources and enables NOMA in the network [77, 78]. The UDNs' spectrum efficiency and overall system throughput are both improved by NOMA. The integration of NOMA in UDNs faces challenges due to the substantial inter-cell interference caused by both intra-tier and inter-tier BSs, as well as intra-cell or inter-user interference arising from NOMA itself. Hence, it is crucial to conduct a comprehensive analysis of inter-cell interference management to address these obstacles and facilitate the successful integration of NOMA in UDN.

6.10.2 NOMA with Simultaneous Wireless Information and Power Transfer (SWIPT)

The utilization of cooperative communication, where relays receive signals from BSs and retransmit them to users at the cell edge, offers an opportunity to enhance coverage for users situated far away from the BSs [83]. In the existing literature, numerous relaying techniques have been proposed to address this objective [84, 85]. Since NOMA's core principle involves users receiving a superimposed signal, combining NOMA with cooperative communication is a logical extension. The cell center user utilizes SIC to decode its own message. Thus, the cell center user already knows the message symbol of the cell edge users. As a result, the user of the cell center can retransmit the message of the cell edge user. Cooperative communication is effective in reducing signal attenuation for users at the cell edge by utilizing relay nodes or users located closer to them. This improvement in signal propagation enhances the throughput performance of cell edge users. However, it is important to consider that

UE and relays are often powered by batteries, and the retransmission process can rapidly deplete their energy resources. Therefore, an advanced approach is needed to ensure an energy-efficient design in such scenarios.

In this context, the concept of energy harvesting, also known as SWIPT, has garnered significant interest among academics in the field of wireless communication. In order to enable the creation of an energy-efficient network, the SWIPT concept, initially put forth in [86], has attracted the interest of many researchers. NOMA used with SWIPT for cooperative communication will provide an energy-efficient method for serving the cell edge user through cooperation in addition to spectrum efficiency. As mentioned in [86], the utilization of SWIPT technology allows radio signals to serve two purposes: information transfer and energy harvesting. The approach described is notably energy-efficient, as it leverages the energy acquired by relays and collaborating users in the first phase to transfer information to the cell edge user in the second phase. Further details regarding this concept can be found in [21, 23]. Further, implementing relaying with SWIPT for cell center users helps mitigate the risk of battery drainage, ensuring sustainable operation and longevity of their battery life.

6.10.3 NOMA with Visible Light Communication (VLC)

VLC, a widely recognized and accepted technology, operates beyond the radio frequency band [87]. VLC plays a pivotal role in enabling light fidelity (LiFi) systems, a technology that has been proposed for diverse applications such as IoT devices, underwater communications, and more. VLC uses light emitting diode (LED) lights as both an optical transmitter and as a source of light. In a standard optical opto-cell, photodiode arrays are utilized at the receiver end [88]. The modulation in VLC is carried out at a speed that the human eye cannot comprehend [89], which enhances the usefulness of LEDs. Illumination is one application, and communication is the other. Furthermore, as compared to the amount needed for radio frequency communication, the power LEDs use to transmit a signal is quite minimal. The candidate multiple access scheme in VLC is suggested to use NOMA as a viable alternative. The main factors to consider using NOMA in VLC are [90]: The limited communication range of VLC results in a dominant line-of-sight component in the channel. This characteristic promotes the need for accurate channel estimation.

Since the SIC performed at the receiver includes decoding and removing messages until the required signal is attained, the accuracy of the channel estimation is crucial in deciding how well NOMA systems operate. The accuracy of the channel estimation has a significant impact on SIC. When SIC is flawed, error spreads throughout each stage and builds up to such a level that it affects throughput performance as a whole. In addition, two more parameters, namely, tuning angle and field of view, provide additional degrees of freedom that can be used to multiplex the signals of several users. Although the integration of NOMA in VLC has been studied in recent literature, the non-linearity caused by the LEDs limits the throughput gains [91]. Using pre-distortion or post-distortion procedures, the non-linearity inherited by the LEDs can be removed. The non-linearity that the LEDs present is taken into account by the authors in [92] when designing a pre-distorter.

6.11 CONCLUSIONS

High system throughput, low latency, and huge connection are three of the primary performance criteria for 5G, and NOMA is a crucial enabling technology for meeting these needs. This study demonstrates how NOMA can use the bandwidth resources more effectively than OMA by taking advantage of the users' varied channel circumstances and QoS requirements. Prior research has also amply proved how NOMA can increase system throughput. Massive connectivity can be realistically achieved with NOMA since multiple users can be served concurrently. The recent incorporation of NOMA into 5G, LTE-A, and digital TV standards reflects the growing industrial interest in making NOMA an essential component of next-generation wireless systems. We believe that this work will help readers in understanding the benefits of NOMA, along with its practical application.

REFERENCES

[1] L. Dai, B. Wang, Y. Yuan, S. Han, I. Chih-Lin, and Z. Wang "Nonorthogonal multiple access for 5G: Solutions, challenges, opportunities, and future research trends," *IEEE Commun. Mag.*, vol. 53, no. 9, pp. 74–81, 2015.

[2] M. Vaezi, Z. Ding, and H. V. Poor "Multiple access techniques for 5G wireless networks and beyond," *Springer*, vol. 159, 2019.

[3] Y. Saito, Y. Kishiyama, A. Benjebbour, T. Nakamura, A. Li, and K. Higuchi "Non-orthogonal multiple access (NOMA) for cellular future radio access," in *Proc. IEEE Veh. Technol. Conf*, Dresden, Germany, pp. 1–5, 2013.

[4] V. Bhatia, P. Swami, S. Sharma, and R. Mitra "Non-orthogonal multiple access: An enabler for massive connectivity," *J. Indian Inst. Sci.*, vol. 100, no. 2, pp. 337–348, 2020.

[5] H. Nikopour and H. Baligh, "Sparse code multiple access," *Proc. IEEE Int. Symp. Pers. Indoor Mobile Radio Commun*, London, U.K., pp. 332–336, 2013.

[6] R. Mitra, S. Sharma, G. Kaddoum and V. Bhatia, "Color-domain SCMA NOMA for visible light communication," *IEEE Commun. Lett.*, vol. 25, no. 1, pp. 200–204, 2021.

[7] S. Sharma, K. Deka, V. Bhatia and A. Gupta, "Joint Power-Domain and SCMA-Based NOMA System for Downlink in 5G and Beyond," *IEEE Commun. Lett.*, vol. 23, no. 6, pp. 971–974, 2019.

[8] M. Taherzadeh, H. Nikopour, A. Bayesteh, and H. Baligh, "SCMA codebook design," in *Proc. IEEE Veh. Technol. Conf.*, Las Vegas, NV, pp. 1–5, 2014.

[9] X. Dai et al, "Successive interference cancelation amenable multiple access (SAMA) for future wireless communications," in *Proc. IEEE Int. Conf. Commun. Syst.*, Coimbatore, India, pp. 222–226, 2014.

[10] M. Al-Imari, M. A. Imran, R. Tafazolli, and D. Chen, "Performance evaluation of low density spreading multiple access," in *Proc. 8th Int. Wireless Commun. Mobile Comput. Conf*, Limassol, Cyprus, pp. 383–388, 2012.

[11] D. Fang, Y.-C. Huang, Z. Ding, G. Geraci, S.-L. Shieh, and H. Claussen, "Lattice partition multiple access: A new method of downlink non-orthogonal multiuser transmissions," in *Proc. IEEE Global Commun. Conf*, Washington, DC, 2016.

[12] L. Ping, L. Liu, K. Wu, and W. K. Leung, "Interleave division multiple access," *IEEE Signal Process. Lett.*, vol. 5, no. 4, pp. 938–947, 2006.

[13] Z. Ding, P. Fan, and H. V. Poor, "Impact of user pairing on 5G nonorthogonal multiple-access downlink transmissions," *IEEE Trans. Veh. Technol.*, vol. 65, no. 08, pp. 6010–6023, 2016.

[14] P. Swami, V. Bhatia, S. Vuppala and T. Ratnarajah, "User fairness in NOMA-HetNet using optimized power allocation and time slotting," *IEEE Sys. J.*, vol. 15, no. 01, pp. 1005–1014, 2021.

[15] "NGMN 5G white paper," NGMN Alliance, Frankfurt, Germany, White Paper, 2015.

[16] "5G innovation opportunities- a discussion paper," techUK, London, U.K., 5G White Paper, 2015.

[17] X. Wang and H. V. Poor, *Wireless Communication Systems: Advanced Techniques for Signal Reception*. New York, NY, USA: Prentice-Hall, 2004.

[18] S. Verdu, *Multiuser Detection*. Cambridge, U.K.: Cambridge University Press, 1998.

[19] *Study on Downlink Multiuser Superposition Transmission for LTE*. document, 3rd Generation Partnership Project (3GPP), Mar. 2015.

[20] L. Zhang et al., "Layered-division-multiplexing: Theory and practice," *IEEE Trans. Broadcast.*, vol. 62, no. 1, pp. 216–232, 2016.

[21] A. S. Parihar, P. Swami, V. Bhatia and Z. Ding, "Performance analysis of SWIPT enabled cooperative-NOMA in heterogeneous networks using carrier sensing," *IEEE Trans. Veh. Technol.*, vol. 70, no. 10, pp. 10646–10656, 2021.

[22] P. Swami, V. Bhatia, S. Vuppala and T. Ratnarajah, "A cooperation scheme for user fairness and performance enhancement in NOMA-HCN," *IEEE Trans. Veh. Technol.*, vol. 67, no. 12, pp. 11965–11978, 2018.

[23] A. S. Parihar, P. Swami and V. Bhatia, "On performance of SWIPT enabled PPP distributed cooperative NOMA networks using stochastic geometry," *IEEE Trans. Veh. Technol.*, vol. 71, no. 05, pp. 5639–5644, 2022.

[24] P. Swami, V. Bhatia, S. Vuppala and T. Ratnarajah, "On user Offloading in NOMA-HetNet using repulsive point process," *IEEE Sys. J.*, vol. 13, no. 02, pp. 1409–1420, 2019.

[25] Y. Sun, D. W. K. Ng, Z. Ding, and R. Schober, "Optimal joint power and subcarrier allocation for full-duplex multicarrier non-orthogonal multiple access systems," *IEEE Trans. Commun.*, vol. 65, no. 03, pp. 1077–1091, 2017.

[26] J. Zhu, J. Wang, Y. Huang, S. He, X. You, and L. Yang, "On optimal power allocation for downlink non-orthogonal multiple access systems," *IEEE J. Sel. Areas Commun.*, vol. 35, no. 12, pp. 2744–2757, 2017.

[27] Z. Ding, M. Peng, and H. V. Poor, "Cooperative non-orthogonal multiple access in 5G systems," *IEEE Commun. Lett*, vol. 19, no. 08, pp. 1462–1465, 2015.

[28] L. Zhang, J. Liu, M. Xiao, G. Wu, Y.-C. Liang, and S. Li, "Performance analysis and optimization in downlink NOMA systems with cooperative full-duplex relaying," *IEEE J. Sel. Areas Commun.*, vol. 35, no. 10, pp. 2398–2412, 2017.

[29] J. Jose, P. Shaik and V. Bhatia, "VFD-NOMA under imperfect SIC and residual inter-relay interference over generalized nakagami-m fading channels," *IEEE Commun. Lett.*, vol. 25, no. 02, pp. 646–650, 2021.

[30] J. Jose, P. Shaik and V. Bhatia, "Performance of Cooperative NOMA with Virtual Full-Duplex based DF Relaying in Nakagami-m Fading," *2021 IEEE 93rd Vehicular Technology Conference (VTC2021-Spring)*, 2021, pp. 1–6.

[31] M. S. ElBamby, M. Bennis, W. Saad, M. Debbah, and M. Latva-Aho, "Resource optimization and power allocation in full duplex non-orthogonal multiple access (FD-NOMA) networks," *IEEE J. Sel. Areas Commun.*, vol. 35, no. 12, pp. 2860–2873, 2017.

[32] S. Luo and K. C. Teh, "Adaptive transmission for cooperative NOMA system with buffer-aided relaying," *IEEE Commun. Lett.*, vol. 21, no. 04, pp. 937–940, 2017.

[33] M. Xu, F. Ji, M. Wen, and W. Duan, "Novel receiver design for the cooperative relaying system with non-orthogonal multiple access," *IEEE Commun. Lett.*, vol. 20, no. 08, pp. 1679–1682, 2016.

[34] Z. Ding, H. Dai, and H. V. Poor, "Relay selection for cooperative NOMA," *IEEE Wireless Commun. Lett.*, vol. 5, no. 4, pp. 416–419, 2016.

[35] T. S. Rappaport et al., "Millimeter wave mobile communications for 5G cellular: It will work!," *IEEE Access*, vol. 1, pp. 335–349, 2013.

[36] *Enable Higher Frequency Spectrum for Future Wireless*, document FCC 16–89, Federal Communications Commission Report, 2016.

[37] E. Nussbaum, "Integrated broadband networks and services of the future," in *Proc. Opt. Fiber Commun. Conf*, Reno, NV, USA, 1987.

[38] B. Wang, L. Dai, Z. Wang, N. Ge, and S. Zhou, "Spectrum and energy efficient beamspace MIMO-NOMA for millimeter- wave communications using lens antenna array," *IEEE J. Sel. Areas Commun.*, vol. 35, no. 10, pp. 2370–2382, 2017.

[39] G. Lee, Y. Sung, and J. Seo, "Randomly-directional beamforming in millimeter-wave multiuser MISO downlink," *IEEE Trans. Wireless Commun.*, vol. 15, no. 2, pp. 1086–1100, 2016.

[40] T. S. Rappaport, E. Ben-Dor, J. N. Murdock, and Y. Qiao, "38 GHz and 60 GHz angle-dependent propagation for cellular & peer-to-peer wireless communications," in *Proc. IEEE Int. Conf. Commun*, Ottawa, ON, Canada, 2012.

[41] Z. Chen, Z. Ding, X. Dai, and G. K. Karagiannidis, "On the application of quasi-degradation to MISO-NOMA downlink," *IEEE Trans. Signal Process.*, vol. 64, no. 23, pp. 6174–6189, 2016.

[42] X. Wang, F. Labeau, and L. Mei, "Closed-form BER expressions of QPSK constellation for uplink non-orthogonal multiple access," *IEEE Commun. Lett.*, vol. 21, no. 10, pp. 2242–2245, 2017.

[43] F. Wei, T. Zhou, T. Xu, and H. Hu, "BER analysis for uplink NOMA in asymmetric channels," *IEEE Commun. Lett.*, vol. 24, no. 11, pp. 2435–2439, 2020.

[44] Q. He, Y. Hu, and A. Schmeink, "Closed-form symbol error rate expressions for non-orthogonal multiple access systems," *IEEE Trans. Veh. Technol.*, vol. 68, no. 7, pp. 6775–6789, May 2019.

[45] T. Assaf, A. Al-Dweik, M. E. Moursi, and H. Zeineldin, "Exact BER performance analysis for downlink NOMA systems over Nakagami-m fading channels," *IEEE Access*, vol. 7, pp. 134 539–134 555, 2019.

[46] I. Lee and J. Kim, "Average symbol error rate analysis for nonorthogonal multiple access with M-ary QAM signals in Rayleigh fading channels," *IEEE Commun. Lett.*, vol. 23, no. 8, pp. 1328–1331, 2019.

[47] L. Bariah, S. Muhaidat, and A. Al-Dweik, "Error probability analysis of non-orthogonal multiple access over Nakagami-m fading channels," *IEEE Trans. Commun.*, vol. 67, no. 2, pp. 1586–1599, 2019.

[48] T. Assaf, A. J. Al-Dweik, M. S. E. Moursi, H. Zeineldin, and M. AlJarrah, "Exact bit error-rate analysis of two-user NOMA using QAM with arbitrary modulation orders," *IEEE Commun. Lett.*, vol. 24, no. 12, pp. 2705–2709, 2020.

[49] S. Bisen, P. Shaik, and V. Bhatia, "On ASER performance of M-ary QAM schemes over DF coordinated-NOMA," in *IEEE Region 10 Conf. (TENCON)*, Nov. 2020, pp. 280–285.

[50] H. Yahya, E. Alsusa, and A. Al-Dweik, "Exact BER analysis of NOMA with arbitrary number of users and modulation orders," *IEEE Trans. Commun.*, vol. 69, no. 9, pp. 6330–6344, 2021.

[51] X. Chen, R. Jia and D. W. K. Ng, "On the Design of Massive Non-Orthogonal Multiple Access With Imperfect Successive Interference Cancellation," *IEEE Trans. Commun.*, vol. 67, no. 3, pp. 2539–2551, 2019.

[52] Y. Wang, J. Wang, D. W. Kwan Ng, R. Schober and X. Gao, "A Minimum Error Probability NOMA Design," *IEEE Trans. Wireless Commun.*, vol. 20, no. 7, pp. 4221–4237, 2021.

[53] F. Kara and H. Kaya, "BER performances of downlink and uplink NOMA in the presence of SIC errors over fading channels," *IET Commun.*, vol. 12, no. 15, pp. 1834–1844, 2018.
[54] Z. Ding, R. Schober, and H. V. Poor, "A new QoS-guarantee strategy for NOMA assisted semi-grant-free transmission," *IEEE Trans. Commun.*, vol. 69, no. 11, pp. 7489–7503, 2021.
[55] Y. Sun, Z. Ding, and X. Dai, "A new design of hybrid SIC for improving transmission robustness in uplink NOMA," *IEEE Trans. Veh. Techno.*, vol. 70, no. 5, pp. 5083–5087, 2021.
[56] J. S. Yeom, H. S. Jang, K. S. Ko, and B. C. Jung, "BER performance of uplink NOMA with joint maximum-likelihood detector," *IEEE Trans. Veh. Technol.*, vol. 68, no. 10, pp. 10 295–10 300, 2019.
[57] R. Campello, G. Carlini, C. E. C. Souza, C. Pimentel, and D. P. B. Chaves, "Successive interference cancellation decoding with adaptable decision regions for NOMA schemes," *IEEE Access*, vol. 10, pp. 2051–2062, 2022.
[58] C. Lin, Q. Chang, X. Li, "A deep learning approach for MIMO-NOMA downlink signal detection," *Sensors 19*, no. 11:2526, 2019.
[59] Narengerile and J. Thompson, "Deep learning for signal detection in nonorthogonal multiple access wireless systems," in *Proc. UK/China Emerg. Technol. (UCET)*, 2019, pp. 1–4.
[60] A. Emir, F. Kara, H. Kaya and H. Yanikomeroglu, "Deep Learning Empowered Semi-Blind Joint Detection in Cooperative NOMA," *IEEE Access*, vol. 9, pp. 61832–61852, 2021.
[61] S.-L. Shieh, C.-H. Lin, Y.-C. Huang, and C.-L. Wang "On gray labeling for downlink non-orthogonal multiple access without SIC," *IEEE Commun. Lett.*, vol. 20, no. 09, pp. 1721–1724, 2016.
[62] S.-L. Shieh and Y.-C. Huang, "A simple scheme for realizing the promised gains of downlink nonorthogonal multiple access," *IEEE Trans. Commun.*, vol. 64, no. 4, pp. 1624–1635, 2016.
[63] Z. Dong, H. Chen, J.-K. Zhang, and L. Huang, "On non-orthogonal multiple access with finite-alphabet inputs in Z- channels," *IEEE J. Sel. Areas Commun.*, vol. 35, no. 12, pp. 2829–2845, 2017.
[64] H. Pan, L. Lu, and S. C. Liew, "Practical power-balanced non-orthogonal multiple access," *IEEE J. Sel. Areas Commun.*, vol. 35, no. 10, pp. 2312–2327, 2017.
[65] S. Bisen, P. Shaik and V. Bhatia, "On performance of energy harvested cooperative NOMA under imperfect CSI and imperfect SIC," *IEEE Trans. Veh. Tehnol.*, vol. 70, no. 09, pp. 8993–9005, 2021.
[66] P. Swami, M. K. Mishra, V. Bhatia, T. Ratnarajah and A. Trivedi, "Performance Analysis of sub-6 GHz/mmWave NOMA Hybrid-HetNets using Partial CSI," *IEEE Trans. Veh. Technol.*, 2022.
[67] F. Fang, H. Zhang, J. Cheng, S. Roy, and V. C. Leung, "Joint user scheduling and power allocation optimization for energy efficient NOMA systems with imperfect CSI," *IEEE J. Sel. Areas Commun.*, vol. 35, no. 12, pp. 2874–2885, 2017.
[68] J. Ma, C. Liang, C. Xu, and L. Ping, "On orthogonal and superimposed pilot schemes in massive MIMO NOMA systems," *IEEE J. Sel. Areas Commun.*, vol. 35, no. 12, pp. 2696–2707, 2017.
[69] Z. Wei, D. W. K. Ng, and J. Yuan, "Power-efficient resource allocation for MC-NOMA with statistical channel state information," in *Proc. IEEE Global Commun. Conf*, Washington, DC, 2017.
[70] Z. Yang, Z. Ding, P. Fan, and G. K. Karagiannidis, "On the performance of non-orthogonal multiple access systems with partial channel information," *IEEE Trans. Commun.*, vol. 64, no. 02, pp. 654–667, 2016.

[71] P. Xu and K. Cumanan, "Optimal power allocation scheme for nonorthogonal multiple access with a-fairness," *IEEE J. Sel. Areas Commun.*, vol. 35, no. 10, pp. 2357–2369, 2017.
[72] P. Xu, Y. Yuan, Z. Ding, X. Dai, and R. Schober, "On the outage performance of non-orthogonal multiple access with 1-bit feedback," *IEEE Trans. Wireless Commun.*, vol. 15, no. 10, pp. 6716–6730, 2016.
[73] P. Swami, M. K. Mishra, V. Bhatia and T. Ratnarajah, "Performance analysis of NOMA enabled hybrid network with limited feedback," *IEEE Trans. Veh. Technol.*, vol. 69, no. 04, pp. 4516–4521, 2020.
[74] Q. Yang, H.-M. Wang, D. W. K. Ng, and M. H. Lee, "NOMA in downlink SDMA with limited feedback: Performance analysis and optimization," *IEEE J. Sel. Areas Commun.*, vol. 35, no. 10, pp. 2281–2294, 2017.
[75] Z. Ding, P. Fan, and H. V. Poor, "Random beamforming in millimeterwave NOMA networks," *IEEE Access*, vol. 5, pp. 7667–7681, 2017.
[76] Z. Ding and H. V. Poor, "Design of massive-MIMO-NOMA with limited feedback," *IEEE Signal Process. Lett.*, vol. 23, no. 5, pp. 629–633, 2016.
[77] Z. Zhang, G. Yang, Z. Ma, M. Xiao, Z. Ding and P. Fan, "Heterogeneous ultradense networks with NOMA: System architecture, coordination framework, and performance evaluation," *IEEE Veh. Technol. Mag.*, vol. 13, no. 2, pp. 110–120, 2018.
[78] Z. Qin, X. Yue, Y. Liu, Z. Ding and A. Nallanathan, "User association and resource allocation in unified NOMA enabled heterogeneous ultra dense networks," *IEEE Veh. Technol. Mag.*, vol. 13, no. 2, pp. 110–120, 2018.
[79] P. Swami, V. Bhatia, S. Vuppala and T. Ratnarajah, "Outage Analysis of NOMA-HCN Using Repulsive Point Process," 2017 *IEEE Globecom Workshops (GC Wkshps)*, 2017, pp. 1–6.
[80] P. Swami, V. Bhatia, S. Vuppala and T. Ratnarajah, "Joint Optimization of Power Allocation and Channel Ratio for Offloading in NOMA-HetNets," 2018 *IEEE Globecom Workshops (GC Wkshps)*, 2018, pp. 1–6.
[81] F. Granelli et al., "Software defined and virtualized wireless access in future wireless networks: Scenarios and standards," *IEEE Veh. Technol. Mag.*, vol. 53, no. 6, pp. 26–34, 2015.
[82] A. Checko et al., "Cloud RAN for mobile networks—A technology overview," *IEEE Commun. Surveys Tuts.*, vol. 07, no. 01, pp. 405–426, 2015.
[83] M. Peng, S. Yan, K. Zhang, and C. Wang, "Fog-computing-based radio access networks: Issues and challenges," *IEEE Netw.*, vol. 30, no. 4, pp. 46–53, 2016.
[84] D. Li, "Opportunistic DF-AF selection for cognitive relay networks," *IEEE Trans. Veh. Tech.*, vol. 65, no. 04, pp. 2790–2796, 2016.
[85] Y. Liu et al., "Hybrid decode-forward & amplify-forward relaying with non-orthogonal multiple access," *IEEE Access*, vol. 4, pp. 4912–4921, 2016.
[86] L. R. Varshney, "Transporting information and energy simultaneously," in *Proc. IEEE Int. Symp. Inf. Theory*, pp. 1612–1616, 2008.
[87] M. Figueiredo, L. N. Alves, and C. Ribeiro, "Lighting the wireless world: The promise and challenges of visible light communication," *IEEE Consum. Electron. Mag.*, vol. 6, no. 4, pp. 28–37, 2017.
[88] L. Yin, W. O. Popoola, X. Wu, and H. Haas, "Performance evaluation of non-orthogonal multiple access in visible light communication," *IEEE Trans. Commun.*, vol. 64, no. 12, pp. 5162–5175, 2016.
[89] B. Inan, S. C. J. Lee, S. Randel, I. Neokosmidis, A. M. J. Koonen, and J. W. Walewski, "Impact of LED nonlinearity on discrete multitone modulation," *IEEE/OSA J. Opt. Commun. Netw.*, vol. 1, no. 5, pp. 439–451, 2009.

[90] H. Marshoud, V. M. Kapinas, G. K. Karagiannidis, and S. Muhaidat, "Non-orthogonal multiple access for visible light communications," *IEEE Photon. Technol. Lett.*, vol. 28, no. 1, pp. 51–54, 2016.

[91] T. P. Lee, "The nonlinearity of double-heterostructure LED's for optical communications," *Proc. IEEE*, vol. 65, no. 9, pp. 1408–1410, 1977.

[92] R. Mitra, and V. Bhatia, "Precoded Chebyshev-NLMS-based pre-distorter for nonlinear LED compensation in NOMA- VLC," *IEEE Trans. Commun.*, vol. 65, no. 11, pp. 4845–4856, 2017.

[93] A. Emir, F. Kara, H. Kaya and H. Yanikomeroglu, "Deep Learning Empowered Semi-Blind Joint Detection in Cooperative NOMA," *IEEE Access*, vol. 9, pp. 61832–61852, 2021.

[94] K. Wang, H. Li, Z. Ding and P. Xiao, "Reinforcement Learning based Latency Minimization in Secure NOMA-MEC Systems with Hybrid SIC," *IEEE Trans. Wireless Commun.*, 2022.

[95] K. B. Letaief, W. Chen, Y. Shi, J. Zhang and Y.-J. A. Zhang, "The Roadmap to 6G: AI Empowered Wireless Networks," *IEEE Commun. Mag.*, vol. 57, no. 8, pp. 84–90, 2019.

[96] H. Huang et al., "Deep Learning for Physical-Layer 5G Wireless Techniques: Opportunities, Challenges and Solutions," *IEEE Wireless Commun.*, vol. 27, no. 1, pp. 214–222, 2020.

[97] A. Emir, F. Kara, H. Kaya and H. Yanikomeroglu, "DeepMuD: Multi-User Detection for Uplink Grant-Free NOMA IoT Networks via Deep Learning," *IEEE Wireless Commun. Lett.*, vol. 10, no. 5, pp. 1133–1137, 2021.

[98] Wang, S., Lv, T., Zhang, X. "Multi-agent reinforcement learning-based user pairing in multi-carrier NOMA systems," in *Proceedings of the 2019 IEEE International Conference on Communications Workshops (ICC Workshops)*, Shanghai, China, 20–24 May 2019; IEEE: Piscataway Township, NJ, USA; pp. 1–6.

[99] Jiang, F., Gu, Z., Sun, C., Ma, R. "Dynamic user pairing and power allocation for NOMA with deep reinforcement learning," in *Proceedings of the 2021 IEEE Wireless Communications and Networking Conference (WCNC)*, Nanjing, China, 29 March–1 April 2021; IEEE: Piscataway Township, NJ, USA; pp. 1–6.

[100] Huang, H.; Guo, S.; Gui, G.; Yang, Z.; Zhang, J.; Sari, H.; Adachi, F. "Deep learning for physical-layer 5G wireless techniques: Opportunities, challenges and solutions," *IEEE Wireless Commun*, vol. 27, no. 1, pp. 214–222, 2019.

[101] Gao, W.; Han, C.; Chen, Z. "DNN-powered SIC-free receiver artificial noise aided terahertz secure communications with randomly distributed eavesdroppers," *IEEE Trans. Wireless Commun.* vol. 21, no. 1, pp. 563–576, 2022.

7 Compensating Inbound Signal Strength for Radio-Controlled Mobile Robots Using ANFIS

M. Bobyr and C. Nolivos
Southwest State University, Kursk, Russia

7.1 INTRODUCTION

Intelligent telecommunications networks (ITN) are communication networks equipped with artificial intelligence technologies that reduce the time for information processing and decision-making, improve the quality of communication, and provide users with new functionality.

Fuzzy logic is an approach to information processing that allows you to work with fuzzy and uncertain data. Using fuzzy logic in ITN can improve system performance and accuracy by allowing the system to process complex data and make decisions based on fuzzy input data. For example, when making decisions about routing traffic, the system can use fuzzy rules to determine the best route based on factors such as time of day, traffic volume, etc. [1].

Automated control systems (ACS) are systems that use computer technology to control processes in various industries such as industry, energy, transportation, etc. ITN can be used in ACS to control and monitor various processes, such as inventory management, product quality control, traffic flow control, etc.

The use of ITN in an ACS can improve productivity and process quality by allowing systems to process data quickly and make decisions based on input parameters. For example, ITN can be used in an ACS to optimize production schedules, taking into account many factors such as order volumes, material availability, etc. [2].

Research in ITN and fuzzy logic aims to develop new technologies and algorithms that improve system performance and decision-making accuracy. They can also be used to create new functionality for users, such as improving voice and video communications, speech recognition and natural language processing, and creating intelligent assistants, etc.

Research in ITN and fuzzy logic also addresses the challenges of processing large amounts of data and creating systems that can adapt to changing conditions in real time. This is especially important in the case of autonomous systems, such as unmanned cars, which must respond quickly to changing road conditions.

DOI: 10.1201/9781003303114-7

The application of ITN and fuzzy logic in ACS and other areas can lead to a significant increase in productivity, reduce the time to make decisions, and improve the quality of system performance. This allows companies to improve their efficiency and competitiveness in the market.

Overall, research in ICS and fuzzy logic has a wide range of applications in industries from industry and energy to transportation and communications. They can lead to the creation of new technologies, improve the performance of existing systems, and create new functionality for users.

Using the Arduino platform, you can create various electronic devices and systems that can use ICS and fuzzy logic. For example, you can create devices that collect data from the environment, such as temperature, humidity, and pressure, and use that data to make decisions, such as automatic control of a heating or air conditioning system [3–6].

One example of using fuzzy logic on an Arduino is a device used to control a robot vacuum cleaner. The device collects data about the distance to obstacles and the angle of the robot and uses fuzzy rules to make decisions about the direction of the robot [7].

One of the advantages of using fuzzy logic on the Arduino is the ability to create systems that can adapt to changing conditions in real time. This allows you to create more efficient and flexible devices. In addition, Arduino is an open platform, allowing users to create their own designs and customize them to their needs.

In general, the use of fuzzy logic and Arduino to implement telecommunication systems can be an effective approach that allows for more flexible, adaptive, and affordable systems. However, before starting to develop such a system, it is necessary to carefully consider and define fuzzy rules and configure the system to work under specific conditions [8–10].

7.2 USING FUZZY LOGIC TO CREATE TELECOMMUNICATION SYSTEMS

Suppose there's a need to create a telecommunication system that automatically detects how much the radio control signal has been disturbed by noise. For this we can use fuzzy logic in conjunction with the Arduino platform.

Let's have two input parameters: noise level $X_1 \in [0; 100]$ and signal quality, which is the ratio of the number of pulses involved in the calculation to the total number of pulses, expressed as a percentage $X_2 \in [0; 100]$; the variables are specified as a percentage. The telecommunication system with help of fuzzy rules must answer the question of how much the signal was disturbed based on these inputs.

This system can be implemented on the Arduino platform using fuzzy logic. To do this, the following sequence of steps must be performed:

Step 1: Define fuzzy sets for each of the input parameters. For the noise level, the linguistic variable is defined by fuzzy sets "low - x_{11}," "medium - x_{12}," and "high - x_{13}." For signal quality, the linguistic variable is defined by the fuzzy sets "bad - x_{21}," "average - x_{22}," and "good - x_{23}." These fuzzy sets are used to build the membership functions shown in Figure 7.1 for noise level and Figure 7.2 for signal quality.

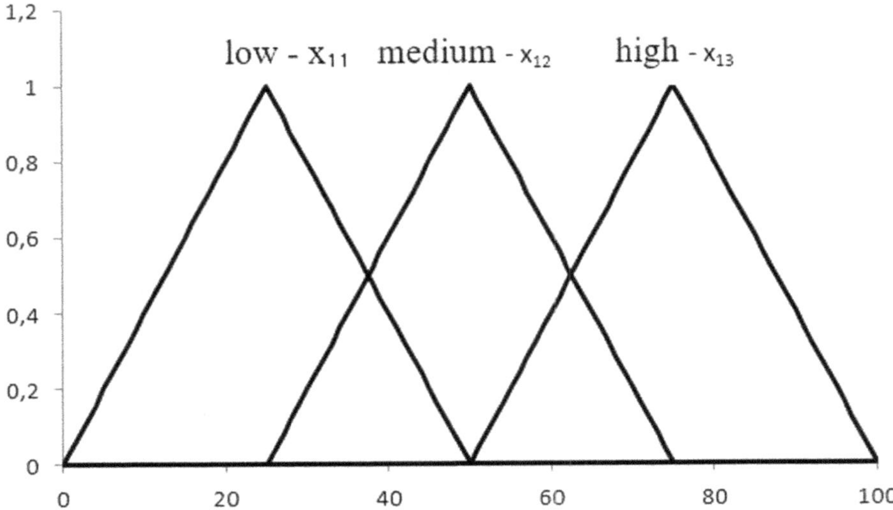

FIGURE 7.1 Membership functions for noise level.

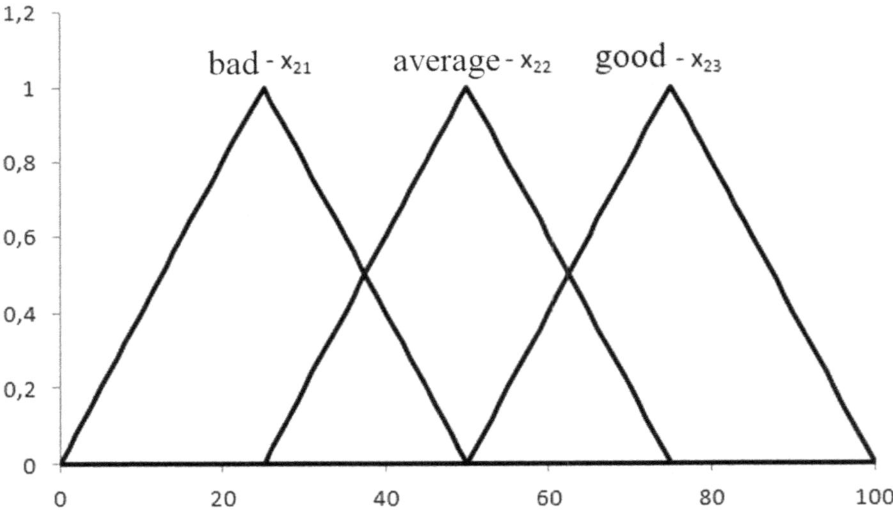

FIGURE 7.2 Membership functions for signal quality.

Step 2: Define fuzzy sets for the output variable. You can use a fuzzy rule that generates an output value between 0 and 10 to determine how badly the signal has been distorted. For the output variable, we define the fuzzy sets "weak signal - y_1" (0 to 8), "weak-moderate signal - y_2" (4 to 12), "moderate signal - y_3" (8 to 16), "moderate-strong signal - y_4" (12 to 20), and "strong signal - y_5" (16 to 24). The resulting membership functions are presented in Figure 7.3.

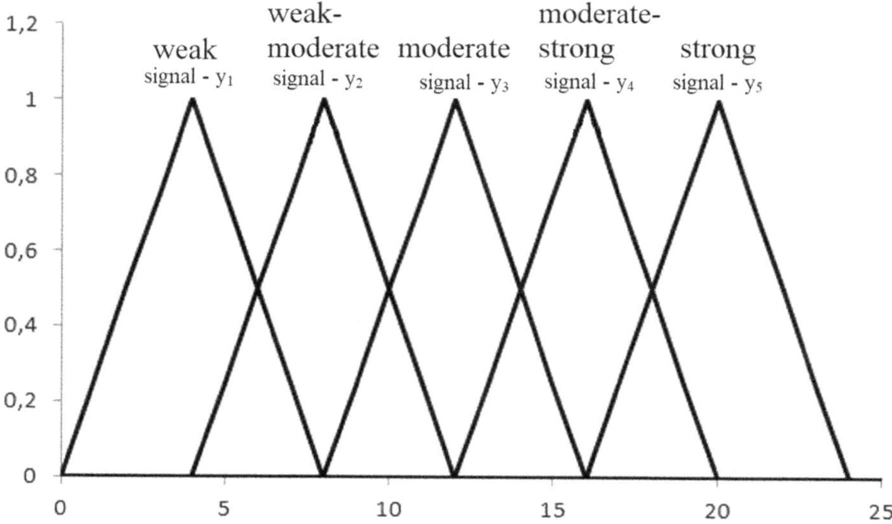

FIGURE 7.3 Membership functions for signal distortion.

Step 3: Define fuzzy rules that determine how badly the signal was disturbed based on the input parameters. For example, if the noise level is "high" and the signal quality is "bad," then the signal has been badly disturbed. This can be represented by the fuzzy rule "IF the noise level is high AND the signal quality is poor, THEN the signal was severely disturbed."

Here is an example of 9 fuzzy rules that can be used to determine how badly a signal has been disturbed based on noise level and signal quality values:

1. **IF** the noise level is "low" **AND** the signal quality is "good", **THEN** the signal is "strong
   IF x_{11} **AND** x_{23} **THEN** y_5.
2. **IF** the noise level is "medium" **AND** the signal quality is "good", **THEN** the signal is "moderately strong
   IF x_{12} **AND** x_{23} **THEN** y_4.
3. **IF** the noise level is "high" **AND** the signal quality is "good" **THEN** the signal is "moderate
   IF x_{13} **AND** x_{23} **THEN** y_3.
4. **IF** the noise level is "low" **AND** the signal quality is "medium" **THEN** the signal is "moderate-strong
   IF x_{11} **AND** x_{22} **THEN** y_4.
5. **IF** the noise level is "medium" **AND** the signal quality is "medium" **THEN** the signal is "moderate
   IF x_{12} **AND** x_{22} **THEN** y_3.

6. **IF** the noise level is "high" **AND** the signal quality is "medium" **THEN** the signal is "slightly moderate
IF x_{13} **AND** x_{22} **THEN** y_2.
7. **IF** the noise level is "low" **AND** the signal quality is "poor", **THEN** the signal is "moderate
IF x_{11} **AND** x_{21} **THEN** y_3.
8. **IF** the noise level is "medium" **AND** the signal quality is "poor", **THEN** the signal is "slightly-moderate
IF x_{12} **AND** x_{21} **THEN** y_2.
9. **IF** the noise level is "high" **AND** the signal quality is "poor", **THEN** the signal is "weak
IF x_{13} **AND** x_{21} **THEN** y_1.

7.3 THE ANFIS METHOD

In order to reduce the computational complexity of traditional fuzzy inference models in some applications, it is recommended to use a simplified model of fuzzy logic inference, which differs from the conclusions of fuzzy control rules using Singleton membership functions (MFs). This algorithm consists of three stages: phasing, fuzzy-logical inference (FLI), and defuzzification. Let us consider an example of its operation. Let the MFs of input and output quantities be given by triangular functions (Figure 7.4), and the control system itself operates on the basis of two fuzzy rules (FRs):

FR_1: IF «a is A_1» AND «b is B_1», THEN «c is C_1»;
FR_2: IF «a is A_2» AND «b is B_2», THEN «c is C_2».

The triangular MFs of the input and output variables A, B, and C are given by the following expressions:

$$A=\{A_1\}+\{A_2\}=\left\{\int_{5}^{15}\left(\frac{a-5}{15-5}\right)\Big/a+\int_{15}^{30}\left(\frac{30-a}{30-15}\right)\Big/a\right\}+$$
$$+\left\{\int_{0}^{10}\left(\frac{a}{10}\right)\Big/a+\int_{10}^{30}\left(\frac{30-a}{30-10}\right)\Big/a\right\},$$

$$B=\{B_1\}+\{B_2\}=\left\{\int_{5}^{15}\left(\frac{b-5}{15-5}\right)\Big/b+\int_{15}^{35}\left(\frac{35-b}{35-15}\right)\Big/b\right\}+$$
$$+\left\{\int_{15}^{40}\left(\frac{b-15}{40-15}\right)\Big/b+\int_{40}^{50}\left(\frac{50-б}{50-40}\right)\Big/b\right\},$$

$$C=\{c_1\}+\{c_2\}=\left\{\int_{0}^{70}25/c\right\}+\left\{\int_{0}^{70}35/c\right\}.$$

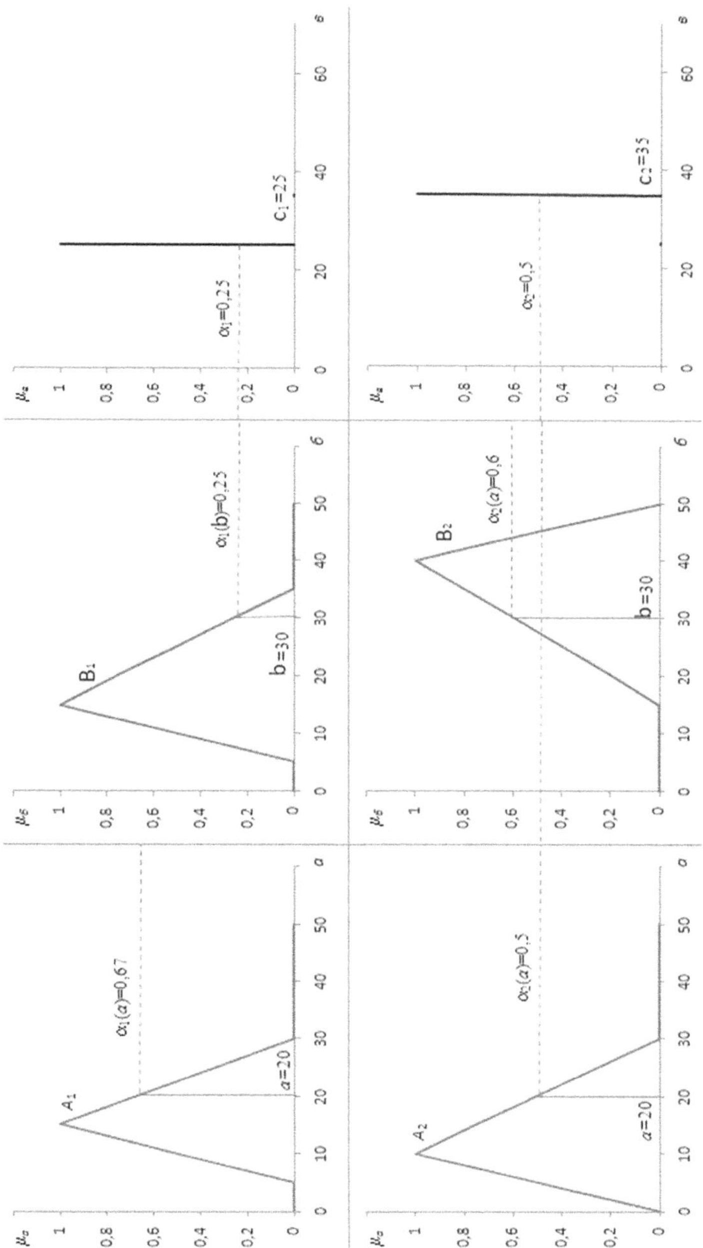

FIGURE 7.4 Structure of a simplified fuzzy inference algorithm.

Let the input of the information-measuring system receive values $a = 20$ and $b = 30$. Then at the first stage the degrees of affiliation are determined:

$$\alpha_1(a) = 0{,}67;\ \alpha_1(b) = 0{,}25;\ \alpha_2(a) = 0{,}5;\ \alpha_2(b) = 0{,}6.$$

At the stage of the logical output are the cutoff levels:

$$\alpha'_1 = \alpha_1(a) \wedge \alpha_1(б) = \min\{0{,}67;\ 0{,}25\} = 0{,}25;$$
$$\alpha'_2 = \alpha_2(a) \wedge \alpha_2(б) = \min\{0{,}5;\ 0{,}6\} = 0{,}5.$$

At realization of the third step by a method of the center of gravity the output value is calculated:

$$y = \frac{\alpha'_1 c_1 + \alpha'_2 c_2}{\alpha'_1 + \alpha'_2} = \frac{0{,}25 \cdot 25 + 0{,}5 \cdot 35}{0{,}25 + 0{,}5} = 31{,}67.$$

For example, if the data $a = 10$ and $b = 40$ come to the input of the information-measuring system, the output value will be the result 35.

Consider the process of training ANFIS, which uses a simplified model of fuzzy inference. Both hard and soft arithmetic operations are used for computation. Let there be the following FRs given by the fuzzy knowledge base (Table 7.1).

Training of the fuzzy system is carried out in several stages:

Step 1: For each of the input variables a_1 and a_2, the degrees of truth are determined as shown in Figure 7.5, according to the data received from the sensors: $a_1 = (\alpha'_{11}, \alpha'_{12}, \alpha'_{13})$; $a_2 = (\alpha'_{21}, \alpha'_{22}, \alpha'_{23})$.

For example, if information $a_1 = 34$ and $a_2 = 163$ is received from the measuring system, the degrees of truth will be equal:

$a_1 = (0.6, 0.4, 0)$; $a_2 = (0.7, 0.3, 0)$.

Step 2: At the stage of logical inference, taking into account the fuzzy knowledge base (Table 7.1) truncated degrees of truth are calculated in Table 7.2 [11]. Next, using the fuzzy relation matrix, the cutoff levels of the output MF are calculated in Table 7.3.

In the final step of the FLI the truncated degrees of truth are merged:

$$\mu(\gamma) = \bigcup_{i=1}^{n} \left(\beta_i;\ \mu(y_i)\right),$$

where n is the number of truncated MFs.

TABLE 7.1
Fuzzy Knowledge Base

FR	If		Then	FR	If		Then	FR	If		Then
r_1	a_{11}	a_{21}	y_{15}	r_4	a_{12}	a_{21}	y_{14}	r_7	a_{13}	a_{21}	y_{13}
r_2	a_{11}	a_{22}	y_{14}	r_5	a_{12}	a_{22}	y_{13}	r_8	a_{13}	a_{22}	y_{12}
r_3	a_{11}	a_{23}	y_{13}	r_6	a_{12}	a_{23}	y_{12}	r_9	a_{13}	a_{23}	y_{11}

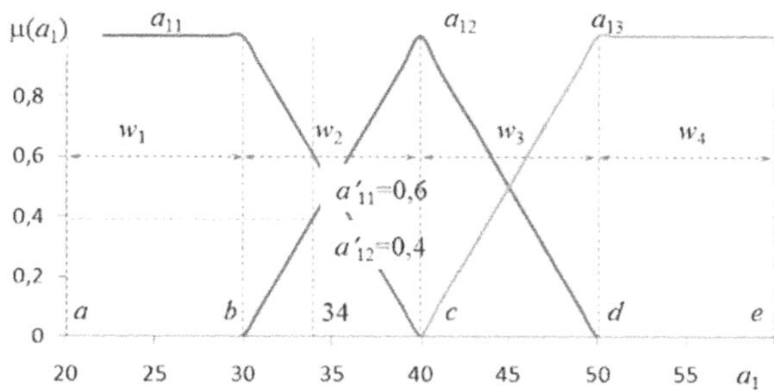

Input MFs

$$w_1 = \begin{cases} 1, x \in if\,[a,b) \\ 0, x \in if\,[a,b) \end{cases}$$

$$w_2 = \begin{cases} 1, x \in if\,[b,c) \\ 0, x \in if\,[b,c) \end{cases}$$

$$w_3 = \begin{cases} 1, x \in if\,[c,d) \\ 0, x \in if\,[c,d) \end{cases}$$

$$w_4 = \begin{cases} 1, x \in if\,[d,e) \\ 0, x \in if\,[d,e) \end{cases}$$

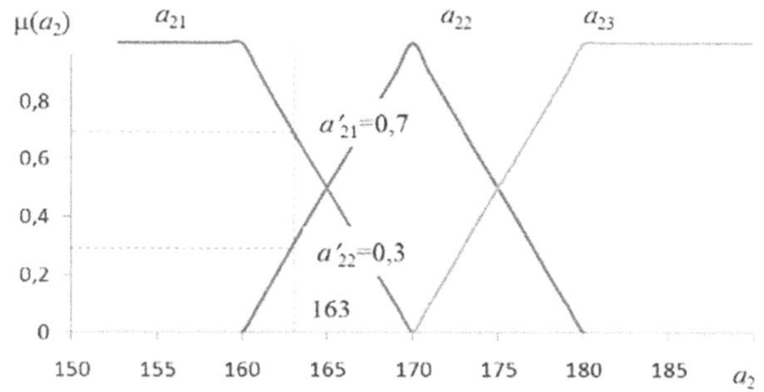

$$f = \begin{cases} w'_1 + w'_2 \cdot \left(\frac{c-x}{c-b} \right) \\ w'_2 \cdot \left(\frac{x-b}{c-b} \right) + w'_3 \cdot \left(\frac{d-x}{d-c} \right) \\ w'_3 \cdot \left(\frac{d-x}{d-c} \right) + w'_4 \end{cases}$$

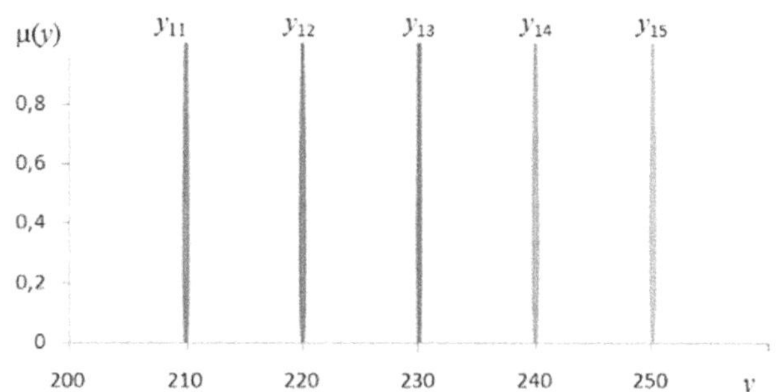

Output MF

$$f(x;a) = \begin{cases} 1, x = y_{nm} \\ 0, x \neq y_{nm} \end{cases}$$

FIGURE 7.5 Membership functions of input and output variables.

TABLE 7.2
Truncated Degrees of Truth

Hard equations	$\alpha''_1 = \min(\alpha'_{11}; \alpha'_{21})$	$\alpha''_2 = \min(\alpha'_{11}; \alpha'_{22})$	$\alpha''_3 = \min(\alpha'_{11}; \alpha'_{23})$
	$\alpha''_4 = \min(\alpha'_{12}; \alpha'_{21})$	$\alpha''_5 = \min(\alpha'_{12}; \alpha'_{22})$	$\alpha''_6 = \min(\alpha'_{12}; \alpha'_{23})$
	$\alpha''_7 = \min(\alpha'_{13}; \alpha'_{21})$	$\alpha''_8 = \min(\alpha'_{13}; \alpha'_{22})$	$\alpha''_9 = \min(\alpha'_{13}; \alpha'_{23})$
Soft equations	$\alpha''_1 = \min_{\delta}(\alpha'_{11}; \alpha'_{21})$	$\alpha''_2 = \min_{\delta}(\alpha'_{11}; \alpha'_{22})$	$\alpha''_3 = \min_{\delta}(\alpha'_{11}; \alpha'_{23})$
	$\alpha''_4 = \min_{\delta}(\alpha'_{12}; \alpha'_{21})$	$\alpha''_5 = \min_{\delta}(\alpha'_{12}; \alpha'_{22})$	$\alpha''_6 = \min_{\delta}(\alpha'_{12}; \alpha'_{23})$
	$\alpha''_7 = \min_{\delta}(\alpha'_{13}; \alpha'_{21})$	$\alpha''_8 = \min_{\delta}(\alpha'_{13}; \alpha'_{22})$	$\alpha''_9 = \min_{\delta}(\alpha'_{13}; \alpha'_{23})$

TABLE 7.3
Cutoff Levels of the Output Membership Functions

Hard equations	$\beta_1 = \alpha''_9$	$\beta_2 = \max(\alpha''_8; \alpha''_6)$	$\beta_3 = \max(\alpha''_7; \alpha''_5; \alpha''_3)$
	$\beta_4 = \max(\alpha''_4; \alpha''_2)$	$\beta_5 = \alpha''_1$	
Soft equations	$\beta_1 = \alpha''_9$	$\beta_2 = \max_{\delta} (\alpha''_8; \alpha''_6)$	$\beta_3 = \max_{\delta} (\alpha''_7; \alpha''_5; \alpha''_3)$
	$\beta_4 = \max_{\delta} (\alpha''_4; \alpha''_2)$	$\beta_5 = \alpha''_1$	

Step 3: Defuzzification of the output quantity is performed using the method of center of gravity:

$$Y = \frac{\sum_{i=1}^{n} \mu(\gamma_i) \cdot y_i}{\sum_{i=1}^{n} \mu(\gamma_i)}.$$

Step 4: After calculation of the actual value of the output variable, the procedure of adaptation of the complex technical system is performed, i.e., adjustment of the terms of the output variable until the obtained value Y becomes as close as possible to the known reference value y_{etal} by the formula:

$$y_{i+1} = y_i + \delta\left(Y - y_{etal}\right),$$

where δ is the learning rate of the neuro-fuzzy inference system (by default $\delta = 0.7$).

Figure 7.6 shows the six-layer structural diagram of ANFIS.

Layer 1. Represents the process of phasing the input variables, each of which has three terms with the membership function shown in Figure 7.5. The inputs of the network are connected only to their terms.
Layer 2. The outputs of this layer are the truth degree values, determined by information from the sensors.
Layer 3. Each node of this layer corresponds to the rules presented in Table 7.1. The outputs of nodes of this layer are values of truncated degrees of truth.
Layer 4. In this layer, the output variable cutoff levels are defined.
Layer 5. It is a process of combining truncated degrees of truth.
Layer 6. Defuzzification of the output parameter is carried out.

In the process of learning the adaptive neuro-fuzzy network, which functions on the basis of a simplified fuzzy inference model, the formation of new FP terms of the output parameter by the iterative formula is performed until the value obtained with the neuro-fuzzy inference system becomes equal to the reference value $y = y_{\text{given}}$.

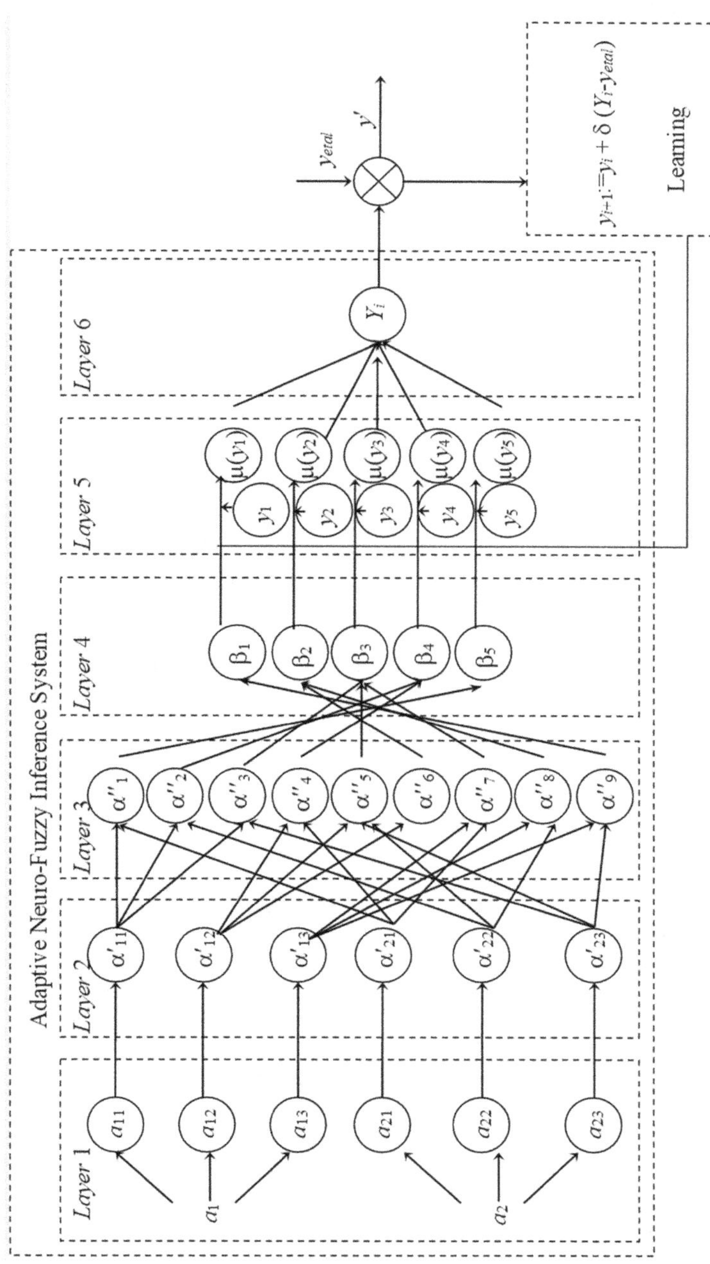

FIGURE 7.6 Adaptive Neuro-Fuzzy inference system.

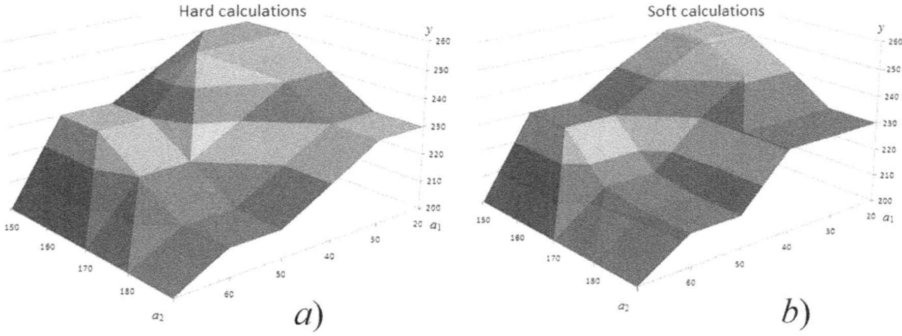

FIGURE 7.7 Graphs simulating the response surface of the ANFIS output variable: (a) based on hard calculations; (b) based on soft calculations.

In order to verify the adequacy and performance of ANFIS, which implements soft and hard arithmetic operations in the conditions of this problem, consider the process of modeling its work.

Let the fuzzy-logical system be given by the rules of the form (Table 7.1) and have two inputs a_1, a_2, and one output variable y. The response surface of the adaptive neuro-fuzzy inference system, based on the simplified fuzzy inference model with soft and hard arithmetic operations, is graphically represented in Figure 7.7.

The calculation data are summarized in Tables 7.4 and 7.5.

Next, consider the ANFIS learning process. Let the sensor data of a complex technical system transmit the following information: $a_1 = 34$, $a_2 = 163$. Given the values of the truth degrees, $a_{1si} = (0.6, 0.4, 0)$, $a_{2si} = (0.7, 0.3, 0)$.

Then the truncated degrees of truth are calculated as shown in Table 7.6.

Using the fuzzy relationship matrix, the cutoff levels are found in Table 7.7.

After the operation of combining truncated degrees of truth, defuzzification of the output value is performed, getting for hard calculations $Y = 242{,}4$, and for soft calculations $Y = 241{,}9$.

Let us assume that the reference value is $y_{etal} = 245$. We see that the difference between the two values (Y-y_{etal}) is 2.6 for hard operations and 3.1 for soft arithmetic operations. Then to achieve the given value, we start the training procedure of the

TABLE 7.4
ANFIS Based on Hard Calculations

	a_2	150	160	170	180	190
a_1						
20		250	250	240	230	230
30		250	250	240	230	230
40		240	250	230	220	220
50		230	230	220	210	210
60		230	230	220	210	210

TABLE 7.5
ANFIS Based on Soft Calculations

a_1 \ a_2	150	160	170	180	190
20	250,8	250,8	241,5	230,1	230,1
30	250,8	250,8	241,5	230,1	230,1
40	241,6	241,6	230,1	230,3	230,3
50	230,1	230,1	214,1	211,9	211,9
60	230,1	230,1	214,1	211,9	211,9

TABLE 7.6
Truncated Degrees of Truth for ANFIS

Hard calculations	$\alpha''_1 = 0{,}6$	$\alpha''_2 = 0{,}3$	$\alpha''_3 = 0$
	$\alpha''_4 = 0{,}4$	$\alpha''_5 = 0{,}3$	$\alpha''_6 = 0$
	$\alpha''_7 = 0$	$\alpha''_8 = 0$	$\alpha''_9 = 0$
Soft calculations	$\alpha''_1 = 0{,}595$	$\alpha''_2 = 0{,}299$	$\alpha''_3 = 0{,}0002$
	$\alpha''_4 = 0{,}399$	$\alpha''_5 = 0{,}295$	$\alpha''_6 = -0{,}024$
	$\alpha''_7 = 0{,}0004$	$\alpha''_8 = -0{,}0008$	$\alpha''_9 = -0{,}024$

TABLE 7.7
Cutoff Levels for ANFIS

Hard calculations	$\beta_1 = 0$	$\beta_2 = 0$	$\beta_3 = 0{,}3$
	$\beta_4 = 0{,}4$	$\beta_5 = 0{,}6$	
Soft calculations	$\beta_1 = -0{,}024$	$\beta_2 = -0{,}002$	$\beta_3 = 0{,}267$
	$\beta_4 = 0{,}394$	$\beta_5 = 0{,}595$	

neuro-fuzzy inference system. Using the iterative equation, we calculate a new value of the coefficient y_{11}, preliminarily taking the learning rate equal to 0.7 as shown in Table 7.8.

where Δ is the offset by which you want to shift the output variable terms.

After shifting the output variable terms by the calculated values, determine the output of ANFIS as $Y = 244{,}2$ for soft operations and $Y = 243{,}9$.

It can be seen that after the first iteration of training, the error between the two values decreased and is 1.1 and 0.8 for soft and hard operations, respectively. Similarly, the bias coefficients for subsequent iterations are calculated. The data from the ANFIS learning process is summarized in Table 7.9.

Table 7.9 shows that ANFIS training for the two methods using soft and hard arithmetic operations was six iterations.

TABLE 7.8
Calculated Values for the Output Variable y_{11}

Soft Operations	Hard Operations
$y_{11} = 210 - 0{,}7(241{,}9 - 245) = 212{,}2,$	$y_{11} = 210 - 0{,}7(242{,}4 - 245) = 211{,}8,$
$\Delta = 212{,}2 - 210 = 2{,}2,$	$\Delta = 212{,}2 - 210 = 1{,}8,$

TABLE 7.9
ANFIS Net Training

Calculation of Output Variable Bias Coefficients

	Soft Operations	Hard Operations
2nd iteration	$y_{12} = 213\Delta = 0{,}8Y = 243{,}9$	$y_{12} = 212{,}4\Delta = 0{,}6Y = 244{,}2$
3rd iteration	$y_{13} = 213{,}3\Delta = 0{,}3Y = 244{,}6$	$y_{13} = 212{,}5\Delta = 0{,}1Y = 244{,}8$
4th iteration	$y_{13} = 213{,}4\Delta = 0{,}1Y = 244{,}9$	$y_{13} = 212{,}6\Delta = 0{,}1Y = 244{,}9$
5th iteration	$y_{13} = 213{,}5\Delta = 0{,}1Y = 245$	$y_{13} = 212{,}7\Delta = 0{,}1Y = 245$

The analysis of the graphs (Figure 7.7) showed that there are no special differences when using soft and hard calculations. However, a peculiarity was noted, which is that in Tables 7.6 and 7.7 the same answers occur. This feature allows us to draw the following conclusion: in order to reduce the number of operations when implementing ANFIS, it is possible to reduce the number of operations calculating the output quantity Y. Thus, Table 7.10 shows for which cells the calculation is necessary, and for which it is sufficient to simply copy the value obtained earlier.

Tables 7.6 and 7.7 show that the calculation of output parameters requires 25 operations, while in Table 7.10 the calculation is reduced to 7 operations, thus the observed feature allows increasing the performance of ANFIS working on the basis of the simplified fuzzy output model by 3.5 times.

An analysis of the learning process using soft and hard operations showed that the same number of iterations is required to train the ANFIS network. Therefore, the

TABLE 7.10
Reducing the Number of Operations in ANFIS

a_1 \ a_2	150	160	170	180	190
20	A	A	D	C	C
30	A	A	D	C	C
40	B	B	E	E	E
50	C	C	F	G	G
60	C	C	F	G	G

simplified ANFIS for automated control systems described in [12–16] will undoubtedly increase their efficiency.

The use of soft arithmetic operations in the simplified fuzzy inference model for training ANFIS network is not always appropriate, since the results show that the gain in this case is not significant, and the complexity of calculations in their use increases many times.

7.4 TRAINING A NEURO-FUZZY SYSTEM BASED ON THE AREA DIFFERENCE METHOD

Consider a fuzzy MISO-system that has input and output variables defined by triangular FPs, and the relationship between them is formed on the basis of fuzzy rules of the form:

Rule$_i$: If $X_1 = x_{i1}$ И $X_2 = x_{i2}$ И … И $X_n = x_{in}$, then $Y = y_i$,

where X_p is the input variable; n is the number of input variables; x_{in} is the linguistic term describing the input MFs; i is the number of terms at the input/output variable; Y is the output variable; xi is the linguistic term describing the output MFs.

A fuzzy MISO system performs a mapping based on one of the traditional defuzzification methods, which include: Middle of Maxima (MM); First of Maxima (FM); Last of Maxima (LM); Center of Gravity (CG); Center of Sums (CS); and Height (Model of Height – MH). For example, when using the Center of Gravity method, the MISO fuzzy system mapping is as follows:

$$y_{defuz} = \frac{\int_{min}^{max} y \cdot \mu'(y)\,dy}{\int_{min}^{max} \mu'(y)\,dy}, \tag{7.1}$$

where min, max – integration limits of the fuzzy set; $\mu(y)$ – MF of the output variable after the implementation based on fuzzy rules of the fuzzy inference procedure.

The problem of estimation of the fuzzy MISO-system is reduced to the search of such a solution at which the root mean square error (RMSE) is minimal [17]:

$$RMSE = \frac{1}{M}\sqrt{\sum_{i=1}^{n}\left(y_{ex} - y_{defuz}\right)^2} \rightarrow \min, \tag{7.2}$$

where u_{ex} – set of trained data; M – number of points in the trained sample.

In traditional fuzzy MISO-systems, the main factors that lead to an increase in RMSE are: imperfect defuzzification models of the resulting fuzzy set; use of rigid formulas for finding the minimum and maximum in the fuzzy inference structure [18], which reduces the additivity of fuzzy models and leads to unsmooth response surface of the resulting variable; large number of fuzzy logic inference, which leads to reduced speed of the control decision when the fuzzy always equal to the number of fuzzy rules [19], so reducing the number of conclusions will increase the speed and minimize the RMSE.

As a rule, the response surface in the trained sample has a complex topography; then the search models based on traditional ANFIS models will not always allow to achieve optimal solutions. It is proposed to use a soft fuzzy logic inference algorithm.

The training method of the fuzzy MISO-system is implemented as the following sequence of steps. It should be noted that a limitation of this method is that the FP terms must be either triangular or trapezoidal.

Step 1: Phasification of input variables. Let the fuzzy MISO system consist of two input variables each having three terms $X_1 = \{x_{11}\}+\{x_{12}\}+\{x_{13}\}$ and $X_2 = \{x_{21}\}+\{x_{22}\}+\{x_{23}\}$ and an output variable having five terms, y $Y = \{y_1\}+\{y_2\}+\{y_3\}+\{y_4\}+\{y_5\}$. In this case, the terms are described by:

Triangular MF

$$f(x;\ a,\ b,\ c) = \begin{cases} 0, & x \le a; \\ \dfrac{x-a}{b-a}, & a \le x \le b; \\ \dfrac{c-x}{c-b}, & b \le x \le c; \\ 0, & c \le x. \end{cases} \tag{7.3}$$

Trapezoidal MF

$$f(x;\ a,\ b,\ c,\ d) = \begin{cases} 0, & x \le a; \\ \dfrac{x-a}{b-a}, & a \le x \le b; \\ 1, & b \le x \le c;\ , \\ \dfrac{d-x}{d-c}, & c \le x \le d; \\ 0, & d \le x. \end{cases} \tag{7.4}$$

where a, b, c, d are MF parameters; x is the quantitative value of the input parameter.

The MF having a triangular form for the fuzzy MISO-system are shown in Figure 7.8.

Step 2: Determine the degrees of affiliation for each prerequisite based on the information coming from the sensors of the active control systems; for example, when $x_1 = 82$ and $x_2 = 256$.

Step 3: Synthesis of the knowledge base containing fuzzy rules of the kind summarized in Table 7.11.

Step 4: Construction of fuzzy relation matrix. It should be noted that the traditional models of Mamdani and Sugheno use the operations of finding the hard minimum and maximum to implement the compositional rule in

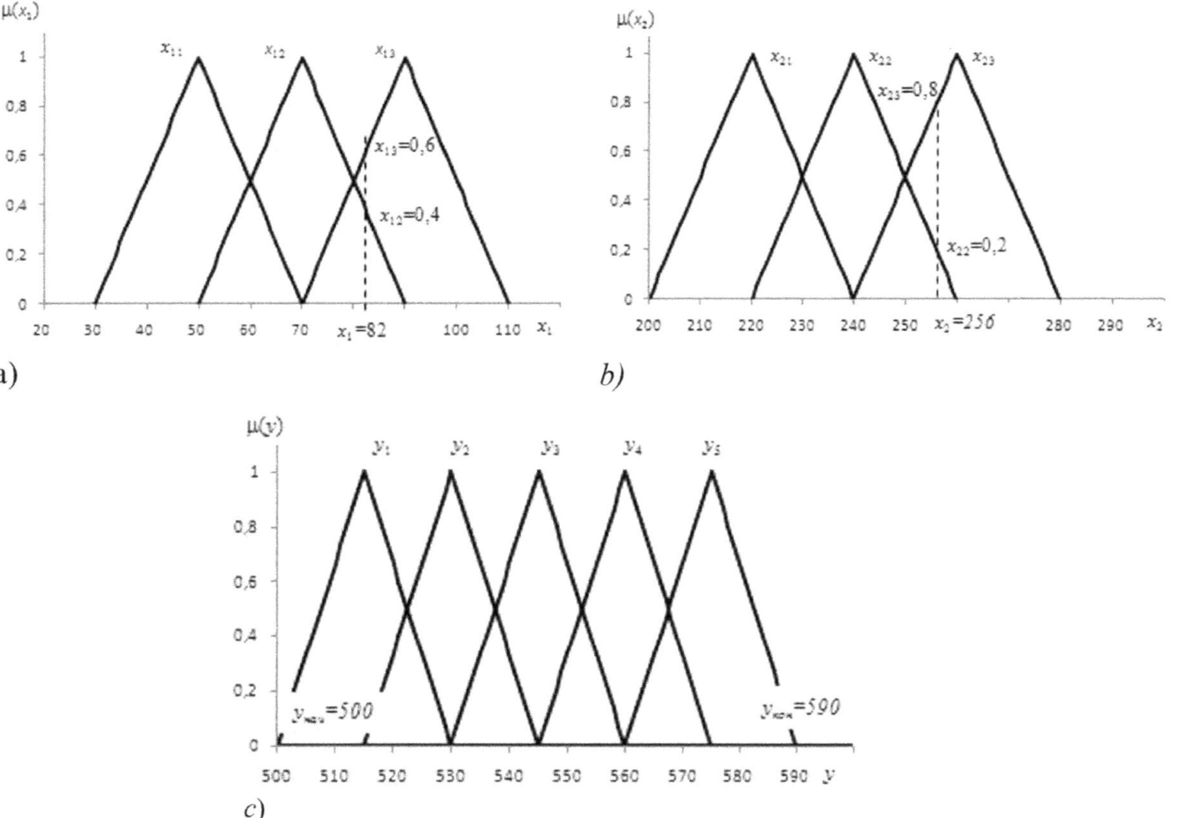

FIGURE 7.8 Plots of MF: (a) first input variable x_1; (b) second input variable x_2; (c) output variable y.

TABLE 7.11
Fuzzy Knowledge Base

FR	If		Then	FR	If		Then	FR	If		Then
FR$_1$	x_{11}	x_{21}	y_5	**FR$_4$**	x_{12}	x_{21}	y_4	**FR$_7$**	x_{13}	x_{21}	y_3
FR$_2$	x_{11}	x_{22}	y_4	**FR$_5$**	x_{12}	x_{22}	y_3	**FR$_8$**	x_{13}	x_{22}	y_2
FR$_3$	x_{11}	x_{23}	y_3	**FR$_6$**	x_{12}	x_{23}	y_2	**FR$_9$**	x_{13}	x_{23}	y_1

TABLE 7.12
Formulas for Soft-minimum and Maximum of Type I

Hard min: $$\min(x_1, x_2) = \frac{x_1 + x_2 - (x_1 - x_2)\cdot \operatorname{sgn}(x_1 - x_2)}{2},$$

$$\operatorname{sgn}(x_1 - x_2) = \begin{cases} -1, & if\ x_1 - x_2 < 0; \\ 0, & if\ x_1 - x_2 = 0; \\ 1, & if\ x_1 - x_2 > 0. \end{cases}$$

Soft min: $$\min_{\delta}(x_1, x_2)_I = \frac{x_1 + x_2 + \delta^2 + \sqrt{(x_1 - x_2)^2 + \delta^2}}{2}.$$

Soft max: $$soft - \max(x_1, x_2) = |\gamma \cdot \max(x_1, x_2) + 0,5(1 - \gamma)(x_1 + x_2)|, \quad where\ \gamma = 0,9$$

fuzzy logic inference. As a consequence, traditional fuzzy logic inference algorithms do not have the property of additivity. Soft-min and soft-max arithmetic operations can eliminate this drawback. Formulas for finding soft-minimum and maximum of type I are presented in Table 7.12.

As shown in [20], the soft minimum formula used to find the soft minimum, at values of input variables equal to 0 gives a negative result, so, for example at $x_1 = 0$ and $x_2 = 0$, the output answer is −0.02375. Negative answers will be obtained until the values of input variables equal to $x_1 = 0.01$ and $x_2 = 0.05$, in which case the answer obtained by using this formula is −0.00077, when $x_2 = 0.06$, the answer will become positive and will be equal to 0.000895. However, this circumstance is a positive property of this formula. When using the well-known formulas for finding the minimum with input values equal to 0, that is, $x_1 = 0$ and $x_2 = 0$, the answer will be 0. At the following fuzzy inference operations, the answer will also be equal to 0, and the fuzzy system itself will become non-additive. Therefore, in the study of fuzzy systems, they have a large value of standard deviation in RMSE.

The following formulae for finding a soft minimum of type II are known (Table 7.13), and this paper gives identical results using the following formulae.

TABLE 7.13
Formulas for Soft Minimum of Type II

Soft min:	$\min\limits_{\delta}\left(x_1,x_2\right)_{II_1}=\dfrac{x_1+x_2-\dfrac{\left(x_1-x_2\right)\cdot\left(x_1-x_2\right)}{\sqrt{\left(x_1-x_2\right)^2+\delta^2}}}{2}=\dfrac{x_1+x_2}{2}-\dfrac{\left(x_1-x_2\right)^2}{2\cdot\sqrt{\left(x_1-x_2\right)^2+\delta^2}}.$
Soft min:	$\min\limits_{\delta}\left(x_1,x_2\right)_{II_2}=0,5\cdot\left[x_1+x_2+\dfrac{\delta^2}{\sqrt{\left(x_1-x_2\right)^2+\delta^2}}-\sqrt{\left(x_1-x_2\right)^2+\delta^2}\right].$

TABLE 7.14
Fuzzy Relationship Matrices for Soft Calculations and Hard Calculations

Terms	Hard Calculations				Soft Calculations			
	Composition			Max	Composition			Max
y'_5	0			0	−0,0238			−0,02375
y'_4	0	0		0	−0,0018	−0,0003		−0,00038
y'_3	0	0,2	0	0,2	0,00047	0,19817	0,0002	0,178883
y'_2	0,2	0,4		0,4	0,19969	0,39969		0,389694
y'_1	0,6			0,6	0,59817			0,598172

Due to the rational arrangement of the elements in the fuzzy relations matrix, the number of conclusions in the fuzzy output is equal to the number of terms in the output FP, in this case, - 5. In traditional fuzzy-logic models, the number of conclusions with the recommendations would be equal to the number of fuzzy rules and should be in the range of 9 to 15. The results of fuzzy relations matrix modeling are shown in Table 7.14.

Step 5: Defuzzification of the output result. This method determines the areas of triangular and/or trapezoidal terms of MFs using the universal formula.

Using this method, there is no need for the operations of truncation of the output MFs terms and their subsequent union, used in traditional fuzzy models.

Step 5.1: Determination of the total area of the figure describing the output variable y, which consists of 5 triangular terms:

$$S_{gen}=n\cdot\frac{b_1}{2}=\frac{5\cdot\left(530-500\right)}{2}=75, \tag{7.5}$$

where n is the number of terms of the output linguistic variable ($n=5$).

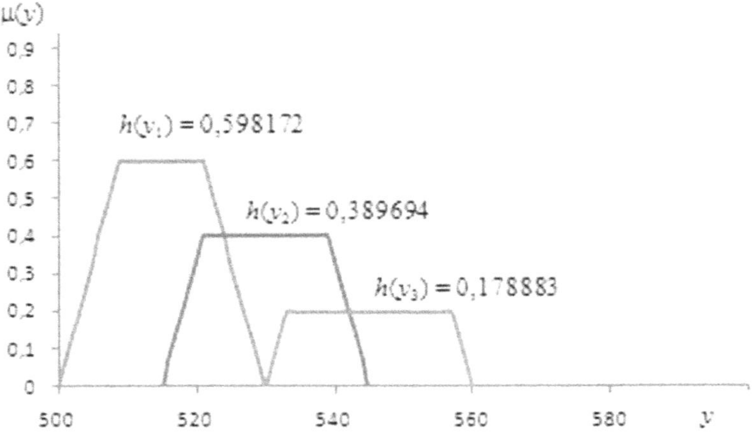

FIGURE 7.9 Determination of the heights of the truncated terms of the output MF.

Step 5.2: Truncate the terms of the output MF. This operation is performed:

$$\mu(y)'_i = soft \overset{n}{\underset{i=1}{-}} \min\left(y'_i;\ \mu(y_i)\right). \tag{7.6}$$

The result of the implementation of this operation in graphical form is shown in Figure 7.9.

Step 5.3: Finding the height of truncated terms of the output membership function. So, in Figure 7.9, the height of the truncated terms is equal to $h(y_1) = 0.598172$, $h(y_2) = 0.389694$, $h(y_3) = 0.178883$, $h(y_4) = 0$, $h(y_5) = 0$.

$$S_n = \begin{cases} S_n = 0, \quad if\ \ h = 0; \\ S_n = \dfrac{b_1}{2}, \quad if\ \ h = 1; \\ S_n = \dfrac{h}{2}\left(b_1 + b_3\right), \quad if\ \ h \in (0,\ 1). \end{cases} \tag{7.7}$$

Then we calculate the areas of the truncated terms:

$$S_1 = \frac{h \cdot (b_1 + b_3)}{2} = \frac{0{,}598172 \cdot \left[(530 - 500) + 13\right]}{2} = 12.86,$$

$$S_2 = \frac{h \cdot (b_1 + b_3)}{2} = \frac{0{,}389694 \cdot \left[(545 - 515) + 19\right]}{2} = 8.38,$$

$$S_3 = \frac{h \cdot (b_1 + b_3)}{2} = \frac{0{,}178883 \cdot \left[(560 - 530) + 25\right]}{2} = 3.85,$$

$$S_4 = 0,$$

$$S_5 = 0.$$

Step 5.4: Determine the total added truncated area of the output variable terms:

$$S_{trunc} = \sum_{i=1}^{n} S_i = 12,86 + 8,38 + 3,85 + 0 + 0 = 25,09.$$

Step 5.5: Calculate the ratio of the total area to the area of the truncated figure:

$$D = \frac{S_{trunc}}{S_{total}} = \frac{25,09}{75} = 0,335.$$

Step 5.6: Find the proportional ratio:

$$y_{defuz} = \left[D \cdot \left(y_{final} - y_{initial}\right)\right] + y_{initial} = \left[0,335 \cdot \left(590 - 500\right)\right] + 500 = 530,15.$$

where $y_{final} - y_{initial}$ are the final and initial values (see Figure 7.9, c) of the range of values of the output variable.

Step 6: Training procedure. At this step, ANFIS is trained using the inverse error propagation method [21–24], i.e., it corrects the truncated areas of the output variable terms until the obtained value of y_{defuz} becomes as close as possible to the known reference value of y_{etal} in accordance with the ratio:

$$y_{out} = \left(w\right)_i + \delta\left(y_{defuz} - y_{etal}\right) \tag{7.8}$$

where δ is the learning step of the neuro-fuzzy inference system (by default $\delta = 0{,}01$).

As a weight parameter w, it is recommended to use the ratio $n/2$, to determine the total area. At the first training step $w = n/2 = 2.5$.

7.5 AGGREGATION OF OUTPUT VALUES

To determine how badly the signal has been disturbed, we can use aggregation of the output values using the minimax method. This means that we select the largest value from all the output values that were derived from the fuzzy rules.

So, formulas implementing this operation have the following form:

$$y_1 = \min\left(x_{13}; x_{21}\right); \tag{7.9}$$

$$y_2 = \max\left(\min\left(x_{12}; x_{21}\right), \min\left(x_{13}; x_{22}\right)\right); \tag{7.10}$$

$$y_3 = \max\left(\min\left(x_{11}; x_{21}\right), \min\left(x_{12}; x_{22}\right), \min\left(x_{13}; x_{23}\right)\right); \tag{7.11}$$

$$y_4 = \max\left(\min\left(x_{11}; x_{22}\right), \min\left(x_{12}; x_{23}\right)\right); \tag{7.12}$$

$$y_5 = \min\left(x_{11}; x_{23}\right); \tag{7.13}$$

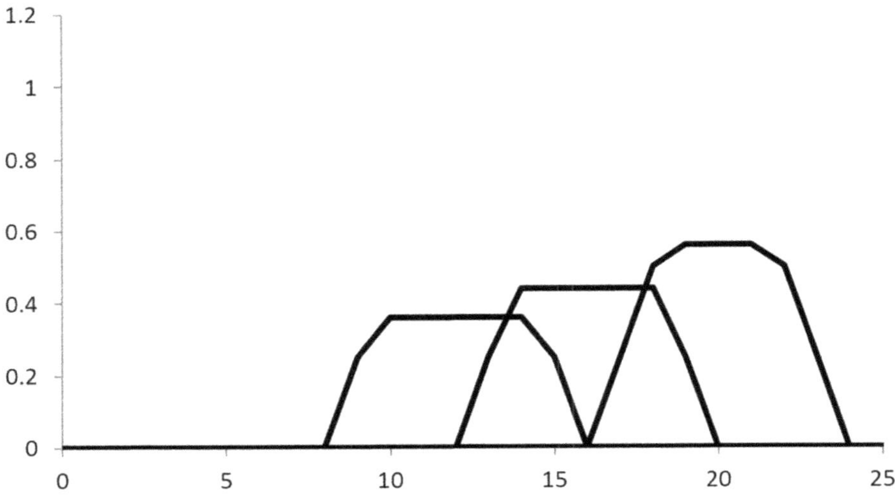

FIGURE 7.10 Aggregation of output variables.

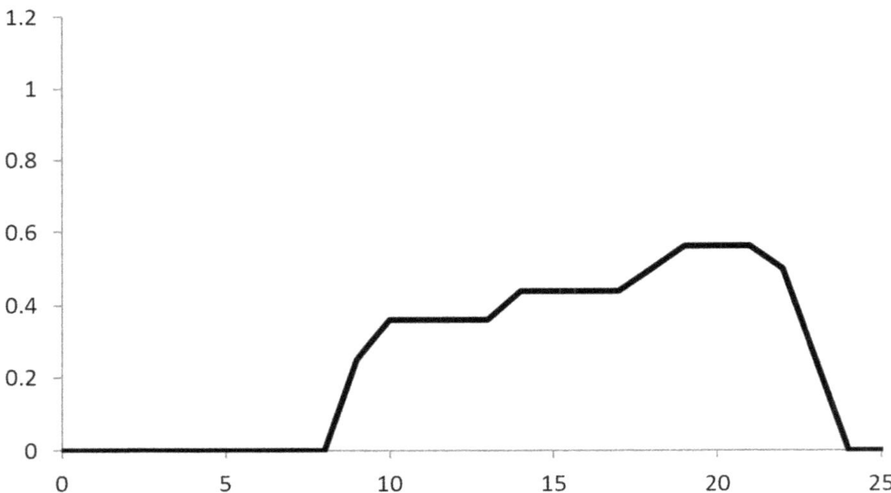

FIGURE 7.11 Global aggregation of output values.

So, when $x_1 = 34$ and $x_2 = 64$, the result of aggregation of output variables is shown in Figure 7.10.

Then the global aggregation of output values is performed. This operation is done using the maximum method. This means that we select the maximum value from the truncated output membership functions. This operation is shown in Figure 7.11.

The defuzzification of output values is needed to obtain a numerical value that will be used for control. To do this, we can use the center of gravity method, which

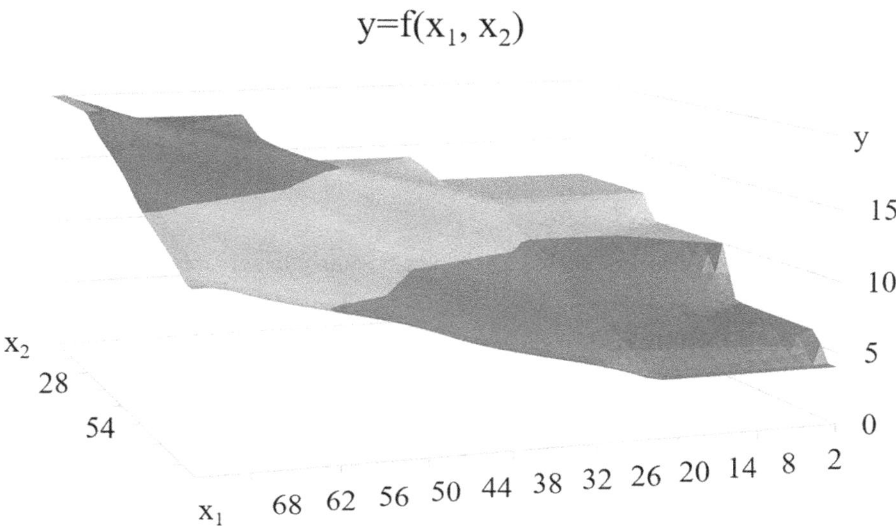

FIGURE 7.12 Result of the fuzzification of output values.

determines the average of the output values weighted by their degree of belonging to fuzzy sets.

To implement this operation, we use the following formula:

$$S = \frac{\sum_{i=1}^{n} d_i \cdot \alpha_i}{\sum_{i=1}^{n} \alpha_i} \tag{7.14}$$

The result of defuzzification is shown in Figure 7.12.

The resulting numerical value can be used to control the telecommunication system. For example, we can use it to adjust the signal gain level to improve signal quality.

7.6 CONCLUSIONS

The advantage of using fuzzy logic and Arduino to implement a telecommunication system is the ability to create more flexible and adaptive systems that can work under different conditions. Fuzzy systems take into account different factors and conditions, which allows to define and solve problems more accurately. In addition, the use of Arduino and other microcontrollers makes it possible to create systems that are more compact and affordable than traditional telecommunications systems.

However, the disadvantage of using fuzzy logic can be the difficulty of defining fuzzy rules, as well as the difficulty of configuring the system to work in different environments. In addition, microcontrollers may have limited ability to process large amounts of data, which can inhibit system performance.

Overall, using fuzzy logic and Arduino to implement telecommunications systems can be an effective approach that allows for more flexible, adaptive, and affordable systems. However, before starting to develop such a system, it is necessary to carefully consider and define fuzzy rules and configure the system to work under specific conditions.

ACKNOWLEDGMENT

This work was supported by the Russian Science Foundation grant No. 23-21-00071.

AUTHOR CONTRIBUTIONS

Conceptualization, methodology, software, and validation program code were handled by Bobyr M.V. Nolivos C.S.A. made formal analysis, wrote the original draft, and implemented data curation and investigation.

REFERENCES

1. Dumitrescu C., Ciotirnae P., Vizitiu C. Fuzzy logic for intelligent control system using soft computing applications. *Sensors*. 2021. 21. 8. C.2617.
2. Khan E. Neural Fuzzy Based Intelligent Systems and //Fusion of neural networks, fuzzy systems and genetic algorithms: industrial applications. – 2020.
3. Wu B. et al. Fuzzy logic based dynamic decision-making system for intelligent navigation strategy within inland traffic separation schemes. *Ocean Engineering*. 2020. 197. C.106909.
4. Lamamra K., Batat F., Mokhtari F. A new technique with improved control quality of nonlinear systems using an optimized fuzzy logic controller. *Expert Systems with Applications*. 2020. 145. C.113148.
5. Nayak J. R. et al. Application of optimized adaptive crow search algorithm based on two degrees of freedom optimal fuzzy PID controller for AGC system. *Engineering Science and Technology, an International Journal*. 2022. 32, 101061.
6. Sain D., Mohan B. M. Simulation and real-time implementation of a nonlinear fuzzy PI/PD controller. *IFAC-PapersOnLine*. 2020. 53. 1. C.673–678.
7. Kamil F. et al. An ANFIS-based optimized Fuzzy-multilayer decision approach for a mobile robotic system in ever-changing environment. *International Journal of Control, Automation and Systems*. 2019. 17. C.253–266.
8. Zhusubaliyev Z. T., Medvedev A., Silva M. M. Bifurcation analysis of PID-controlled neuromuscular blockade in closed-loop anesthesia. *Journal of Process Control*. 2015. 25. 152–163.
9. Bobyr M. V., Yakushev A. S., Dorodnykh A. A. (2020). Fuzzy devices for cooling the cutting tool of the CNC machine implemented on FPGA. *Measurement: Journal of the International Measurement Confederation*, 152. https://doi.org/10.1016/j.measurement.2019.107378
10. Bobyr M., Arkhipov A., Emelyanov S., Milostnaya N. A method for creating a depth map based on a three-level fuzzy model. *Engineering Applications of Artificial Intelligence*, 2023. 117. https://doi.org/10.1016/j.engappai.2022.105629
11. Bobyr M. V., Milostnaya N. A., Bulatnikov V. A. The fuzzy filter based on the method of areas' ratio. *Applied Soft Computing*. 2022. C.108449.

12. Bobyr M. V., Arkhipov A. E., Yakushev A. S. (2021). Shade recognition of the color label based on the fuzzy clustering. *Informatics and Automation*, 20. 2. 407–434. https://doi.org/10.15622/IA.2021.20.2.6
13. Cummings L., Russkikh M. The use of non-standard thermistors with a family of temperature converters to digital code. *Components and Technologies*. 2021. 5. C.17–22.
14. Zhu W., Nikafshan Rad H., Hasanipanah M. (2021). A chaos recurrent ANFIS optimized by PSO to predict ground vibration generated in rock blasting. *Applied Soft Computing*. 108. https://doi.org/10.1016/j.asoc.2021.107434
15. Alaei M., Khorsand R., Ramezanpour M. (2021). An adaptive fault detector strategy for scientific workflow scheduling based on improved differential evolution algorithm in cloud. *Applied Soft Computing*. 99. https://doi.org/10.1016/j.asoc.2020.106895
16. Ani K. A., Agu C. M. (2022). Predictive comparison and assessment of ANFIS and ANN, as efficient tools in modeling degradation of total petroleum hydrocarbon (TPH). *Cleaner Waste Systems*. 3. 100052. https://doi.org/10.1016/j.clwas.2022.100052
17. Denga X., Wang X. Incremental learning of dynamic fuzzy neural networks for accurate system modeling. *Fuzzy Sets and Systems*. 2009. 160. 7. 972–987.
18. Khodashinsky I.A., Dudin P.A. Identification of fuzzy systems based on direct ant colony algorithm *Artificial Intelligence and Decision Making*. 2011. 3. C.26–33.
19. Jayaram M.A., Chandana M. (2023). Design of flexible pavements through fuzzy inference system with genetic algorithm optimized rule base. *International Journal of Transportation Science and Technology*. https://doi.org/10.1016/j.ijtst.2023.03.001
20. Bobyr M.V., Titov D.V., Kulabukhov S.A. Evaluation of forecasting of control decisions under uncertainty // *Telecommunications*. 2015. 11. C.39–44.
21. Leekwijck W.V., Kerre E.E. Continuity focused choice of maxima: Yet another defuzzification method // *Fuzzy Sets and Systems*. 2001. 122. 303–314.
22. Perelman Y.I. Amusing Physics. Book 1 // Rimmis. 2015.
23. A.V. Kravchenko Robot Labyrinth // Radioamator. *Practical Radioelectronics*. 2010, 2. C.30–32.
24. Morishita T, Tojo O. Integer inverse kinematics method using Fuzzy logic *Intelligent Service Robotics*. 2013. 6. 2. 101–108.

8 Optimizing Wireless Sensor Networks Using Machine Learning

Sandhya Armoogum and Doorgesh Sookarah
University of Technology Mauritius, Port Louis, Mauritius

8.1 INTRODUCTION

Sensors identify physical phenomenon by measuring physical input from its environment such as temperature, motion, light, heat, proximity, and sound, and they output a signal that represents data captured that can be interpreted and analyzed. A sensor network consists of multiple sensor nodes that are often used to cooperatively sense and control an environment. The sensor network often connects to a computer network or a cloud via the Internet to transfer collected data for storage and analysis. Typically, the sensor nodes can communicate to each other as well as a base station over short distances, while the base station (edge node) connects to the Internet. Wired sensor networks consist of sensors physically connected by wires, whereas wireless sensor networks communicate using wireless technologies such as Bluetooth/Bluetooth Low-Energy (BLE), Near Field Communication (NFC) protocols, Zigbee, Z-Wave, 6LoWPAN, WiFi, and cellular. Data transmitted or received by the sensor network can trigger processing and analysis for gathering insights about the monitored environment and take actions if required.

Since the turn of the century, sensor networks and by extension, wireless sensor networks (WSNs), have been at the forefront of several research endeavors. These pertain mostly to industry-based applications where the interest in this specific segment of technology has only been accentuated over the past few years. When compared to conventional data capture and data dissemination techniques, WSNs present an added value as they can be implemented at a relatively low cost without completely disrupting existing systems and through near seamless integration.

Moreover, managing WSNs, especially those that involve a significant number of sensor nodes deployed over a wide area, can be tedious and complex. Such WSNs should be efficiently scalable and adapted to the environments and application requirements while minimizing the strain exerted on the underlying sensor nodes and infrastructure. This can be challenging, as any change in the already intricate network ecosystem will inevitably require a redesign activity, as WSNs have traditionally been explicitly programmed.

The promise of an easily deployable wireless sensor network is consequently substantial to system designers who may adopt varying configurations of their sensor networks to meet technical and operational requirements. As such, intrinsic factors

 DOI: 10.1201/9781003303114-8

and, more commonly, external ones can be accounted for providing dynamic adaptability to WSNs in the ever-evolving environments. This consideration would certainly help to better manage routing strategies, network latency, multidimensional design, localization, network span and coverage, network quality assurance, and overall link reliability. In addition, faults and shortcomings of a WSN would be more easily identifiable and remedied using a flexible and less taxing approach.

Machine Learning (ML) is a technique that emphasizes the autonomous learning and performance improvement of a computer system when dealing with a certain problem. In other words, machine learning enables learning from experience and focuses on self-improvement without being explicitly programmed. Using machine learning techniques, systems can be made computationally efficient, reliable, and cheaper. Such ML models support the analysis of high volumes of data in a better structured, faster, and more dependable manner. This ultimately helps in developing adaptable, quick, yet accurate sensor networks.

With the recent advent of increased computational power and the liberalization of digital technologies, ML has been applied to circumvent problems in several domains including WSNs. ML can augment performance of WSNs, while barely requiring any human intervention. Thus, redesign and periodic program updates are not required, as would have previously been the case. Another factor that has precipitated the involvement of ML in sensor networks is the massive volume of data that has now become widely available using cheap and widespread sensors, or alternatively higher storage capacities. ML provides the right framework and acts as a facilitator to extract, preprocess, analyze, and utilize the extensive amounts of data captured by sensors. Moreover, it can easily accommodate Internet of things (IoT) components.

Section 8.2 describes wireless sensor networks. Section 8.3 presents different machine learning techniques. In Section 8.4, the ML techniques for optimizing WSN design is presented. Section 8.5 covers the use of ML in WSN applications, and finally, Section 8.6 concludes the chapter.

8.2 SENSOR NETWORK

A sensor network consists of networked sensor nodes placed at various levels (e.g., ground level, underground level, undersea level, etc.). The sensor nodes are often small, cheap, battery powered, and having sensing, communication, computation, as well as limited storage capabilities. Sensor nodes are set out in large numbers, often to monitor various environmental conditions such as temperature, vibration, pressure, motion, and heat. Sensor nodes cooperatively transfer their data to a base station or sink where the data can be further transmitted to a server for processing and analysis. A WSN is made of numerous sensor nodes that have ***wireless*** communication capabilities. WSN are more practical in unattended, difficult-to-access and/or hostile environments such as undersea to detect seismic activities or military target tracking. WSN constitute an ad hoc network, and the structure of a WSN can follow a star, tree, mesh, or a hybrid star-mesh topology. Moreover, the topology of WSN may often change dynamically due to environmental factors in the region where the sensor nodes are positioned. Figure 8.1 depicts the typical WSN architecture.

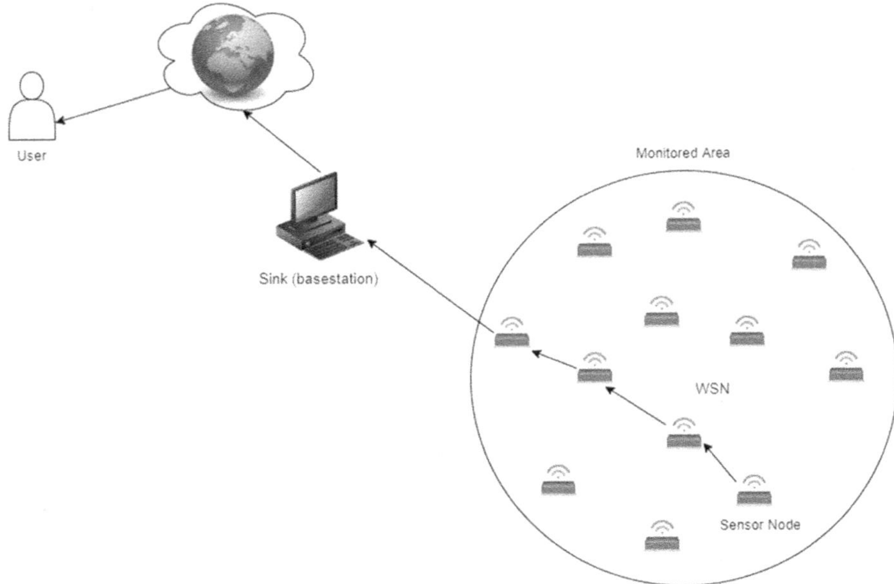

FIGURE 8.1 WSN Architecture.

8.2.1 Types of WSN

The main types of WSN include overground, underground, undersea, multimedia, and mobile WSNs [1]:

Overground WSNs consist of numerous wireless sensor nodes deployed either in an ad hoc (randomly dropped via airplane) or in a structured optimal placement (if accessibility is not an issue). Such sensor nodes communicate efficiently with the base stations.

Underground WSNs consist of sensor nodes located underground, for example, to monitor tunnels or mines. Such underground sensor nodes relay data to sink nodes located above ground. Underground WSN involves more expensive infrastructure (sensor nodes capable of reliable communication through layers of soil, rock, or concrete), complex deployment, and high maintenance. It requires careful planning due to signal losses and inherently elevated attenuation. Moreover, it is challenging to recharge or replace a sensor node's battery.

Undersea WSN consists of sensor nodes deployed undersea for exploring oceans, submarine detections, oil rigs monitoring, and detection of earthquake and tsunamis. Such sensor nodes communicate via acoustics and should be able to self-organize and adjust to hostile environment. Their batteries cannot be easily replaced or recharged. Furthermore, underwater WSN may experience longer propagation delay and limited bandwidth as well as occasional sensor failures. Often, underwater sensor nodes capture and send data to a base station on ground via satellite communications or underwater cables. Autonomous underwater vehicles may also be used for receiving data from these sensor nodes.

Multimedia WSNs are mainly used for collection and onwards transmission of images, audio, and video to enable monitoring of events. Sensor nodes are outfitted with cameras and microphones. Multimedia WSNs are often deployed in an organized manner to ensure sufficient coverage of the monitored environment. They usually have high bandwidth requirement, high energy consumption, and complex data processing/compressions operations, which makes quality of service (QoS) provisioning a challenge.

Mobile WSNs consist of sensors that can travel autonomously within the deployed environment for sensing and collecting data. Such a WSN may not require numerous sensor nodes, which results in decreased cost of the WSN. Mobile WSNs are characterized as having more reliable communication, better coverage, and being more energy-efficient. Mobile WSN is popular for real-time tracking and monitoring, for example, of hazardous material.

8.2.2 Applications of WSN

WSNs are the culmination of different areas, such as communications, electronics, and control systems. With the recent evolution in Internet of things (IoT), engineers and researchers find it more and more important to collect data from the surroundings to learn and respond to changes and threats; and WSNs play a fundamental role in such IoT systems. With the increasing number of different sensors, WSNs can sense the environments they are deployed in very accurately. WSNs are most suitable in harsh or hostile environments where they can be released from an airplane or drone into the target area. They can self-configure in a network arrangement for effective communication. WSNs can be easily scaled up by adding additional nodes to communicate on the same network without the need to change or modify existing sensor nodes. Furthermore, they are highly adaptive as they support dynamically changing connections between nodes (addition, removal, failure of nodes), and changes in their positioning and resulting change in topology. Long distance data collection is enabled as the sensor nodes can cooperate and relay data to other sensor nodes for the data to be retrieved by the base station. In many cases, some sensor nodes are not within the connectivity range of the base station. Data captured in dense WSNs results in data redundancy, which ensures data availability. Thus, WSNs have great potential for numerous types of applications.

As such, WSNs are heavily used for environmental monitoring [2]. WSNs are being used to monitor our natural environment for natural disaster detection/prevention to save lives, i.e., detection of earthquake and tsunamis, landslides, forest fires, volcanic activity, or flood due to rising water levels in rivers or lakes. Greenhouse effect monitoring allows scientists to better forecast weather and climate changes using WSN. Precise monitoring of agricultural environments provides important knowledge to farmers through WSNs that enable them to take informed decisions, for example, to determine harvest time, water saving approach for irrigation [3, 4].

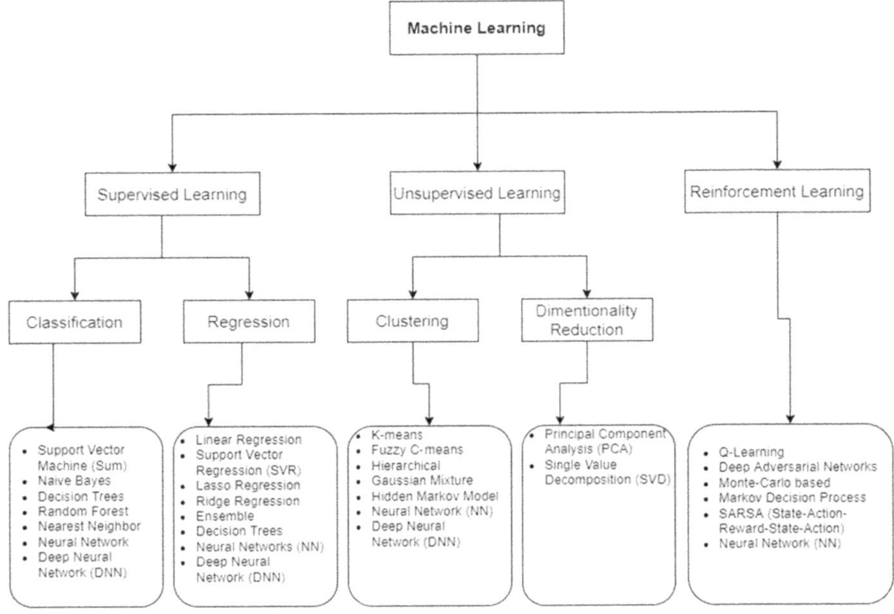

FIGURE 8.2 WSN Applications.

WSNs are also used for greenhouse monitoring and livestock tracking. WSN plays a fundamental role in Industry 4.0 for monitoring industrial processes in manufacturing plants [5]. In the area of healthcare, WSN can help monitor vital patient information, i.e., implanted sensors can help monitor patients' health [6]. Smart homes and smart cities also rely on WSNs and IoT, for example, for air pollution, building structure monitoring, and temperature control [7]. Figure 8.2 depicts the various WSN applications.

8.2.3 Challenges in WSN and Design Objectives

The design and implementation of WSNs are hampered due to their limited resources, namely energy, computing, and memory/storage, which have to be overcome. Given that sensor nodes are battery powered, a fundamental concern is energy conservation in WSNs by optimization of algorithms that lead to energy-efficient processing and communications. Energy conservation will increase the WSN lifetime. Often, sensor nodes on the ground have a secondary power source such as power cells to supplement them with energy. However, WSNs are often deployed in harsh environments inflicting damage on sensor nodes, which can result in sensor node failures.

Similarly, data latency is a major issue in real-time WSNs monitoring applications where delay can lead to wrong decisions or delayed action/feedback. WSNs are highly unreliable networks offering variable link capacity and prone to errors (depending on interference level and obstructions). Thus, bandwidth and transmission delay are location-dependent and fluctuate constantly, making it a challenge for real-time WSN applications. Adaptive and scalable communications, error control, as well as time-synchronization protocols, are needed for WSNs to be more efficient.

Due to the high density of the sensor nodes deployed, there is data redundancy. Data aggregation involves some pre-processing done by sensor nodes, which can lead to a smaller amount of data to be sent to the base station by relaying sensor nodes. This can decrease the amount of energy utilization related to transmission and eventually aid in increasing the lifetime of the WSN.

Moreover, sensor nodes often being deployed in open or adverse environments and are also subject to security attacks. Due to their limited computing power, sensor nodes cannot implement traditional encryption techniques and lack robust security mechanisms, and are thus vulnerable. Security is an important requirement in sensor networks to protect data and prevent malicious attacks such as DDoS and Sybil attacks.

WSNs involve large numbers of sensor nodes that have to self-configure autonomously to establish connections using a different routing strategy. Scalable architecture and efficient routing protocols are required, which also enhances the flexibility, robustness, and reliability of the WSN. Moreover, load balancing such as cluster head rotation can provide optimum usage of sensor nodes and augment the overall life span of the WSN. Tracking the location of sensor nodes is also important in WSNs, as the collected data does not have much value if its location is unknown.

These WSN challenges can be addressed using machine learning, such that WSNs become more efficient and useful. In the next section, different machine learning techniques are described.

8.3 MACHINE LEARNING (ML)

ML is based on systems that can learn without relying on explicit programming. ML is applied widely in various industries such as bioinformatics, healthcare, agriculture, financial services, environment, commerce, energy, telecommunications, automobiles (e.g., self-driving cars), entertainment, and social media [8]. ML is popularly used for data analytics and automation, especially network automation, to improve operational efficiency. The biggest benefit of ML is the possibilities it offers at performing generalization tasks; that is, it can provide solutions to an array of problems and is not necessarily bound by situational or time-based factors. This allows the architecture to sustain learning and ameliorate its performance over successive experiences. Thus, there is no one distinct research or applicative area where ML can be, but strictly speaking, ML is an interdisciplinary area, and it plays a fundamental role in a multitude of sectors.

A dataset is typically used, which often undergoes some preprocessing. The dataset is then divided into training data and testing data. Suitable ML algorithms are applied to the training data to 'learn,' whereby ML identifies patterns from the training data, and then the 'trained model' is applied on the new set of testing data to determine its performance. In general, an ML model is developed such that it can receive input data and make predictions within an acceptable range. There are different categories of ML algorithms, namely supervised learning, unsupervised learning, and reinforcement learning. Deep learning involves advanced techniques such as

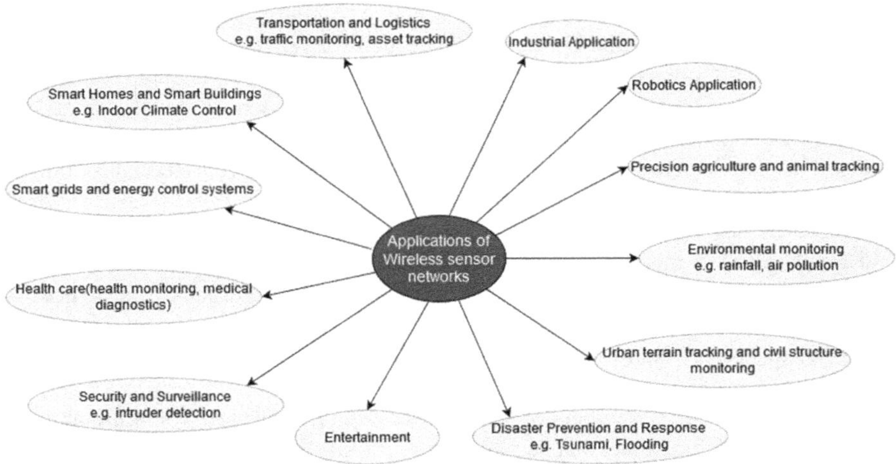

FIGURE 8.3 Machine Learning Taxonomy.

Artificial Neural Networks (ANN). The taxonomy of the most common machine leaning models is presented in Figure 8.3.

8.3.1 Supervised Machine Learning

Supervised learning involves the machine being taught by example using a dataset that consists of inputs as well as their corresponding outputs (labeled data). The ML algorithms then identify patterns in the training data and find the relationship between the inputs and the output(s) while training the system. The algorithm improves the model iteratively until the performance reaches an acceptable level. Supervised learning consists of two main types of tasks: Classification and Regression. Classification tasks involve predicting output to be one among two or more possible output classes, i.e., the ML model predicts only discrete output values. Examples of classification include analyzing input emails and to classify each of them as being spam or not, or analyzing images of cats and dogs and classifying them accordingly. Some common ML classification algorithms are Logistic Regression, K-Nearest Neighbor (KNN), Naïve Bayes (NB), Support Vector Machine (SVM), Decision Tree (DT), and Random Forest (RF).

Tasks that require the computation of a numerical output are known as regression tasks. The ML algorithm must estimate the output after understanding the relationships between the input variables and the output variable. An example of regression task is determining the price of a plot of land by leveraging features such as plot area, location, and weather. Popular regression algorithms include Linear Regression, Regression trees, and Support Vector Regression (SVR).

8.3.2 Unsupervised Learning

Supervised learning relies on labeled data to infer the ML model, whereas unsupervised ML involves analyzing the data to identify likeness, patterns, and correlations to establish relationships within the unlabeled data. Unsupervised algorithms can be used for a wide range of tasks where there is no labeled data. As the algorithm trains on more data, its ability to make decisions on that data gradually progresses and it becomes more fine-tuned. Unsupervised ML algorithms often fall under two categories: Clustering and Dimension Reduction.

8.3.2.1 Clustering

Clustering deals with finding meaningful structure and groupings inherent in a set of data points (based on defined criteria) in a collection of unlabeled data such that data points in the same groups are more related/close to other data points in the same group and dissimilar to the data points in other groups. Clustering algorithms can be applied in many fields; for instance, in marketing to find groups of customers with comparable behavior to help companies target their customers better to get better return on their investment. Some popular clustering algorithms include k-means clustering, fuzzy k-means clustering, hierarchical clustering, and probabilistic clustering such as Gaussian Mixture Models.

8.3.2.2 Dimension Reduction

Raw data contains a lot of information (curse of dimensionality), and it is quite challenging to identify significant patterns and variables—and also more difficult to visualize datasets. Dimension reduction is often used when the number of features, or dimensions, in a given dataset is too high. Dimension reduction allows to decrease the number of data inputs while preserving the integrity of the dataset, i.e., simplifies the dataset without losing too much information. Two popular dimension reduction methods are Principal Component Analysis (PCA) and Singular Value Decomposition (SVD). PCA consists of extracting the most *interesting* features and reducing redundancies in the dataset. Data with high variance contribute to more information. PCA allows identifying new dimensions (features) that capture maximum variance (i.e., information).

8.3.3 Reinforcement Learning

Reinforcement learning (RL) is an adaptive ML technique where the ML algorithm learns by interacting with the environment and collects information to take certain actions by trial and error. The RL system consists of an agent, the environment the agent interacts with, the policy that the agent applies to take actions, and the reward signal (also value) the agent observes upon taking actions. The general aim is to discover the best actions to maximize rewards. Two popular RL algorithms are Q-Learning and SARSA (State-Action-Reward-State-Action), which are model free value-based. RL is popular for developing games, e.g., AlphaGo, and robotics, industrial automation, and autonomous driving.

8.3.4 Deep Learning

Deep learning is a subset of ML. It mainly uses ANNs inspired by the configuration of the human brain to imitate a human's thinking and learning process. Thus, deep learning allows solving more complex tasks. Deep learning often uses neural networks consisting of at least three layers. The first layer is the input layer, which receives the input, while the last layer is called the output layer, where the final prediction or classification is made. Deep learning does not require too much preprocessing and feature extraction, as is the case of machine learning. Feature analysis is inherently inbuilt within the neural network.

The middle layers are called hidden layers. Simple neural networks consist of a few hidden layers, but some deep neural networks can be built with multiple layers (e.g., 100) of interconnected nodes. Additional hidden layers can optimize and improve the accuracy of the output. Typically, each layer builds upon the previous layer to refine and optimize the prediction—forward propagation. Additionally, back-propagation using algorithms such as gradient descent allows the network to make predictions and correct for any errors to become more accurate with time.

Two popular complex neural networks are Convolutional Neural Networks (CNNs) and Recurrent Neural Networks (RNNs). Deep learning applications include self-driving cars, intelligent chat bots, and credit card fraud detection, among others.

In the next section, we present the applications of ML to optimize various WSN design challenges.

8.4 ML FOR OPTIMIZING WSNs

8.4.1 Security

Security is an important challenge in WSN, given that it is deployed in an open environment. Some common attacks perpetrated in WSN include Denial of Service (DoS); Distributed Denial of Service attack (DDoS); Man in the Middle Attack (MITM); Sybil attacks; Blackhole, Sinkhole, Wormhole attacks; jamming; and spoofing [9]. However, the limited computing power and energy in WSN prevent the adoption of traditional security protocols, as they are battery-draining and not suitable for the ad hoc communication and the dynamic WSN topology. Security mechanisms for WSN have to be energy-efficient, i.e., low computational and communication complexity, but effective. In [10], a review of ML techniques for WSN Security is presented.

8.4.1.1 Anomaly Detection

ML is applied to detect anomalous data from faulty or damaged sensor nodes (hardware failure or malicious attacks), as well as noise/interference to reduce false alarms or erroneous values when using such data. The authors in [11] identified the following techniques used for classifying anomalous (outlier) sensor data: Statistical, Nearest Neighbor, Clustering, and Classification. Ye Yuan et al. [12] compared the performance of SVM, Naïve Bayes (NB), and Gradient Lifting Decision Tree (GBDT) for anomalous data detection with (1) noise fault injection, (2) short-term fault injection, and (3) fixed fault injection. The detection accuracy for noise fault injection was 87% for SVM, 82% for NB, and 92% for GBDT. For short-term fault injection, the detection accuracy was 80% for SVM, 84% for NB, and 91% for GBDT, while for fixed fault injection it was 78% for SVM, 82% for NB, and 88% for GBDT. Overall, the experiments showed that the GBDT algorithm performs better fault detection.

Similarly, Gil et al. [13] evaluated two novel online outlier detection mechanisms, namely the Support Vector Novelty Detection (SVND) and the robust OPAST with rank-1 modification (R-OPASTr). SVND uses a Least Squares-Support Vector Machine (LS-SVM) implemented with a sliding window-based learning algorithm, while R-OPASTr is PCA-based using the orthonormal projection approximation subspace tracking (OPAST) algorithm. Though SVND had a better detection rate, R-OPASTr had lower energy consumption. In [14], the authors used Discrete Wavelet Transform (DWT) together with the self-organizing map (SOM) neural network to accurately detect anomalous data from WSNs. Support Vector Data Description (SVDD) evolved from Support Vector Classifier, and it is popularly used for outlier detection [15]. Feng et al. [16] proposed a new method based on SVDD to detect anomalies in sensor data captured from WSN with an accuracy of 85% to 95% across the different variations of SVDD examined.

Cheng and Zhu [17] proposed two lightweight anomaly detection (LAD) algorithms that are based on the one-class quarter-sphere support vector machine (QSSVM), which provides detection accuracy of 94% while maintaining low computational complexity. In [18], two online and distributed anomaly detection techniques are presented, which are based on the hyperellipsoidal one-class support vector machine, namely Ellipsoidal SVM-based online outlier detection technique (EOOD), Ellipsoidal SVM-based adaptive outlier detection technique (EAOD), SVM-based adaptive outlier detection technique (SAOD), and SVM-based batch outlier detection technique (SBOD). The detection algorithms were tested using synthetic and real datasets. Testing results demonstrate that EAOD has better detection accuracy and lower false alarm compared to the other algorithms. Zheng et al. [19] proposed the "Improved Distributed PCA-based Outlier Detection (IDPCA)," which is based on PCA for detecting anomalous data. The IDPCA achieves a detection accuracy of 96% and a low false alarm rate of 1.5%. In [20], the authors use the unsupervised ML approach for anomaly detection in WSN data. They adopt the one-class SVM (OCSVM) together with Unsupervised PCA to reducing the computation and communication overhead on the sensor nodes. Gethzi and Paramasivan [21] propose a mechanism to perform online detection of anomalous sensor data in real time using the Linear Weighted Progression Regression (LWPR) for making predictions using only a subset of data to reduce the computation complexity. The LWPR also performs dimensionality reduction using PCA, and a dynamic threshold value is determined to find anomalous sensor values.

8.4.1.2 Intrusion Detection

Intrusion detection system (IDS) involves one or more nodes/units called IDS agent(s), which continuously analyze the traffic for suspicious/malicious packets to detect attacks occurring in the WSN. IDS methods are signature-based analysis and anomaly-based analysis. In signature-based IDS, a database of known attack signatures is used. The IDS compares packets in the network with the database to identify known attacks. Anomaly-based IDS are trained to detect suspicious traffic in the network after familiarizing with the normal baseline behavior of the network. Such IDS does not rely on the patterns of known attacks and can detect any attack in the WSN. Anomaly behavior detection in WSN, however, poses certain challenges due to the dynamic topology, ad hoc routing characteristic of the WSN, resource limitations of sensor nodes, and high probability of sensor node failure in harsh environments [22]. From the survey in [23], several intrusion detection techniques are presented. Recently, more elaborate ML-based methods are increasingly being adopted for intrusion detection in WSNs.

Tan et al. [24] uses the Random Forest (RF) algorithm for intrusion detection after using the synthetic minority oversampling technique (SMOTE) for balancing the dataset. When using RF only, the accuracy was 92.39%, while the accuracy of the RF combined with SMOTE (RF+SMOTE) was 92.57%, depicting a slight improvement in the accuracy. Belavagi and Balachandra [25] compared the performance of SVM, RF, Logistic Regression (LR), and Gaussian Naïve Bayes (GNB) for intrusion detection, where they concluded that the RF was the most accurate classifier. Elbahadır and Erdem [26] adopted a hybrid model using ML algorithms BayesNet, J48, and RF to classify normal and abnormal traffic.

Ma et al. [27] proposes a novel approach that uses both spectral clustering and DNN algorithms (SCDNN). Experimental results show that the SCDNN algorithm has improved accuracy compared to back-propagation NN (BPNN), SVM, Random Forest, and Bayes tree models for the six datasets created from KDD'99 and NSL-KDD. For datasets 1, 2, and 3, the precision was 92% for SCDNN; for dataset 4 and 5, the precision with SCDNN was 72 and 71%, respectively. For dataset 6, the precision was low (45%) for SCDNN, though still better than for the other algorithms. In [28] the authors use the unsupervised learning Mean Shift Clustering Algorithm followed by SVM to enhance the detection accuracy using the proposed knowledge-based intrusion detection strategy (KBIDS). The proposed algorithm was tested using simulated dataset, real data set NSL-KDD, and tested in a real system. The average detection rate was 97.9% and higher than the other models (K-Means, Decision Tree, Logistic Regression, etc.), and the average false alarm rate was around 2.1%. For the real dataset, the detection rate of KBIDS was 97.49% and higher than other models with the lowest false alarm rate of 2.26%. When tested in the real environment, KBIDS had a detection rate of 96.5% and a false alarm rate of 2.7%. In [29] an edge intelligence framework for intrusion detection is proposed, which uses the supervised ML kNN, as well as the parallel levy and arithmetic optimization algorithm (AOA). The proposed Parallel Lévy AOA (PL-AOA) achieves an accuracy of 99%, compared to 88% with kNN, which can greatly improve intrusion detection in WSNs.

Wazid and Das [30] attempt to detect hybrid anomaly in the WSN (i.e., anomaly resulting from the infiltration of different types of attacker nodes on the WSN) using k-means clustering. The machine learning model has a detection rate of 98.6% and was successful in detecting blackhole and misdirection nodes in the WSN. In [31] the authors propose the energy-efficient Advanced Hybrid Intrusion Detection System (AHIDS), which is based on fuzzy rule sets and the Multiplayer Perceptron NN for detecting different attack nodes such as Sybil attack, wormhole, and hello flood. The detection rate of Sybil, hello flood, and wormhole attack was 99.4%, 98.2%, and 99.2%, respectively. Ifzarne et al. [32] proposes the ID-GOPA detection algorithm, which is based on information gain ratio and online passive aggressive classifier for detection of scheduling, grayhole, flooding, and blackhole attacks based on the

WSN-DS dataset. The detection rate for scheduling, grayhole, flooding, and blackhole attacks were 86%, 68%, 63%, and 46%, respectively.

The Random Support Vector Regression (RSVR) has been proposed in [33] to detect DDoS attacks in WSN with a detection accuracy of 91% compared to 82% with SVM and 87% with KNN. Elsaid and Albatati [34] proposed the optimized collaborative IDS, which fine-tunes the weighted support vector machine (WSVM) classifier using improved artificial bee colony such that the IDS is resource-aware. The model was trained and tested using the NSL-KDD dataset and achieved a detection rate of 97.9% with a false alarm rate of 1.8%. In [35] the authors use enhanced empirical-based component analysis for the selection of relevant features and Long Short-Term Memory (LSTM) recurrent neural network for classification of WSN traffic using three datasets (NSLKDD, UNSWNB15, and CICIDS2017) with an accuracy of 99.76% with the CICIDS dataset and 99.95% with the UNSW-NB15 and NSLKDD datasets. Similarly, several research works adopt Deep Neural Networks for intrusion detection in networks [36–38]. Premkumar and Sundararajan [39] present a lightweight DoS detection method called "Deep Learning-based Defense Mechanism (DLDM)," which uses a Radial basis function (RBF)-based neural deep learning. However, much of the research work used the NSL-KDD dataset, which does not mainly consist of WSN traffic.

8.4.2 Routing

Routing principles and protocols are fundamental considerations in the design, implementation, and operation of a WSN. The different types of WSNs have extensively been used in numerous practical applications based on context-specific and environmental factors. In some instances, the deployment of sensory networks is required for probabilistic events such as product defect occurrence in large industrial settings, prediction of natural catastrophes such as tsunamis or floods, and traffic congestion in metropolitan cities. This diversified use of WSNs often involves an element of uncertainty, which accentuates the challenges encountered in the gathering, processing, and delivery of data packets across the whole network architecture. Efficient routing, therefore, involves the transmission of these data packets from a source node to the sink node using the optimal path in the WSN [40]. In addition to guaranteeing faster functional operation, such approaches ensure that the WSN is energy-efficient, thus extending its overall lifetime. ML has successfully been used for both static and adaptive routing in WSNs. The main advantages that ML presents when it comes to sensory networks is its ability to minimize communication overhead and to find the ideal number of transmission nodes for a specific routing cycle [41].

Conventional ML algorithms such as Naïve Bayes (NB) classifiers have previously been used to perform routing functions in WSNs, and these typically involve the generation of projection vectors and the implementation of a goal-seeking behavior in the form of dedicated target nodes. This procedure is repeated for each transmission iteration, i.e., every sensing and reaction activity in the network. The authors in [42] have built up on this approach and devised a learning process that is suited for both centralized and decentralized infrastructures. The novel aspect of this research is the inclusion of a scheduling methodology to quantify energy consumption in the network and balance out losses that would bring up the long-term efficiency of the WSN. K-means is another ML algorithm, and it has been used for clustering approaches for WSN routing. The main benefits using K-means for routing include an enhanced Packet Delivery Ratio (PDR), low energy usage, higher throughputs, and reduced communication overhead.

EKMT, an energy-efficient k-means clustering algorithm, is one of the highest performing clustering-based routing protocols for sensor networks, as it operates in a dynamic manner to determine nodes that are the closest to the optimal transmission path towards the sink node [43]. It achieves improvement in residual energy, throughput, and delay while having less jitter in the WSN. ANN has been used for routing purposes under numerous declinations; for example, ELDC, an ANN-based energy-efficient and robust routing scheme [44] that uses the concept of backpropagation neural networks (BPNNs) to balance out the expenditure of energy and loss of data in WSNs. Additionally, more in-depth research has been conducted using both competitive and corrective learning procedures. The latter has generally been favored by academics and can be implemented using techniques such as Self-Organizing Maps (SOMs), which are widely discussed in literature pertaining to routing problems in WSNs. The work of [45] makes use of SOM for the creation of an *Energy-based Clustering* routing protocol, where the object is to reduce the energy utilization of nodes by performing network load balancing and in turn achieving better longevity of the WSN. Another technique that has been explored in recent times for the elaboration of routing protocols for sensor networks is Adaptive Resonance Theory (ART). The most commonly used mechanisms under ART include minimum node separation distance, node rotation, and cost functions. Other ANN-based protocols include PEGASIS and RZ LEACH, which have proven to be better performing at finding optimal routes in WSNs while sustaining their operational lifetime [46].

In [47], the authors were able to develop a routing protocol that utilizes unlicensed ultrawideband (UWB) technology and Q-Learning, which is one of the key concepts in Reinforcement Learning to produce node mapping for data packet transfer based on geographical factors. This was a significant achievement, as it allows large scale and even nationwide control of WSN traffic. Aside standard ML models, Reinforcement Learning, and ANNs, other techniques such as Genetic Algorithms and Fuzzy Logic systems have also been used to augment the routing capabilities of WSNs.

8.4.3 Data Aggregation

Sensor nodes cooperate to send data to the base station. Data transmission in WSN is more energy-depleting than computation on the sensor nodes. Data aggregation techniques are recommended to decrease the amount of data sent. Instead of each node sending data to the base station, especially if the data consist of redundancy, the redundancy can be reduced by aggregating the data, i.e., collecting and combining useful data, thereby decreasing the communication overhead and improving the lifetime of the WSN. Randhawa and Jain [48] present a survey of data aggregation in WSN.

Lee and Chung [49] propose a method based on the Kohonen Self-Organizing Map (SOM) method for aggregating data in a cluster in an energy-efficient manner.

Chen et al. [50] propose a data compression algorithm based on PCA, and in [53] an adaptive-PCA data aggregation technique is proposed for WSNs. Similarly, in [51, 52] the authors propose a data aggregation technique based on PCA, whereas in [54, 55, 56, 57] data aggregation is based on Independent Component Analysis (ICA). Sudha et al. [54] propose the EML-DA architecture, which uses a hybrid cluster head selection based on ANN and ICA for data aggregation for reducing energy consumption. In Strong Clustering Algorithm and Data Aggregation (SCADA) [56], an ANN is used to optimize cluster head formation, and ICA is used for data aggregation. In [58] a three-layer back-propagation neural network (BPNN) is proposed for data aggregation. Sung [59] proposes a BPNN with associative memory for data fusion. Ullah and Youn [60] propose a data aggregation scheme based on clustering and extreme learning machine (ELM); ELM is a single feed forward neural network. The authors in [61] compared three data reduction approaches, namely ANN, ICA, and deep learning-based regression. Best data reduction was observed with the ANN. Sudha et al. [62] proposes the ACNM technique for accurate data aggregation created by neural network and data classification. The ACNM demonstrates better accuracy and less delay as compared to the ELM neural network.

Yousefpoor et al. further recommends that data security is provided during the data aggregation process by means of secure data aggregation methods [63]. Similarly, Gavel et al. propose a data aggregation technique for fault detection, as well as data sensing in WSNs [64].

8.4.4 Coverage and Connectivity

The span of a WSN is generally constrained by either geographical or regulatory aspects, i.e., there may be physical boundaries that prevent further scalability of the network or alternatively, local authorities may not permit the installation of necessary equipment in certain areas. Such limitations imply that a WSN will be prone to coverage and connectivity issues. Connectivity refers to the communication link that exists between individual or clustered sensor nodes and their Base Station either through a direct channel or via intermediary relay nodes. On the other hand, coverage describes the area that can be monitored at a time by a specific sensing unit. An impaired coverage or unreliable connectivity will most likely result in gaps within the network architecture, thus heavily impacting the overall performance of the WSN [65, 66].

For many years, the standard practice to remedy such issues was to use traditional network planning methods and perform aggregated on-site testing to identify any of these gaps in the network. Academic research has even been done on the subject [67], whereby end-to-end data delivery reliability was assessed and investigated. However, with the emergence of newer research avenues such as Artificial Intelligence (AI) and ML, efforts have been redirected to these segments of technology.

In their paper, authors of [68] investigated anomalies that occur in the coverage radius of WSN and devised a method using Reinforcement Learning (RL), which allows sustained network performance by proactively detecting coverage holes and by implementing node repositioning and the automatic adjustment of the WSN sensing range using game theory based on Q-learning algorithm. Similar research was conducted by [69], where other Reinforcement Learning techniques such as Nash

Q-Learning were used to develop a scheduling algorithm (CCM-RL) to automatically modify the current status of nodes throughout the network. The accuracy of this algorithm was contrasted with other comparable ones, but the ability to set the nodes to active, sleep, or hibernate proved to be highly beneficial for performance optimization. In [112], the authors adopt a decision-making algorithm based on fuzzy logic for self-healing sensor networks in cases of sensors deployed randomly and sensor malfunction to improve network coverage in WSNs. In [113] Reinforcement Learning-based independent learner (IL) algorithm is proposed for optimizing coverage and energy consumption in WSN, which achieves better accuracy in high-density WSN, while a learning automata-based algorithm was proposed in [114]. Chen, Li, and Zhao [115] propose a reinforcement learning-based sleep scheduling for coverage (RLSSC) algorithm for improving network coverage.

8.4.5 Localization

The spatial coordinates or indoor location of the sensor nodes is crucial to network operations (i.e., location-based routing) and in most WSN applications for the proper analysis and interpretation of the captured data. In mobile devices location is available via the global positioning system (GPS). However, GPS-equipped sensor nodes can be costly and consume more energy. Furthermore, GPS locations in indoor environments (multistory building) may not be very precise. In WSN, sensor nodes often discover their relative locations based on their neighbors, i.e., location of a beacon/anchor node [70]. The range-based localization approach determines the distance between the sensor node and the neighboring anchor nodes by measuring the received signal strength indicator (RSSI), time of arrival (TOA), time difference of arrival (TDOA), or angle of arrival (AOA) [71]. The later measures the angle at which the signal arrives and requires compasses or an array of antennas [72].

Bayesian algorithm [73], BPNN [116], and DNN [75] have been applied for sensor node location WSNs. The authors in [74] explore the localization techniques in WSN based on Neural Fuzzy Inference System (ANFIS) and a hybrid ANN with 3 optimization algorithms, namely Particle Swarm Optimization (PSO), Gravitational Search Algorithm (GSA), and Backtracking Search Algorithm (BSA). The best location accuracy was observed with GSA-ANN. The authors in [76] compared the results of ANN, SVM, Decision Tree, and Naïve Bayes and determined that Naïve Bayes performed better. Bhatti [77] adopted a multivariate regression model and SVR model with radial basis function (RBF) kernel for localization in large-scale WSNs. Similarly, Singh et al. [78] adopt an SVR approach for reducing the average location error in WSN. The Kernel Extreme Learning Machines based on Hop-count Quantization (KELM-HQ) is proposed in [79], which improve the accuracy of determining the node location. In [80] a node localization algorithm based on fuzzy logic, extreme learning machines, and PSO is proposed.

8.4.6 Quality of Service (QoS)

WSNs are often an integral part of IoT systems. In real-time WSN applications such as industrial systems and environmental disaster monitoring require guaranteed QoS.

QoS embodies reliability, robustness, low latency, privacy, and security levels, as well as efficient energy consumption. In [81], the authors focus on different aspects of routing protocols such as metric selection, state propagation and maintenance, scalability, and reliability for improving performance in WSNs. Additionally, specific routing protocols were considered for a qualitative comparison of their use in WSNs. These protocols include Sequential Assignment Routing (SAR), Multi-Path Multi-Speed Protocol (MMSPEED), Reliable Information Forwarding using Multipath (ReInForM), and Directed Alternative Spanning Tree (DAST), among others. Mbowe and Oreku [82] recommend that reliability, availability, and serviceability are important factors to attain QoS in WSNs. Pundir and Sandhu [83] present a review of ML for QoS in WSNs. Highlights of the latter's work include a comprehensive analysis of conventional ML techniques when applied to various QoS metrics such as security, confidentiality, integrity, safety, maintainability, bandwidth, and connectivity. For instance, it was established that Decision Trees are significantly better at link cost estimation with a maximum accuracy of 94%, as opposed to other techniques such as Multilayer Perceptrons, which could only achieve 56% accuracy. Sujanthi and Nithya Kalyani [84] propose a novel Crossover-based Fitted Deep Neural Network (Co-FitDNN) for optimal route selection for improving QoS. Recently, Ateeq et al. [85] showed there is a relationship between the communication parameters of a WSN and performance metrics; using these correlations, deep learning models were trained to predict QoS metrics. Such a model can be used for designing adaptive WSN for different QoS requirements related to communication performance.

In this context, a Deep Neural Network (DNN) was successfully devised by the authors, and this specific model was able to predict the Signal to Noise Ratio (SNR) and the Packet Delivery Ratio (PDR) in a simulated environment at 96% and 98% accuracy, respectively, while only using 10% or less of the originally consolidated dataset. Similarly, Mazloomi et al. [86] investigate the relationship between communication parameters configuration in WSN for low energy consumption; a machine learning model based on SVR and genetic algorithm is presented for optimizing communication parameters for energy efficiency and QoS.

8.4.7 Energy Harvesting

A recurrent theme in the operation of WSNs in open environments is their dependence on battery-powered units. The power limitation of a WSN directly impacts the performance of the network, as every aspect of data gathering, processing, and transmission involves energy consumption. It is, therefore, critical that the operational availability of WSNs and their continuity of service are extended via all possible means. Energy harvesting is one of these countermeasures, and it has been heavily explored by the academic community over the past decade or so. The process of harvesting energy for WSN usage involves capturing ambient energy in different forms from the sensor network's immediate environment and converting this to electrical energy. The latter can then be used as a supplementary power source to offset the energy expenditure when high performance standards are being implemented.

Energy harvesting is inherently a multidisciplinary field, as it combines elements from electrical theory, acoustics, electromagnetism, and sustainable energy management [87]. ML in the context of energy harvesting presents interesting prospects, as it allows an unprecedented opportunity to ameliorate the conventional processes and techniques used to capture energy. For instance, ML can be used for forecasting the peaks in energy availability in the WSN's surroundings either due to climatic conditions or other environmental factors. Alternatively, ML can be beneficial for automated and smart switching of energy harvesting tools. These switching mechanisms are also relevant to the interoperability of primary and secondary power sources for the WSN.

Dynamic Power Management (DPM) is an energy utilization technique that was proposed by [88], which makes use of Reinforcement Learning for the optimization of energy harvesting in solar-powered WSNs. This optimal operation of the energy capture units is achieved by ensuring smart switching, designating idle times, and minimizing transition delays. Another promising ML-based solution for energy harvesting is Reinforcement Learning Energy Management (RLMan), whereby an in-depth analysis of the time-varying behavior of energy sources is carried out [89]. The biggest benefit of RLMan is its use of linear function approximations and its minute computational and memory requirements that make it ideal for resource-limited applications such as WSNs. A comparative study of RLMan versus other similar algorithms demonstrated that it outperforms the other models by accomplishing an approximate gain of 70% on the packet rate. More recently, other techniques such as Fuzzy Logic and Q-Learning (FQL) have been used to dynamically manage energy considerations for WSNs. For example, the authors of [90] proposed a new fuzzy Q-learning-based dynamic energy management (FQLDEM) model capable of resolving the work duty cycle of sensing components, which in turn maintains the charge level of associated battery units at higher levels than other methods. More precisely, FQLDEM was proven to be able to meet a target 65% of initial quality control requirements devised by the research team. Lastly, [91] presents a concise list and descriptions of state-of-the-art technologies currently being utilized for energy harvesting in WSNs.

8.4.8 Congestion Control

Communication channels are used for the exchange of data in various forms, and they are an essential component for the dissemination of information in network systems. One of the main hindrances in achieving high reliability and high performance for a communication channel is poor traffic flow control. Inadequately estimating the volume of traffic on a communication link will either result in congestion of the overall network or, alternatively, in low resource utilization at the broader level of the system. WSNs are not safe from these phenomena, and congestion control is in fact a major focus of research in the field of sensor networks [92].

Due to the versatile nature of WSNs and their frequent application in sensitive or critical settings, they can easily become prone to traffic surges due to the uncertainty surrounding the trigger for such events. In these scenarios, the communication channel may engage in excessive transmission of data without any concern for its maximum bandwidth. The immediate effect will be a high packet arrival rate, a lower Bit

Error Rate (BER) for node pairs in the network, and collision of packets at multiple nodes. The repercussions of network congestion on WSNs are critical as they impact roundtrip communication times, increase the end-to-end delay of routes, decrease the packet delivery rate of the system, reduce the Quality of Service (QoS) metrics, and significantly increase the energy consumption of the WSN, which can result in a shortened lifetime.

The use of ML algorithms is therefore a powerful tool in combating the problem of congestion in WSNs. These learning models assist in reducing end-to-end system delays while adjusting the transmission range in a dynamic and automated manner. Moreover, ML can facilitate the recognition of communication patterns and proactively safeguard the network against traffic surges. Over the years, different ML techniques have been used and numerous models have been developed to tackle the issue of congestion in WSNs. One of these models was developed by [93], and the rationale for congestion control was to use ANNs for compressing the transmitted data to reduce the traffic load on the network. The operating principle of the model, Rate-distortion Balanced Data Compression for WSNs (RBDC), was based on spatio-temporal correlation of data packets for successive sensing phases of the individual nodes. By utilizing an adaptive rate-distortion feature, the authors were able to balance the actual data size per transmission and the resulting error rate or distortion level. Subsequent analysis of the energy expenditures in the investigated WSN showed considerable improvements with the system lifetime increasing multiple times its original value.

Other research avenues for congestion control and avoidance in WSNs have not explored data compression but instead investigated the hopping mechanism that occurs in sensor networks. The work of [94] is a prime example of this, whereby multi-classification using Support Vector Machines (SVMs) was implemented and optimized to calculate the occupancy rates of specific nodes in a sensor network. Their proposed model was named Genetic Algorithm Support Vector Machine (GASVM). Through strategic inference on the availability of nodes and their ability to accommodate traffic, an effective control method was devised, and a buffer mechanism was designed to better distribute traffic throughout a whole WSN. Consequently, notable improvements were observed on end-to-end delays, packet loss, and aggregated energy consumption for high traffic applications. In [95] the effects of congestion in WSNs are analyzed. However, the concept of Multi Agent Reinforcement Learning (MRL) was coupled with energy-aware models such as the convex-hull algorithm for self-configuration and self-optimization of unattended WSNs. This research endeavor gave rise to the MRL-SCSO model, which prioritizes specific communication channels based on factors such as residual energy and buffer dimensions. Lastly, the survey article [96] presents a comprehensive listing of recent methodologies adopted for congestion control and avoidance in WSNs.

Table 8.1 summarizes the different machine learning techniques identified in this chapter for optimizing WSNs.

8.4.9 Challenges of ML for WSN

Most of the research work involves machine learning training performed using some existing dataset(s), often a dataset consisting of network traffic that is not WSN.

TABLE 8.1
ML Techniques for Optimizing WSN

WSN Area	Machine Learning Techniques
Anomaly Detection	SVM, NB, GBDT, SVND, R-OPASTr, DWT+SOM, SVDD, LAD, EOOD, EAOD, SAOP, SBOD, IPPCA, OCSVM, LWPR
Intrusion Detection	RF, NB, RF+SMOTE, SVM, LR, GNB, SCDNN, KBIDS, PLAOA, BPNN, K-means, AHIDS, ID-GOPA, RSVR, WSVM, LSTM RNN, DNN, DLDM
Routing	NB, RL, K-means, EKMT, ANN, ELDC, BPNN, SOM, ART, PEGASIS, RZ LEACH, Genetic Algorithm, Fuzzy Logic
Data Aggregation	SOM, PCA, Adaptive-PCA, ICA, BPNN, ANN, ELM, ACNM
Coverage and Connectivity	RL, Nash Q Learning, CCM-RL, Fuzzy Logic, IL, RLSSC
Localization	NB, DT, SVR, SVM, SVR+RBF, ANN, DNN, ANFIS, GSA-ANN, PSO-ANN, GSA-ANN, KELM-HQ
QoS	DT, SVR, DNN, CoFitDNN, Genetic Algorithm
Energy Harvesting	RL, RLMan, FQL, FQLDEM
Congestion Control	RBSC, SVM, GASVM, ANN, MRL, MRL-SCSO

It would be preferable to have more WSN-specific datasets for training. In practice, it would require time for sufficient data to be captured for training the ML model before it can actually be useful. The more data available, the better the ML model. However, if the ML training is done online by sensor nodes, this can become an overhead on the resource-limited sensor nodes and result in increased energy depletion of the node. Often, data captured must be preprocessed before it can be used by ML algorithms; this can add further burden on the resource constrained nodes. The issues of coverage, localization, QoS, and congestion control are more challenging to address and require more ML techniques.

8.5 ML IN WSN APPLICATIONS

Machine learning algorithms are also used to improve the efficiency of WSN for intrusion detection. WSNs are commonly deployed on frontiers for border control as well as military applications to quickly detect the presence of intruders [97]. Singh et al. [98] have proposed a WSN for rapid intruder detection using a ML approach based on Gaussian Process Regression (GPR) model. The authors in [99] proposed a log-transformed feature learning and feature scaling-based ML algorithm and a tuned SVR model for swift intruder detection using WSNs. In [100] a deep learning-based intruder detection system using WSN, the Restricted Boltzmann-based Clustered IDS (RBC-IDS) model, is proposed for monitoring and protection of critical infrastructures. Singh et al. [101] further proposed an automated ML model for intruder detection, which automatically selects the ML model (e.g., SVR, Gaussian process regression, binary DT, bagging ensemble learning, boosting ensemble learning, kernel regression, and LR model) and the hyperparameters optimization using Bayesian optimization for intruder detection.

WSN and IoT are popularly adopted for precision agriculture to optimize cultivation of crops and increase yield by closely monitoring weather and soil parameters;

to monitor pests, weed, and diseases; and control fertilizer usage, among others. Rahaman and Azharuddin [102] survey the application of WSN and ML in agriculture, where they identified the importance of ML in WSN in several areas such as clustering, event detection, and query processing to extend the life span of WSNs. Machine learning is used for monitoring soil moisture and smart irrigation systems [103]. In [104], the authors investigate the use of SVM, Multiple LR, NN, and Bayesian Network-based techniques for crop pest prediction. Wani and Ashtankar [105] proposed utilizing the Naive Bayes Kernel Algorithm for predicting pest/diseases of crops using a WSN. Rozario and Vasanthi [106] proposed a solution for crop growth supervision and leaf area index estimation using WSN and CNN. Marković et al. [107] uses sensor networks and ML to help farmers identify the occurrence of pests, thereby allowing farmers to prioritize the plots to be treated; the ML algorithms considered were k-NN, SVM, DT, RF, Multi-layer Perceptron classifier (NN), Ada Boost, Gaussian Naive Bayes, and Quadratic Discriminant Analysis (QDA). The authors in [108] proposed an ANN for fast and accurate pest identification and categorization using WSNs deployed in the field. Nitrogen can improve crop growth and development, thereby improving crop yield and crop produce quality. Chlingaryan et al. [109] discuss the ML approach for crop yield prediction and nitrogen status estimation in precision agriculture.

In the area of healthcare, WSNs are typically implemented in the form of a Wireless Body Area Network (WBAN), whereby wireless sensor devices are placed on the body to examine patient vital information such as heart rate, blood pressure, and pulse oxygen saturation (SpO2). Measurements from such sensors are subject to anomalies due to flawed calibration, electromagnetic interference, patient sweat, etc. Such anomalies impact on the diagnosis, which can endanger patient's lives. Pachauri and Sharma [110] use RF algorithm and Additive Regression techniques for anomaly detection in medical WSNs. Vidya and Sasikumar [111] captured multi-sensor activity data from the (WSN) nodes and inertial sensor embedded in a smartphone to perform Human Activity Recognition (HAR). HAR can be useful in healthcare, e.g., for fall detection, as well as in other areas such as smart city applications. The data were first preprocessed and then trained using the following four ML algorithms—SVM, KNN, Ensemble classifier, and DT—to classify human activities with an accuracy of 99.63%.

8.6 CONCLUSIONS

Current technology advances in IoT promise to transform our industries and the society. IoT consist of interconnected devices that are able to gather, share, analyze information, and to take decisions accordingly, thereby enhancing control and productivity. WSN is a vital element of IoT. WSNs allow to conveniently capture data in a range of environments. WSN have numerous applications, as they are low cost, flexible, and easy to maintain. However, WSNs face certain challenges, which machine learning techniques can address. In this chapter, the applications of ML to tackle security, QoS, routing, data aggregation, coverage and connectivity, localization, energy harvesting, congestion control, as well as application specific issues, are presented to enable the WSNs to fulfill their role despite their limited resources.

REFERENCES

[1] Yick, J., Mukherjee, B., & Ghosal, D. (2008). Wireless sensor network survey. *Computer Networks* (Elsevier), *52*, 2292–2330.

[2] Mohd Fauzi Othman, Khairunnisa Shazali, Wireless Sensor Network Applications: A Study in Environment Monitoring System, *Procedia Engineering*, Volume 41, 2012, Pages 1204–1210, ISSN 1877-7058, https://doi.org/10.1016/j.proeng.2012.07.302

[3] Loubna Hamami, Bouchaib Nassereddine, Application of wireless sensor networks in the field of irrigation: A review, Computers and Electronics in Agriculture, Volume 179, 2020, 105782,ISSN 0168-1699, https://doi.org/10.1016/j.compag.2020.105782

[4] Zirong Wang, Greenhouse data acquisition system based on ZigBee wireless sensor network to promote the development of agricultural economy, *Environmental Technology & Innovation*, Volume 24, 2021, 101689, ISSN 2352–1864, https://doi.org/10.1016/j.eti.2021.101689

[5] R. Teti, D. Mourtzis, D.M. D'Addona, A. Caggiano, Process monitoring of machining, CIRP Annals, 2022, ISSN 0007-8506, https://doi.org/10.1016/j.cirp.2022.05.009

[6] Gouse Baig Mohammad, S. Shitharth, Wireless sensor network and IoT based systems for healthcare application, Materials Today: Proceedings, 2021, ISSN 2214-7853, https://doi.org/10.1016/j.matpr.2020.11.801

[7] Shengbo Chen, Lanxue Zhang, Yuanmin Tang, Cong Shen, Roshan Kumar, Keping Yu, Usman Tariq, Ali Kashif Bashir, Indoor temperature monitoring using wireless sensor networks: A SMAC application in smart cities, *Sustainable Cities and Society*, Volume 61, 2020, 102333, ISSN 2210-6707, https://doi.org/10.1016/j.scs.2020.102333

[8] Sandhya Armoogum, Xiao Ming Li, Chapter 2 - Big Data Analytics and Deep Learning in Bioinformatics With Hadoop, Editor(s): Arun Kumar Sangaiah, *Deep Learning and Parallel Computing Environment for Bioengineering Systems*, Academic Press, 2019, Pages 17–36, ISBN 9780128167182, https://doi.org/10.1016/B978-0-12-816718-2.00009-9

[9] M. Keerthika, D. Shanmugapriya, Wireless Sensor Networks: Active and Passive attacks - Vulnerabilities and Countermeasures, Global Transitions Proceedings, Volume 2, Issue 2, 2021, Pages 362–367, ISSN 2666-285X, https://doi.org/10.1016/j.gltp.2021.08.045

[10] Ahmad, Rami, Raniyah Wazirali, and Tarik Abu-Ain. 2022. "Machine Learning for Wireless Sensor Networks Security: An Overview of Challenges and Issues" *Sensors* 22, no. 13: 4730. https://doi.org/10.3390/s22134730

[11] Shahid, Nauman, Ijaz Haider Naqvi, and Saad Bin Qaisar. "Characteristics and classification of outlier detection techniques for wireless sensor networks in harsh environments: a survey." *Artificial Intelligence Review* 43, no. 2 (2015): 193–228.

[12] Y. Yuan, S. Li, X. Zhang and J. Sun, "A Comparative Analysis of SVM, Naive Bayes and GBDT for Data Faults Detection in WSNs," *2018 IEEE International Conference on Software Quality, Reliability and Security Companion (QRS-C)*, 2018, pp. 394–399, https://doi.org/10.1109/QRS-C.2018.00075

[13] Gil, P., Martins, H. & Januário, F. Outliers detection methods in wireless sensor networks. *Artif Intell Rev* **52**, 2411–2436 (2019). https://doi.org/10.1007/s10462-018-9618-2

[14] Supakit Siripanadorn, Wipawee Hattagam, Neung Teaumroong, Anomaly detection in wireless sensor networks using self-organizing map and wavelets, ACS'10: *Proceedings of the 10th WSEAS international conference on Applied computer science*, October 2010 Pages 381–387.

[15] Tax, D.M., Duin, R.P. Support Vector Data Description. *Machine Learning* **54**, 45–66 (2004). https://doi.org/10.1023/B:MACH.0000008084.60811.49

[16] Feng, Zhen, Jingqi Fu, Dajun Du, Fuqiang Li, and Sizhou Sun. "A new approach of anomaly detection in wireless sensor networks using support vector data description." *International Journal of Distributed Sensor Networks* 13, no. 1 (2017): 1550147716686161.

[17] Cheng, Pu, and Minghua Zhu. "Lightweight Anomaly Detection for Wireless Sensor Networks." *International Journal of Distributed Sensor Networks*, (August 2015). https://doi.org/10.1155/2015/653232

[18] Zhang, Yang, Nirvana Meratnia, and Paul JM Havinga. "Distributed online outlier detection in wireless sensor networks using ellipsoidal support vector machine." *Ad hoc networks* 11, no. 3 (2013): 1062–1074.

[19] Zheng, W., Yang, L., Wu, M. (2018). An Improved Distributed PCA-Based Outlier Detection in Wireless Sensor Network. In: Sun, X., Pan, Z., Bertino, E. (eds) Cloud Computing and Security. ICCCS 2018. Lecture Notes in Computer Science(), vol. 11067. Springer, Cham. https://doi.org/10.1007/978-3-030-00018-9_4

[20] Zamry, Nurfazrina Mohd, Anazida Zainal, and Murad A. Rassam. "Unsupervised anomaly detection for unlabelled wireless sensor networks data." *International Journal of Advances in Soft Computing & Its Applications* 10, no. 2 (2018).

[21] I. Gethzi Ahila Poornima, B. Paramasivan, Anomaly detection in wireless sensor network using machine learning algorithm, Computer Communications, Volume 151, 2020, Pages 331–337, ISSN 0140-3664, https://doi.org/10.1016/j.comcom.2020.01.005

[22] Ghosal, A., Halder, S. (2013). Intrusion Detection in Wireless Sensor Networks: Issues, Challenges and Approaches. In: Khan, S., Khan Pathan, AS. (eds) Wireless Networks and Security. Signals and Communication Technology. Springer, Berlin, Heidelberg. https://doi.org/10.1007/978-3-642-36169-2_10

[23] Farooqi, A.H., Khan, F.A. (2009). Intrusion Detection Systems for Wireless Sensor Networks: A Survey. In: Ślęzak, D., Kim, Th., Chang, A.C.C., Vasilakos, T., Li, M., Sakurai, K. (eds) Communication and Networking. FGCN 2009. Communications in Computer and Information Science, vol 56. Springer, Berlin, Heidelberg. https://doi.org/10.1007/978-3-642-10844-0_29

[24] Tan, Xiaopeng, Shaojing Su, Zhiping Huang, Xiaojun Guo, Zhen Zuo, Xiaoyong Sun, and Longqing Li. 2019. "Wireless Sensor Networks Intrusion Detection Based on SMOTE and the Random Forest Algorithm" *Sensors* 19, no. 1: 203. https://doi.org/10.3390/s19010203

[25] Belavagi, Manjula C., and Balachandra Muniyal. "Performance evaluation of supervised machine learning algorithms for intrusion detection." *Procedia Computer Science* 89 (2016): 117–123.

[26] H. Elbahadır and E. Erdem, "Modeling Intrusion Detection System Using Machine Learning Algorithms in Wireless Sensor Networks," *2021 6th International Conference on Computer Science and Engineering (UBMK)*, 2021, pp. 401–406, https://doi.org/10.1109/UBMK52708.2021.9558928

[27] Ma, Tao, Fen Wang, Jianjun Cheng, Yang Yu, and Xiaoyun Chen. "A hybrid spectral clustering and deep neural network ensemble algorithm for intrusion detection in sensor networks." *Sensors* 16, no. 10 (2016): 1701.

[28] Hongchun Qu, Zeliang Qiu, Xiaoming Tang, Min Xiang, Ping Wang, Incorporating unsupervised learning into intrusion detection for wireless sensor networks with structural co-evolvability, Applied Soft Computing, Volume 71, 2018, Pages 939–951, ISSN 1568-4946, https://doi.org/10.1016/j.asoc.2018.07.044

[29] Gaoyuan Liu, Huiqi Zhao, Fang Fan, Gang Liu, Qiang Xu and Shah Nazir, An Enhanced Intrusion Detection Model Based on Improved kNN in WSNs, *Sensors* 2022, *22*(4), 1407; https://doi.org/10.3390/s22041407

[30] Wazid, M., Das, A.K. An Efficient Hybrid Anomaly Detection Scheme Using K-Means Clustering for Wireless Sensor Networks. *Wireless Pers Commun* **90**, 1971–2000 (2016). https://doi.org/10.1007/s11277-016-3433-3

[31] Rupinder Singh, Jatinder Singh, Ravinder Singh, "Fuzzy Based Advanced Hybrid Intrusion Detection System to Detect Malicious Nodes in Wireless Sensor Networks", *Wireless Communications and Mobile Computing*, vol. 2017, Article ID 3548607, 14 pages, 2017. https://doi.org/10.1155/2017/3548607

[32] Samir Ifzarne, Hiba Tabbaa, Imad Hafidi and Nidal Lamghari, Anomaly Detection using Machine Learning Techniques in Wireless Sensor Networks, *Journal of Physics: Conference Series*, Volume 1743, The International Conference on Mathematics & Data Science (ICMDS) 2020 29–30 June 2020 Khouribga, Morocco.

[33] Gang Xu, Allemar Jhone P. Delima, Ivy Kim D. Machica, Jan Carlo T. Arroyo, Zhengfang He, Weibin Su. (2022). Improvement of Wireless Sensor Networks Against Service Attacks Based on Machine Learning. International Journal of Engineering Trends and Technology, 70(5), 74–79. https://doi.org/10.14445/22315381/IJETT-V70I5P209

[34] Elsaid, S.A., Albatati, N.S. An optimized collaborative intrusion detection system for wireless sensor networks. *Soft Comput* **24**, 12553–12567 (2020). https://doi.org/10.1007/s00500-020-04695-0

[35] Liu Zhiqiang, Ghulam Mohiuddin, Zheng Jiangbin, Muhammad Asim, Wang Sifei, Intrusion detection in wireless sensor network using enhanced empirical based component analysis, Future Generation Computer Systems, Volume 135, 2022, Pages 181–193, ISSN 0167-739X, https://doi.org/10.1016/j.future.2022.04.024

[36] Folino, Francesco, Gianluigi Folino, Massimo Guarascio, Francesco Sergio Pisani, and Luigi Pontieri. "On learning effective ensembles of deep neural networks for intrusion detection." *Information Fusion* 72 (2021): 48–69.

[37] Yao, Ruizhe, Ning Wang, Zhihui Liu, Peng Chen, and Xianjun Sheng. 2021. "Intrusion Detection System in the Advanced Metering Infrastructure: A Cross-Layer Feature-Fusion CNN-LSTM-Based Approach" *Sensors* 21, no. 2: 626. https://doi.org/10.3390/s21020626

[38] Yang, Wencheng, Song Wang, and Michael Johnstone. "A comparative study of ML-ELM and DNN for intrusion detection." In *2021 Australasian Computer Science Week Multiconference*, pp. 1–7. 2021.

[39] M. Premkumar, T.V.P. Sundararajan, DLDM: Deep learning-based defense mechanism for denial of service attacks in wireless sensor networks, *Microprocessors and Microsystems*, Volume 79, 2020, 103278, ISSN 0141–9331, https://doi.org/10.1016/j.micpro.2020.103278

[40] Shafiq, M., Ashraf, H., Ullah, A., & Tahira, S. (2020). Systematic literature review on energy efficient routing schemes in WSN–A survey. Mobile Networks and Applications, 25(3), 882–895.

[41] Alsheikh, M.A., Lin, S., Niyato, D. and Tan, H.P., 2014. Machine learning in wireless sensor networks: Algorithms, strategies, and applications. *IEEE Communications Surveys & Tutorials*, *16*(4), pp. 1996–2018.

[42] F. Kazemeyni, O. Owe, E.B. Johnsen, I. Balasingham, Formal modeling and analysis of learning-based routing in mobile wireless sensor networks, in: Integration of Reusable Systems, Springer, 2014, pp. 127–150.

[43] B. Jain, G. Brar, J. Malhotra, EKMT-k-means clustering algorithmic solution for low energy consumption for wireless sensor networks based on minimum mean distance from base station, in: Networking Communication and Data Knowledge Engineering, Springer, 2018, pp. 113–123.

[44] Mehmood, A., Lv, Z., Lloret, J., & Umar, M. M. (2017). An Artificial neural network based energy-efficient and robust routing scheme for pollution monitoring in WSNs. IEEE Trans. Emerg. Top Comput., No. PP, 1.

[45] Neda Enami, Reza Askari Moghadam, Energy-based clustering self organizing map protocol for extending wireless sensor networks lifetime and coverage, *Can. J. Multimedia Wirel. Networks* 1 (4) (2010).

[46] Deepshikha, Priyanka Arora, Varsha, Enhanced NN based RZ LEACH using hybrid ACO/PSO based routing for WSNs, in; *IEEE 2017 8th International Conference on Computing, Communication and Networking Technologies (ICCCNT).*

[47] S. Dong, P. Agrawal, K. Sivalingam, Reinforcement learning-based geographic routing protocol for UWB wireless sensor network, in: Global Telecommunications Conference, IEEE, 2007, pp. 652–656.

[48] Randhawa, S., Jain, S. Data Aggregation in Wireless Sensor Networks: Previous Research, Current Status and Future Directions. *Wireless Pers Commun* **97**, 3355–3425 (2017). https://doi.org/10.1007/s11277-017-4674-5

[49] Lee, S., Chung, T. (2005). Data Aggregation for Wireless Sensor Networks Using Self-organizing Map. In: Kim, T.G. (eds) Artificial Intelligence and Simulation. AIS 2004. Lecture Notes in Computer Science, vol 3397. Springer, Berlin, Heidelberg. https://doi.org/10.1007/978-3-540-30583-5_54

[50] Chen, F. & Li, M. & Wang, D. & Tian, B. (2013). Data compression through principal component analysis over wireless sensor networks. *Journal of Computational Information Systems*. 9. 1809–1816.

[51] A. Morell, A. Correa, M. Barceló and J. L. Vicario, "Data Aggregation and Principal Component Analysis in WSNs," in *IEEE Transactions on Wireless Communications*, vol. 15, no. 6, pp. 3908–3919, June 2016, https://doi.org/10.1109/TWC.2016.2531041

[52] J. Li, S. Guo, Y. Yang and J. He, "Data Aggregation with Principal Component Analysis in Big Data Wireless Sensor Networks," *2016 12th International Conference on Mobile Ad-Hoc and Sensor Networks (MSN)*, 2016, pp. 45–51, https://doi.org/10.1109/MSN.2016.015

[53] P. Poekaew and P. Champrasert, "Adaptive-PCA: An event-based data aggregation using principal component analysis for WSNs," *2015 International Conference on Smart Sensors and Application (ICSSA)*, 2015, pp. 50–55, https://doi.org/10.1109/ICSSA.2015.7322509

[54] C. Sudha, D. Suresh and A. Nagesh, "An Enhanced Machine learning data aggregation model for Efficient data processing in wireless sensor networks," *2021 6th International Conference on Communication and Electronics Systems (ICCES)*, 2021, pp. 1748–1753, https://doi.org/10.1109/ICCES51350.2021.9489124

[55] Shahina, K., and T. S. Pradeep Kumar. "Similarity-based clustering and data aggregation with independent component analysis in wireless sensor networks." *Transactions on Emerging Telecommunications Technologies*: e4462. https://doi.org/10.1002/ett.4462

[56] William, P., Badholia, A., Verma, V., Sharma, A., Verma, A. (2022). Analysis of Data Aggregation and Clustering Protocol in Wireless Sensor Networks Using Machine Learning. In: Suma, V., Fernando, X., Du, K.L., Wang, H. (eds) Evolutionary Computing and Mobile Sustainable Networks. Lecture Notes on Data Engineering and Communications Technologies, vol. 116. Springer, Singapore. https://doi.org/10.1007/978-981-16-9605-3_65

[57] Shahina, K. and V. Vaidehi, "Clustering and Data Aggregation in Wireless Sensor Networks Using Machine Learning Algorithms," *2018 International Conference on Recent Trends in Advance Computing (ICRTAC)*, 2018, pp. 109–115, https://doi.org/10.1109/ICRTAC.2018.8679318

[58] L. Sun, W. Cai and X. Huang, "Data aggregation scheme using neural networks in wireless sensor networks," *2010 2nd International Conference on Future Computer and Communication*, 2010, pp. V1-725-V1-729, https://doi.org/10.1109/ICFCC.2010.5497335

[59] Sung, Wen-Tsai. "Employed BPN to multi-sensors data fusion for environment monitoring services." In *International conference on autonomic and trusted computing*, pp. 149–163. Springer, Berlin, Heidelberg, 2009.

[60] Ullah, I., Youn, H.Y. Efficient data aggregation with node clustering and extreme learning machine for WSN. *J Supercomput* **76**, 10009–10035 (2020). https://doi.org/10.1007/s11227-020-03236-8

[61] J. Abdullah, M. K. Hussien, N. A. M. Alduais, M. I. Husni and A. Jamil, "Data Reduction Algorithms based on Computational Intelligence for Wireless Sensor Networks Applications," *2019 IEEE 9th Symposium on Computer Applications & Industrial Electronics (ISCAIE)*, 2019, pp. 166–171, https://doi.org/10.1109/ISCAIE.2019.8743665

[62] C. Sudha, D. Suresh, A. Nagesh, Accurate data aggregation created by neural network and data classification processed through machine learning in wireless sensor networks, Theoretical Computer Science, Volume 925, 2022, Pages 25–36, ISSN 0304-3975, https://doi.org/10.1016/j.tcs.2022.04.020

[63] Mohammad Sadegh Yousefpoor, Efat Yousefpoor, Hamid Barati, Ali Barati, Ali Movaghar, Mehdi Hosseinzadeh, Secure data aggregation methods and countermeasures against various attacks in wireless sensor networks: A comprehensive review, *Journal of Network and Computer Applications*, Volume 190, 2021, 103118, ISSN 1084-8045, https://doi.org/10.1016/j.jnca.2021.103118

[64] Gavel, S., Charitha, R., Biswas, P. et al. A data fusion based data aggregation and sensing technique for fault detection in wireless sensor networks. *Computing* **103**, 2597–2618 (2021). https://doi.org/10.1007/s00607-021-01011-y

[65] J. Qin, W. Fu, H. Gao, W.X. Zheng, Distributed k-means algorithm and fuzzy c -means algorithm for sensor networks based on multiagent consensus theory, *IEEE Trans. Cybern.* 47 (3) (2017) 772–783.

[66] X. Chang, J. Huang, S. Liu, G. Xing, H. Zhang, J. Wang, L. Huang, Y. Zhuang, Accuracy-aware interference modeling and measurement in wireless sensor networks, *IEEE Trans. Mob. Comput.* 15 (2) (2016) 278–291.

[67] W. Sun, X. Yuan, J. Wang, Q. Li, L. Chen, D. Mu, End-to-end data delivery reliability model for estimating and optimizing the link quality of industrial WSNs, *IEEE Trans. Autom. Sci. Eng.* 15 (3) (2018) 1127–1137.

[68] Hajjej, F., Hamdi, M., Ejbali, R. and Zaied, M., 2020. A distributed coverage hole recovery approach based on reinforcement learning for Wireless Sensor Networks. *Ad Hoc Networks*, 101, p. 102082. https://doi.org/10.1016/j.adhoc.2020.102082

[69] Sharma, A. and Chauhan, S., 2020. A distributed reinforcement learning based sensor node scheduling algorithm for coverage and connectivity maintenance in wireless sensor network. *Wireless Networks*, 26(6), pp. 4411–4429. https://doi.org/10.1007/s11276-020-02350-y

[70] J. Kuriakose, V. Amruth and N. S. Nandhini, "A survey on localization of Wireless Sensor nodes," *International Conference on Information Communication and Embedded Systems (ICICES2014)*, 2014, pp. 1–6, https://doi.org/10.1109/ICICES.2014.7033874

[71] Wang, J., Ghosh, R.K. & Das, S.K. A survey on sensor localization. *J. Control Theory Appl.* **8**, 2–11 (2010). https://doi.org/10.1007/s11768-010-9187-7

[72] Paul, Anup Kumar, and Takuro Sato. 2017. "Localization in Wireless Sensor Networks: A Survey on Algorithms, Measurement Techniques, Applications and Challenges," *Journal of Sensor and Actuator Networks* 6, no. 4: 24. https://doi.org/10.3390/jsan6040024

[73] M. R. Morelande, B. Moran and M. Brazil, "Bayesian node localisation in wireless sensor networks," *2008 IEEE International Conference on Acoustics, Speech and Signal Processing*, Las Vegas, NV, USA, 2008, pp. 2545–2548, https://doi.org/10.1109/ICASSP.2008.4518167

[74] S. Gharghan, R. Nordin and M. Ismail, "A wireless sensor network with soft computing localization techniques for track cycling applications", *Sensors*, vol. 16, no. 8, pp. 1043, Aug. 2016.
[75] Shivakumar Kagi, Basavaraj S. Mathapati, Optimal Trained Deep Neural Network for Localization in Wireless Sensor Network, International Journal of Computational Methods, https://doi.org/10.1142/S0219876221420068
[76] H. Ahmadi and R. Bouallegue, "Exploiting machine learning strategies and RSSI for localization in wireless sensor networks: A survey", *Proc. 13th Int. Wireless Commun. Mobile Comput. Conf. (IWCMC)*, pp. 1150–1154, Jun. 2017.
[77] Bhatti, Ghulam. (2018). Machine Learning Based Localization in Large-Scale Wireless Sensor Networks. *Sensors*. 18. 4179. https://doi.org/10.3390/s18124179
[78] A. Singh, V. Kotiyal, S. Sharma, J. Nagar and C.-C. Lee, "A Machine Learning Approach to Predict the Average Localization Error With Applications to Wireless Sensor Networks," in *IEEE Access*, vol. 8, pp. 208253–208263, 2020, https://doi.org/10.1109/ACCESS.2020.3038645
[79] L. Wang, M. J. Er and S. Zhang, "A Kernel Extreme Learning Machines Algorithm for Node Localization in Wireless Sensor Networks," in *IEEE Communications Letters*, vol. 24, no. 7, pp. 1433–1436, July 2020, https://doi.org/10.1109/LCOMM.2020.2986676
[80] Songyut Phoemphon, Chakchai So-In, Dusit (Tao) Niyato, A hybrid model using fuzzy logic and an extreme learning machine with vector particle swarm optimization for wireless sensor network localization, *Applied Soft Computing*, Volume 65, 2018, Pages 101–120, ISSN 1568-4946, https://doi.org/10.1016/j.asoc.2018.01.004
[81] B. Bhuyan, H. Sarma, N. Sarma, A. Kar and R. Mall, "Quality of Service (QoS) Provisions in Wireless Sensor Networks and Related Challenges," *Wireless Sensor Network*, Vol. 2 No. 11, 2010, pp. 861–868. https://doi.org/10.4236/wsn.2010.211104
[82] J. E. Mbowe and G. S. Oreku, "Quality of Service in Wireless Sensor Networks," *Wireless Sensor Network*, Vol. 6 No. 2, 2014, pp. 19–26. https://doi.org/10.4236/wsn.2014.62003
[83] Meena Pundir, Jasminder Kaur Sandhu, A Systematic Review of Quality of Service in Wireless Sensor Networks using Machine Learning: Recent Trend and Future Vision, *Journal of Network and Computer Applications*, Volume 188, 2021, 103084, ISSN 1084-8045, https://doi.org/10.1016/j.jnca.2021.103084
[84] Sujanthi, S., Nithya Kalyani, S. SecDL: QoS-Aware Secure Deep Learning Approach for Dynamic Cluster-Based Routing in WSN Assisted IoT. *Wireless Pers Commun* **114**, 2135–2169 (2020). https://doi.org/10.1007/s11277-020-07469-x
[85] Muhammad Ateeq, Muhammad Khalil Afzal, Sheraz Anjum, Byung-Seo Kim, Cognitive quality of service predictions in multi-node wireless sensor networks, *Computer Communications*, Volume 193, 2022, Pages 155–167, ISSN 0140-3664, https://doi.org/10.1016/j.comcom.2022.06.042
[86] Neda Mazloomi, Majid Gholipour, Arash Zaretalab, Efficient configuration for multi-objective QoS optimization in wireless sensor network, Ad Hoc Networks, Volume 125, 2022, 102730, ISSN 1570-8705, https://doi.org/10.1016/j.adhoc.2021.102730
[87] Shaikh, F.K. and Zeadally, S., 2016. Energy harvesting in wireless sensor networks: A comprehensive review. *Renewable and Sustainable Energy Reviews*, 55, pp. 1041–1054.
[88] Chaoming Hsu, R., Liu, C.T. and Lee, W.M., 2009, June. Reinforcement learning-based dynamic power management for energy harvesting wireless sensor network. In International Conference on Industrial, Engineering and Other Applications of Applied Intelligent Systems (pp. 399–408). Springer, Berlin, Heidelberg.
[89] Aoudia, F.A., Gautier, M. and Berder, O., 2018. RLMan: An energy manager based on reinforcement learning for energy harvesting wireless sensor networks. *IEEE Transactions on Green Communications and Networking*, 2(2), pp. 408–417.

[90] Hsu, R.C., Lin, T.H. and Su, P.C., 2022. Dynamic energy management for perpetual operation of energy harvesting wireless sensor node using fuzzy Q-learning. *Energies*, 15(9), p. 3117.

[91] Ehlali, S. and Sayah, A., 2022. Towards Improved Lifespan for Wireless Sensor Networks: A Review of Energy Harvesting Technologies and Strategies. *European Journal of Electrical Engineering and Computer Science*, 6(1), pp. 32–38.

[92] Yadav, S.L., Ujjwal, R.L., Kumar, S., Kaiwartya, O., Kumar, M. and Kashyap, P.K., 2021. Traffic and energy aware optimization for congestion control in next generation wireless sensor networks. Journal of Sensors, 2021.

[93] M. Abu Alsheikh, S. Lin, D. Niyato and H.-P. Tan, "Rate-Distortion Balanced Data Compression for Wireless Sensor Networks," in IEEE Sensors Journal, vol. 16, no. 12, pp. 5072–5083, June 15, 2016, https://doi.org/10.1109/JSEN.2016.2550599

[94] Gholipour, M., Haghighat, A.T. and Meybodi, M.R., 2017. Hop-by-Hop Congestion Avoidance in wireless sensor networks based on genetic support vector machine. Neurocomputing, 223, pp. 63–76.

[95] MRL-SCSO: Multi-agent Reinforcement Learning-Based Self-Configuration and Self-Optimization Protocol for Unattended Wireless Sensor Networks.

[96] Renold, A.P. and Chandrakala, S., 2017. MRL-SCSO: multi-agent reinforcement learning-based self-configuration and self-optimization protocol for unattended wireless sensor networks. Wireless Personal Communications, 96(4), pp. 5061–5079.

[97] Sharma, M. & Kumar, C. Machine learning-based smart surveillance and intrusion detection system for national geographic borders. In *Artificial Intelligence and Technologies* 165–176 (Springer, 2022).

[98] Singh, A., Nagar, J., Sharma, S. & Kotiyal, V. A gaussian process regression approach to predict the k-barrier coverage probability for intrusion detection in wireless sensor networks. *Expert Syst. Appl.* **172**, 114603 (2021).

[99] Singh, A., Amutha, J., Nagar, J., Sharma, S. & Lee, C.-C. Lt-fs-id: Log-transformed feature learning and feature-scaling-based machine learning algorithms to predict the k-barriers for intrusion detection using wireless sensor network. *Sensors* https://doi.org/10.3390/s22031070 (2022).

[100] Otoum, S., Kantarci, B. & Mouftah, H. T. On the feasibility of deep learning in sensor network intrusion detection. *IEEE Netw. Lett.* **1**, 68–71 (2019).

[101] Singh, A., Amutha, J., Nagar, J. et al. AutoML-ID: automated machine learning model for intrusion detection using wireless sensor network. *Sci Rep* **12**, 9074 (2022). https://doi.org/10.1038/s41598-022-13061-z

[102] Md Mohinur Rahaman, Md Azharuddin, Wireless sensor networks in agriculture through machine learning: A survey, Computers and Electronics in Agriculture, Volume 197, 2022, 106928, ISSN 0168-1699, https://doi.org/10.1016/j.compag.2022.106928

[103] Sami, Maira, Saad Qasim Khan, Muhammad Khurram, Muhammad Umar Farooq, Rukhshanda Anjum, Saddam Aziz, Rizwan Qureshi, and Ferhat Sadak. 2022. "A Deep Learning-Based Sensor Modeling for Smart Irrigation System" *Agronomy* 12, no. 1: 212. https://doi.org/10.3390/agronomy12010212

[104] Yun Hwan Kim, Seong Joon Yoo, Yeong Hyeon Gu, Jin Hee Lim, Dongil Han, Sung Wook Baik, Crop Pests Prediction Method Using Regression and Machine Learning Technology: Survey, *IERI Procedia*, Volume 6, 2014, Pages 52–56, ISSN 2212-6678, https://doi.org/10.1016/j.ieri.2014.03.009

[105] H. Wani and N. Ashtankar, "An appropriate model predicting pest/diseases of crops using machine learning algorithms," *2017 4th International Conference on Advanced Computing and Communication Systems (ICACCS)*, 2017, pp. 1–4, https://doi.org/10.1109/ICACCS.2017.8014714

[106] G. R. S and V. Vasanthi, "Crop Growth Monitoring and Leaf Area Index Estimation Using Wireless Sensor Network and CNN," *2021 Third International Conference on Inventive Research in Computing Applications (ICIRCA)*, 2021, pp. 1031–1036, https://doi.org/10.1109/ICIRCA51532.2021.9545062

[107] Marković D, Vujičić D, Tanasković S, Đorđević B, Ranđić S, Stamenković Z. Prediction of Pest Insect Appearance Using Sensors and Machine Learning. Sensors (Basel). 2021 Jul 16;21(14):4846. https://doi.org/10.3390/s21144846

[108] Kamred Udham Singh, Ankit Kumar, Linesh Raja, Vikas Kumar, Alok Kumar Singh Kushwaha, Neeraj Vashney, Manoj Chhetri, "An Artificial Neural Network-Based Pest Identification and Control in Smart Agriculture Using Wireless Sensor Networks", *Journal of Food Quality*, vol. 2022, Article ID 5801206, 12 pages, 2022. https://doi.org/10.1155/2022/5801206

[109] Anna Chlingaryan, Salah Sukkarieh, and Brett Whelan. 2018. Machine learning approaches for crop yield prediction and nitrogen status estimation in precision agriculture: A review. Comput. Electron. Agric. 151, C (Aug 2018), 61–69. https://doi.org/10.1016/j.compag.2018.05.012

[110] Girik Pachauri, Sandeep Sharma, Anomaly Detection in Medical Wireless Sensor Networks using Machine Learning Algorithms, Procedia Computer Science, Volume 70, 2015, Pages 325–333, ISSN 1877-0509, https://doi.org/10.1016/j.procs.2015.10.026

[111] B Vidya, Sasikumar P, Wearable multi-sensor data fusion approach for human activity recognition using machine learning algorithms, Sensors and Actuators A: Physical, Volume 341, 2022, 113557, ISSN 0924-4247, https://doi.org/10.1016/j.sna.2022.113557

[112] X. Du, M. Zhang, K. Nygard, M. Guizani and H.H. Chen, "Distributed Decision Making Algorithm for Self-Healing Sensor Networks," *2006 IEEE International Conference on Communications*, Istanbul, Turkey, 2006, pp. 3402–3407, https://doi.org/10.1109/ICC.2006.255598

[113] M. W. M. Seah, C.-K. Tham, V. Srinivasan and A. Xin, "Achieving Coverage through Distributed Reinforcement Learning in Wireless Sensor Networks," *2007 3rd International Conference on Intelligent Sensors, Sensor Networks and Information*, Melbourne, VIC, Australia, 2007, pp. 425–430, https://doi.org/10.1109/ISSNIP.2007.4496881

[114] Hosein Mohamadi, Shaharuddin Salleh, Mohd Norsyarizad Razali, Sara Marouf, A new learning automata-based approach for maximizing network lifetime in wireless sensor networks with adjustable sensing ranges, Neurocomputing, Volume 153, 2015, Pages 11–19, ISSN 0925-2312, https://doi.org/10.1016/j.neucom.2014.11.056

[115] H. Chen, X. Li and F. Zhao, "A Reinforcement Learning-Based Sleep Scheduling Algorithm for Desired Area Coverage in Solar-Powered Wireless Sensor Networks," in *IEEE Sensors Journal*, vol. 16, no. 8, pp. 2763–2774, April 15, 2016, https://doi.org/10.1109/JSEN.2016.2517084

[116] L. Zhao, X. Wen and Dan Li, "Amorphous localization algorithm based on BP artificial neural network," *International Conference on Software Intelligence Technologies and Applications & International Conference on Frontiers of Internet of Things 2014*, Hsinchu, 2014, pp. 178–183, https://doi.org/10.1049/cp.2014.1556

9 Machine Learning-Assisted Interference Management in the 6G UAV Networks with Soft Frequency Reuse

Md. Sakir Hossain
Bangabandhu Sheikh Mujibur Rahman Aviation and Aerospace University, Lalmonirhat, Bangladesh

Md. Imran Ahmed, Rayhan Khan Ridoy, and Mirza Hasibul Hasan
American International University-Bangladesh, Dhaka, Bangladesh

Md. Shakhawat Hossain
Independent University, Dhaka, Bangladesh

9.1 INTRODUCTION

Our way of life has greatly changed in the last two decades due to the rapid development of communication systems. Cellular communication, in particular, plays a vital role in this rapid transformation of our societies by bringing communication facilities to everyone's door. Cellular communication services are easier to get and cheaper compared to traditional wired-based telephone systems. People are performing a wide variety of tasks including education, financial transactions, recreation, and so on, over communications systems. However, ubiquitous and seamless connectivity is still elusive even if these are the requirements of the 6G wireless communication systems.

There are some scenarios where the ubiquity of connectivity cannot be ensured; for example, communication systems may collapse due to natural disasters. For quick search and rescue operations, the presence of communication systems is very important. If communication systems are destroyed due to disasters, reinstalling them may take several days. Similarly, when a huge number of people gather in a stadium for sports or concert events, traditional cellular networks cannot provide services to the suddenly increased number of users. However, high data rate connectivity is highly demanded in such scenarios. To provide the rapid deployment of

 DOI: 10.1201/9781003303114-9

communication systems and high data rate services, the concept of UAV-assisted wireless networks has recently become attractive to academia and industries. In such networks, UAVs can be used as aerial base stations for user equipment (UE) in cellular networks, the mobile relay in flying ad hoc networks, and are promising for wireless backhauling in smart cities where UAVs provide seamless network coverage for both indoor and outdoor scenarios [1]. UAV base stations can effectively improve existing cellular networks because of their mobility, flexibility, and adaptive altitude by providing additional capacity to hotspot areas and can deliver better network coverage for rural areas [2]. However, many challenges remain to make UAV-assisted mobile networks practical, including deployment, interference management, channel modeling, energy efficiency, and so on. In this chapter, we deal with the interference management issue of UAV-assisted wireless networks.

As UAV technology has advanced significantly in last the two decades, many researchers anticipate using UAVs to improve existing cellular networks for better user experience and quality of service. UAV-assisted wireless networks are used for a wide variety of applications such as wireless sensor networks (WSN) [3], vehicular networks [8, 10], mobile networks [5–7, 9, 11, 12], disaster management [4, 5], and many others. In most cases [3, 4, 6, 9, 11], the UAV deployment is an important issue. A significant effort is devoted to the optimization of radio resource management [5, 7]. Most of the resource management optimization methods are computationally expensive; due to the limited computational capability of UAVs, such solutions are not practically viable for UAV-assisted wireless networks. To overcome this limitation, some low-complex power and bandwidth allocation methods are proposed in [16, 17], where the soft frequency reuse (SFR) schemes are exploited in resource management. In [16], the power and bandwidth are allocated among UAVs using graph theory. Although this method can significantly reduce the computational overhead, more efforts are required to further reduce computational complexity. A deep learning method is proposed in [17] to approximate the graph theory algorithm proposed in [16]. This method reduces the computational complexity dramatically. However, the imperfect approximation causes a degradation of the throughput.

In this chapter, we address the low accuracy issue of the existing method in allocating resource plans among UAVs in a cellular network with SFR. The contribution of this chapter is outlined below:

- We present a comprehensive literature review on the progress of research in interference management in mobile networks and UAV-assisted mobile networks.
- A machine learning model is proposed to allocate resource plans among the UAVs in a multi-UAV network.
- We investigate whether the positions of all UAVs should be considered in allocating resource plans to a specific UAV.

The rest of the chapter is organized as follows. Section 9.2 highlights the latest development in UAV deployment and resource management. The concept of UAV-assisted wireless communications is briefly described in Section 9.3. Then, the need for SFR in UAV-assisted networks is explained in Section 9.4. Thereafter, in Section 9.5, we describe the proposed machine learning method for allocating radio resources among UAVs. The performance of the proposed method is analyzed in Section 9.6, before concluding the chapter in Section 9.7.

9.2 RELATED WORKS

In [3], a design and deployment technique of UAV-assisted WSN is proposed to minimize the impact of natural disasters through proper warning and monitoring of the post-disaster recovery procedures. The efficiency of localization and navigation performance of the proposed system is mainly evaluated based on the Kalman filter. It is shown that a UAV can maintain a predetermined trajectory successfully to deploy WSNs for post-disaster monitoring. A novel method is proposed in [3] for optimizing the deployment delay of UAVs for emergency coverage of wireless networks. Specifically, two deployment scenarios are considered: a small number of UAVs for small areas and a large number for large areas. The proposed model is used to reduce the data traffic and delay of UAV deployment to provide emergency coverage of wireless networks. The authors in [4] solve a UAV-aided emergency rescue problem.

Additionally, each UAV acts as a base station, where the locations of a large number of users are covered by the minimum number of UAVs. Besides, to achieve a minimum data rate and overcome the transmission power limitation of each UAV, a heuristic approach using the genetic algorithm is proposed and evaluated. In [5], a scheme is proposed to deploy UAVs for providing network coverage to the disaster-affected area. To maximize the sum spectral efficiency of disaster-affected wireless networks, the k-means clustering algorithm is used to determine the two-dimensional (2D) placement of the UAVs. Two approaches, an exhaustive search and particle swarm optimization (PSO), are used to determine the accurate altitude of the UAVs.

The authors in [6] proposed a novel framework that enables a predictive deployment of UAVs to complement ground cellular systems in downlink traffic offload. In addition, a novel learning method based on weighted expectation maximization (WEM) for estimating the user distribution and downlink traffic demand is also proposed. This WEM approach achieves a significant improvement in prediction accuracy compared to the expectation minimization and k-mean approaches. However, in this approach, the energy consumption is relatively higher. Various optimization techniques are exploited in [7] to find which one provides a higher network throughput and reduces the transmit power. Among all solutions, the heuristic algorithm increases the network throughput and minimizes transmission power, and better performance is achieved by stochastic analysis in lower-dense ground stations connected with UAVs. In [8], a UAV-assisted vehicular network architecture is proposed to integrate UAVs with ground vehicular networks to efficiently improve system performance. In addition, an integrated simulation platform is built to validate the performance. Besides, the simulation results demonstrate that the performance of vehicular networks can be significantly enhanced with the proposed UAV network architecture.

The authors in [9] propose a UAV deployment method using a bisecting k-means clustering algorithm based on the current user's QoS requirements. In addition, two fractional frequency reuse (FFR) schemes are proposed to significantly reduce the outage probability. A mobility model is presented in [10] with the goal of characterizing the fundamental metrics such as the UAV arrival process, speed, and density, including delay analysis and performance. In addition, this proposed model has higher improvement in terms of path availability and average data delivery delay. However, it does not cover all mobility patterns.

The authors in [11] propose a cache-enabling UAV cellular network with a large access capability in non-orthogonal multiple access (NOMA) systems. For content delivery minimization, the authors in [11] formulate the long-term caching placement and resource allocation optimization as a Markov decision process (MDP), where MDP is used for modeling decision-making in dynamic systems. It decides which action to take depending on the state of the environment the MDP is working on to attain the goal. Each action changes the state of the environment, and a reward is provided based on each action. MDP is widely used in optimization, manufacturing, economics, robotics, and so on. In [11], a Q-learning-based caching algorithm and resource allocation algorithm are used. Although it can improve performance for large networks, it is inappropriate for small networks where there are limited numbers of users. A coverage-aware three-dimensional (3D) UAV deployment algorithm is proposed in [12] where the UAV trajectories are planned in such a way that the coverage hole can be reduced to the minimum. A dueling double deep Q network (dueling DDQN) is trained based on the signal measurement for UAV navigation. Although this method improves the network coverage, it incurs an increased computational burden on the network.

The works described above deal with the deployment of UAVs. However, the unpredictable positions of UAVs pose a serious threat to user experiences due to increased interference. The users located in the cell-edge areas suffer from low throughput and deep null due to the high interference from the neighboring cells. To alleviate a similar problem in traditional cellular networks, SFR is a proven solution. For example, an algorithm called adaptive soft frequency reuse (ASFR) is used in [13] to extract subcarrier and power allocations followed by exhaustive search and greedy decent methods. The authors in [14] propose a novel soft frequency reuse scheme for LTE-A networks based on three frequency segments including two-band SFR (2B-SFR), three-band SFR (3B-SFR), and three-band improved SFR (3B-ISFR). Additionally, the frequency reuse pattern allocation scheme for microcells is centralized at each level of each macrocell, and the scheme is based on the exchange of measurements made by the microcells with their macrocell. In terms of outage probability and average cell capacity among three frequency segments, the 3B-ISFR has shown significant improvement in resource reuse and spectrum efficiency. A multilayer SFR for 5G heterogeneous cellular networks (HetNet) scheme is proposed in [15], which improves network throughput and outage probability.

In [13–15], we see that the SFR is an effective method of resource management in cellular networks for interference management. Inspired by this, in [16], the SFR is applied in UAV-assisted mobile networks. To deploy UAVs, k-means clustering is used. To tackle the unpredictable positions of UAVs, a graph theory-based resource plan allocation scheme is proposed where the research plan is a specific combination of the bandwidth and transmit power. Through simulations, it is shown that the graph theory-based solution can effectively reduce interference. However, such a solution suffers from high computational complexity. As UAVs are highly resource-limited vehicles, such high complexity is a bottleneck in such a network. To alleviate this shortcoming, a low-complex deep learning-based resource plan allocation method is proposed in [17]. A dataset in [17] is created by artificially deploying UAVs using k-means clustering and allocating resource plans using the graph theory-based algorithm proposed. Then, a deep learning model is trained by the dataset to predict the resource plans for various UAVs by giving the positions of the UAVs as inputs to the model. However, the accuracy of the deep learning-based method is very limited.

9.3 UAV-ASSISTED WIRELESS NETWORKS

In some scenarios traditional cellular networks cannot ensure ubiquitous and seamless connectivity. For example, the communication systems may collapse when the cell tower gets broken down due to a natural disaster. In this case, it may take several days to weeks to restore the communication systems. In a normal situation, when people gather in a location far away from the cell tower, it is not economical to set up base stations to connect the people to the cellular networks, as they gather in the place for a short period and seldomly. In such scenarios, deploying a portable base station (or flying base station) can be a good solution. In such networks, a small base station is mounted to a UAV. Then, the base station-mounted UAVs are deployed to provide connectivity to the seldom gathering. Such networks are called UAV-assisted wireless networks. In such networks, the flying base station connects people to the nearby base stations. Figures 9.1 through 9.3 show some scenarios where UAV-assisted wireless networks can be used to ensure ubiquitous and seamless connectivity. In Figure 9.1, a disaster area is shown where the cellular base station is destroyed. In this case, a flying base station can be deployed to connect the people to the cellular networks. In Figure 9.2, multiple UAVs are used to relay communication signals to connect users located in a place where there are no cellular network base stations. Another scenario is shown in Figure 9.3, where a flying base station is deployed to provide connectivity to users avoiding the impact of large obstacles between the cell tower and users.

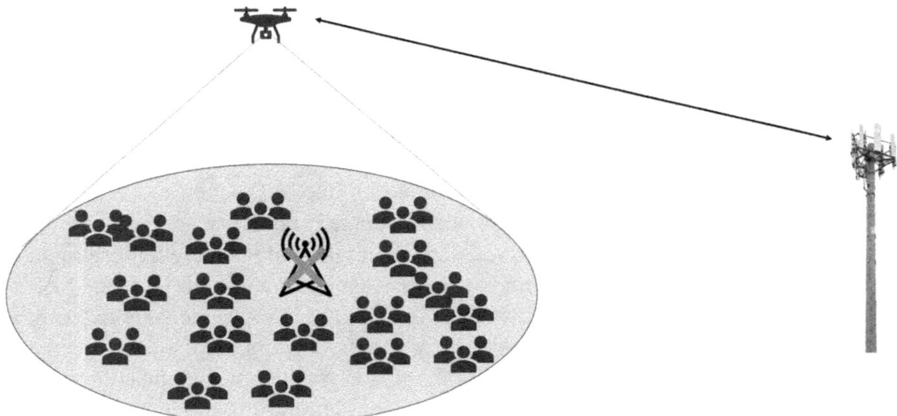

FIGURE 9.1 UAV-assisted network in a disaster area.

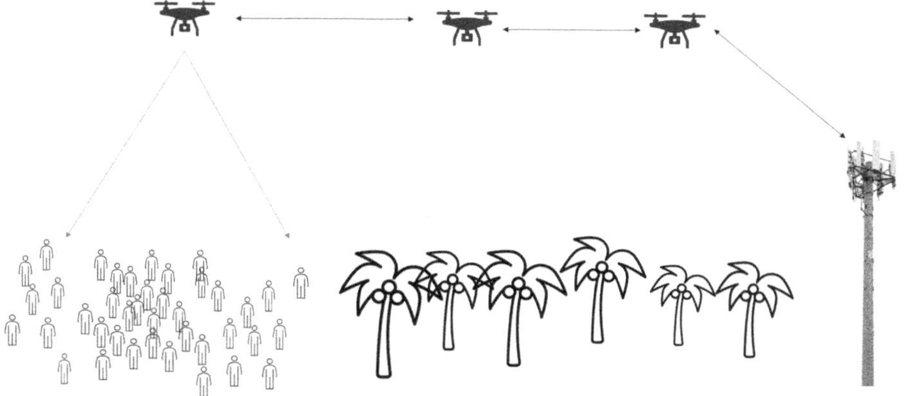

FIGURE 9.2 UAV-assisted networks for relaying.

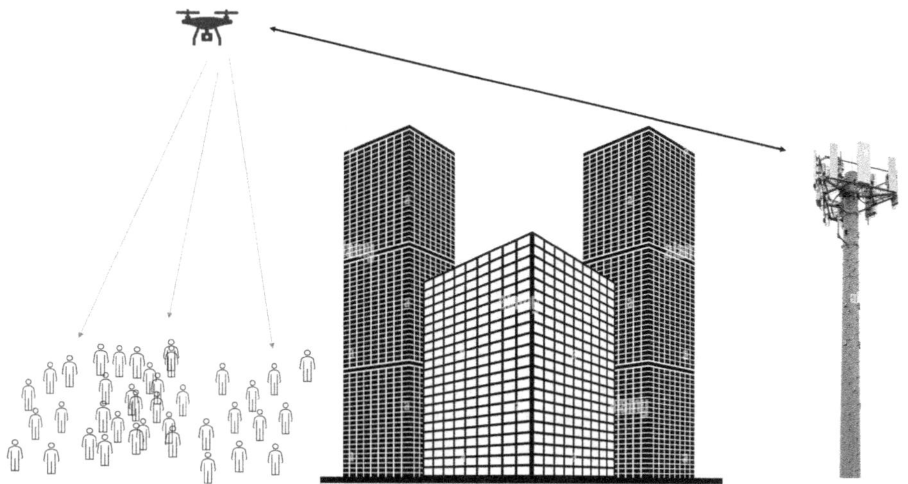

FIGURE 9.3 UAV-assisted networks to avoid big obstacles in communication paths.

Another benefit of the UAV-assisted wireless network is that it also enables increased spectrum reuse to ensure a higher network throughput. Since the possibility of getting a line-of-sight (LOS) path between the users and UAV is higher, such networks can be a good application for higher frequency bands such as millimeter wave bands and terahertz bands.

9.4 SOFT FREQUENCY REUSE IN UAV-ASSISTED NETWORKS

Soft frequency reuse is a popular approach that has been applied to wireless systems over the last two decades. It was first introduced in the GSM network system and also brought under the 3GPP LTE infrastructure with the goal of giving a higher

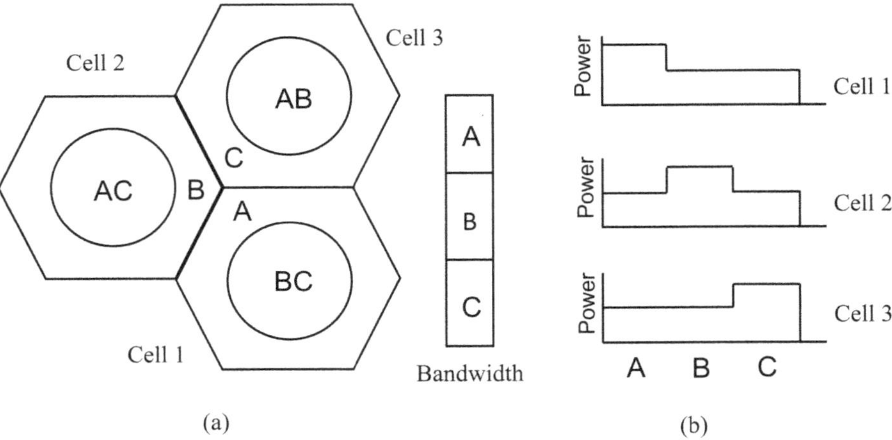

FIGURE 9.4 The frequency planning and power allocation for the SFR scheme: (a) bandwidth (b) power allocation.

performance for users near the cell edge area [18]. It is considered as an effective frequency reuse method for inter-cell interference coordination by following spectrum efficiency [19]. Users within each cell are divided into two groups—cell-center users and cell-edge users—based on their distance to the base station. Cell-edge users are confined to the reserved cell-edge band, while cell-center users have exclusive access to the cell-center band and can also have access to the cell-edge band but with lower precedence. A frequency planning of SFR can be performed in a three-cell cluster, as shown in Figure 9.4(a), where the cell-center users can use two-thirds of the bandwidth, but the cell-edge users only use a third of the bandwidth. The downlink transmitting power for the cell-edge users is higher to enhance their data rates, whereas the transmit power for the cell center should be lower compared to the power for the cell-edge area. The available bandwidth is divided into three sub-bands: A, B, and C. The sub-band allocated to the cell-edge area of a cell is used in the cell center of the other two cells. The bandwidth allocated for the cell-edge region may also be used in the central region if it is not being used at the cell-edge region. The power distribution of SFR is in Figure 9.4(b). A particular combination of bandwidth and power is termed a resource plan. For example, in cell 1, a higher power is allocated to the sub-band A, while a comparatively lower power is allocated to the sub-bands B and C.

Recently, the concept of a multi-level soft frequency reuse (MLSFR) scheme is proposed in [20], where each cell is divided into more than two circular regions. The number of circular regions is 2^n, where n is an integer. The innermost and the outermost region form an SFR, while the second innermost region and the second outermost regions form another SFR. In this way, a cell can have more than one SFR at a time. This technique can ensure more throughput to the cell-edge users reducing the throughput to the cell-center users.

In conventional mobile networks, the network layout is predesigned. Thus, the area of each cell is predetermined. A radio frequency engineer can allocate frequencies and power to various cells before implementing the network in practice. However,

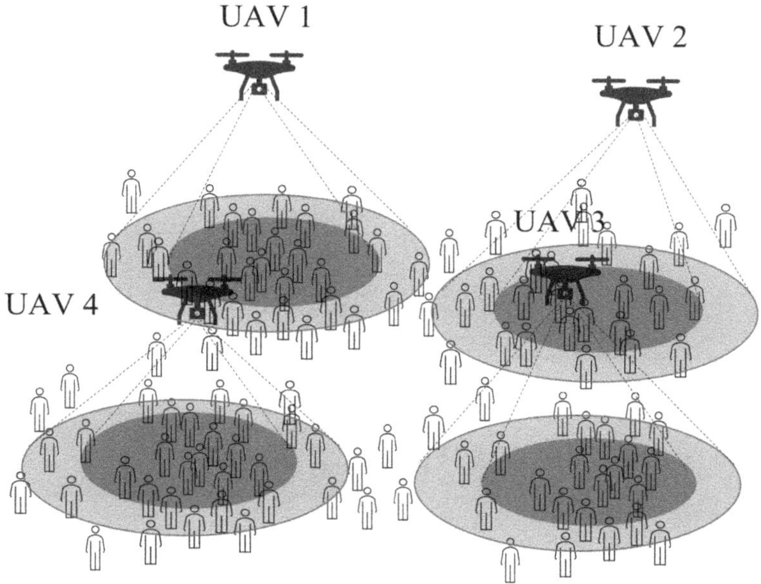

FIGURE 9.5 SFR in the UAV-assisted wireless networks.

the network layout is not known a priori in the UAV-assisted mobile networks, as the UAVs change their positions over time based on the positions of users. For this reason, a predetermined allocation of the radio resource plans is not possible, as the positions of various UAVs cannot be known a priori, and their positions change over time. For this reason, a dynamic resource plan allocation is required for UAV-assisted wireless networks. Figure 9.5 shows the scenario. In this scenario, we require an intelligent system to allocate the resource plans among the UAV in such a way that the distance between the UAVs with the same resource plans is maximized. The maximized distance leads to minimized interference.

9.5 PROPOSED MACHINE LEARNING MODEL

In this section, we present the proposed machine learning model. The purpose of this section is to allocate resource plans among the UAVs deployed in a service area. The resource plans should be allocated in such a way that the distance between any two UAVs with the same resource plan is maximized. We input the positions of the UAVs to the machine learning classifiers, while it would be the task of the classifier to allocate a resource plan to each UAV.

The proposed system model is shown in Figure 9.6. First, the input dataset is split into training and test datasets. We use 80% samples of the dataset as the training dataset, while the rest of the 20% is used as the test dataset. Then, we use two methods for allocating the resource plans: without and with feature selection. In the first method, we directly input the positions of all UAVs to a classifier to train it. Then, the performance of the trained classifier is evaluated. In the second method, we first

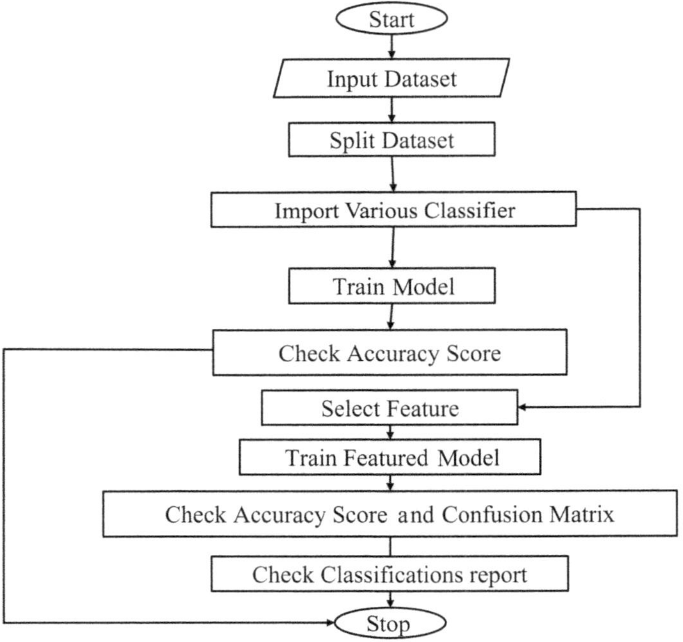

FIGURE 9.6 Proposed system model.

perform feature selection. In this case, the feature means the positions of the UAVs. We investigate if the positions of all UAVs are required to allocate a resource plan to a specific UAV. We use the recursive feature elimination (RFE) [21] method for feature selection. After selecting a subset of the features (i.e., positions of the UAVs), we train a machine learning model with only those selected features. Then, the performance is evaluated with respect to the test dataset. The purpose of the feature selection is to devise a low-complex training phase. The low-complexity feature will enable us to retrain the model whenever we require it within a comparatively shorter time.

9.6 SIMULATION RESULTS AND ANALYSIS

In this section, we use the CTU_UAV_SFR resource plan allocation dataset [17]. The dataset is created by the Czech Technical University in Prague, Czech Republic. It is an artificially generated dataset using MATLAB. The data generation procedure is illustrated in Figure 9.7. To generate the data, first, the users are uniformly distributed in the service area. Then, UAVs are deployed using k-means clustering algorithm in such a way that the distance between users and their serving UAV is minimized. Thereafter, the graph theory algorithm proposed in [16] is exploited to allocate resources among the UAVs. Finally, the UAV positions and resource plans allocated to each UAV are stored in a CSV file.

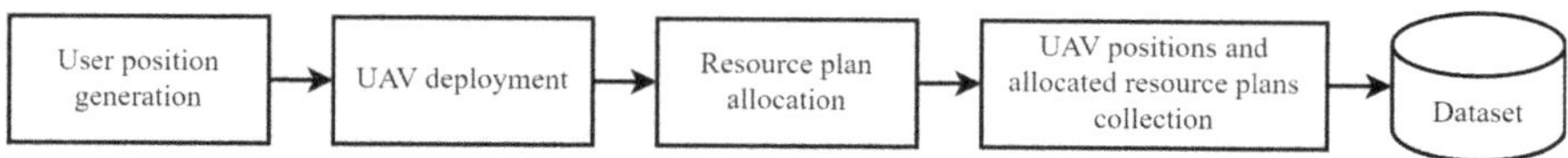

FIGURE 9.7 Data generation process.

The dataset consists of the positions of the UAVs, and the resource plan allocated to the UAV whose position is mentioned at the beginning among all UAVs (i.e., the first two elements of a row). Every two columns in the dataset represent the position of a UAV. Besides, the last column represents the target class (that is, the resource plan). The three classes are denoted as 1, 2, and 3. This dataset consists of data for 3, 4, 5, 6, 7, 8, 9, 10, 11, 12, 13, 14, 15, 18, 21, 24 UAVs. In addition, each dataset has a different number of samples. For example, the dataset UAV_8 has 10,40,000 rows and 16 features, where rows represent the position of the UAVs in numerical values. In addition, the UAV-12 has approximately 15,60,000 rows and 24 features. Due to the time and resource constraints, we work on the UAV_8 and UAV_12 cases only.

The performance of the proposed model will be evaluated next. The 8 and 12 UAVs cases will be evaluated separately. First, we consider the 8 UAVs case. The performance of the classifiers without any feature selection is shown in Table 9.1. The notations are defined as follows: GNB: Gaussian Naïve Bayes, SVC: support vector classifier, LR: logistic regression, KNN: k-nearest neighbor, DT: decision tree, RF: random forest, XGB: extreme gradient boosting, and GB: gradient boosting. The performance of the classifiers varies from 40% to 78%. The maximum accuracy of 78.3% is obtained using the random forest classifier. The accuracy rate declines to 75.2% for XGBoost and goes further down to 74.6% while using KNN. The accuracies of the GNB, GB, LR, and Adaboost classifiers are around 40%. The performance shown in Table 9.1 is obtained using the full dataset. Since the dataset contains more than a million samples, the training of the classifier requires considerable time. To investigate the impact of the amount of the dataset, we use a partial dataset to train the classifiers. Table 9.2 shows the performance of the classifiers for various amounts of data. We consider 20%, 50%, and 75% of the data for 8 UAV cases. It is found that almost 73% accuracy can be obtained using only 20% of the whole data. For the classifier LR, the increase in the amount of data slightly reduces the accuracy. On the other hand, a 2% to 4% improvement in accuracy is obtained for SVC, KNN, XGB, and RF.

Next, we investigate whether the positions of all UAVs are required to allocate a resource plan to a UAV. For this reason, we use the recursive feature selection method. As we find the highest accuracy for the random forest classifier (see Table 9.1), we

TABLE 9.1
Performance of the Proposed Models for 8 UAVs

GNB	SVC	LR	KNN	DT	RF	XGB	GB	ADA
40.12	68.50	39.42	74.60	68.24	78.30	75.20	43.66	41.26

TABLE 9.2
Impact of the Amount of Data on the Percentage of Accuracy of the Classifiers for 8 UAVs

Data	SVC	LR	KNN	DT	RF	XGB	GB	ADA
20%	63.84	39.26	69.94	64.85	70.48	72.92	43.95	42.10
50%	66.72	39.13	72.18	66.74	72.52	74.14	43.69	41.68
75%	68.05	39.12	73.13	67.91	73.93	74.96	43.54	41.14

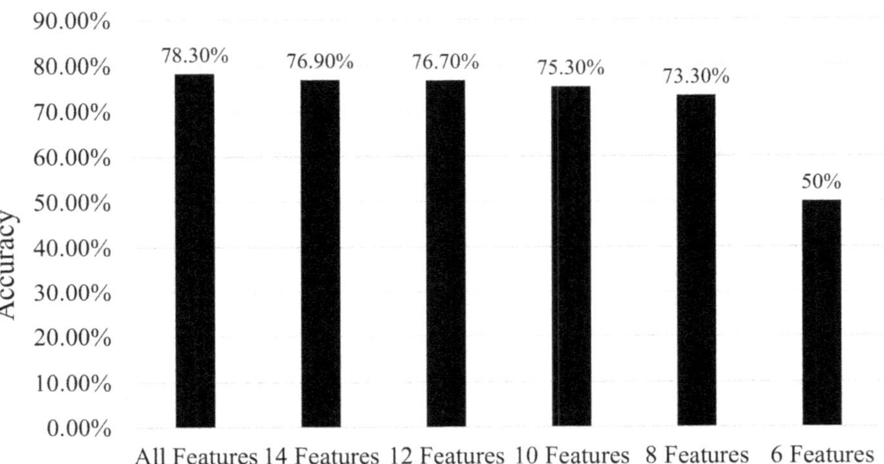

FIGURE 9.8 Impact of feature selection on the accuracy for 8 UAVs.

consider the random forest classifier for the feature selection. The impact of using a varying number of features is shown in Figure 9.8. The highest accuracy of 78.3% is obtained considering all features. When we use feature selections, the accuracy decreases; for 14 features the accuracy is 76.9%, which means the accuracy decreases by almost 1.4%. The accuracy continues to decrease as we reduce the number of features. However, the decrease in the accuracy is not significant as long as the number of features is more than or equal to 50% of the total number of features. That is, the reduction of the 8 features decreases the accuracy by 5%. However, the accuracy goes down by 23% due to the reduction of 2 more features.

Next, we consider the 12 UAV case. Table 9.3 shows the accuracy of the various classifiers for the 12-UAV deployment case. The maximum accuracy of 60.60% is obtained using the random forest classifier, a reduction of almost 18% compared to the 8 UAVs case. Similar to the 8 UAV case, the XGBoost is the second best performer; it attains just 0.5% less accuracy compared to the random forest. However, the difference between the highest and lowest achiever has reduced to about 20%, where the accuracy is 38% for the 8 UAV case. The UAV_12 dataset contains more than 1.5 million samples. Table 9.4 shows that the improvement in accuracy is not

TABLE 9.3
Performance of the Proposed Model for 12 UAVs

GNB	SVC	LR	KNN	DT	RF	XGB	GB	ADA
39.10	56.66	36.52	55.76	53.89	60.60	60.01	44.30	38.27

TABLE 9.4
Impact of the Amount of the Data on the Percentage of Accuracy of the Classifiers for 12 UAVs

Data	SVC	LR	KNN	DT	RF	XGB	GB	ADA
20%	52.57	36.78	49.80	51.54	52.93	58.02	45.81	38.58
50%	54.65	36.62	52.65	52.71	54.91	59.34	46.04	38.32
75%	55.35	36.54	53.76	52.89	55.60	59.75	46.28	37.44

noteworthy for increasing the amount of data by 2.5%. For some classifiers such as the LR and Adaboost, the accuracy reaches its pick at 20% of the dataset. Further increasing the data does not improve performance. However, for the classifier such as the random forest and XGBoost, the accuracy steadily increases with the increase of the amount of data.

Finally, the impact of feature selection on the resource plan allocation accuracy is shown in Figure 9.9 for 12 UAV cases. In contrast to the 8 UAV case, we find an improvement in the accuracy due to the feature selection. The accuracy is 60.6% when we consider the positions of all UAVs. Then, the accuracy increases with the decrease of the number of features before it reaches its pick for 18 features. Then, it starts going down. That is, for 12 UAV cases, we do not have to consider the positions of all UAVs for allocating a resource plan to a specific UAV. The positions of the 9 UAVs give us the best performance. To allocate a resource plan to a UAV, we do not need to consider the positions of those UAVs that are located considerably far from the UAV to which we are going to allocate a resource plan.

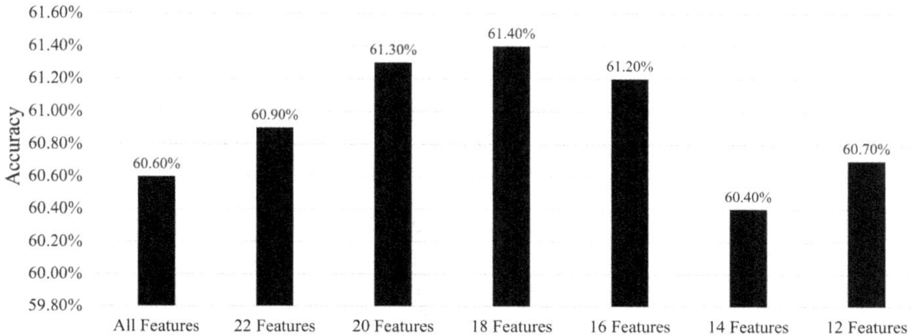

FIGURE 9.9 Impact of feature selection on accuracy for 12 UAVs.

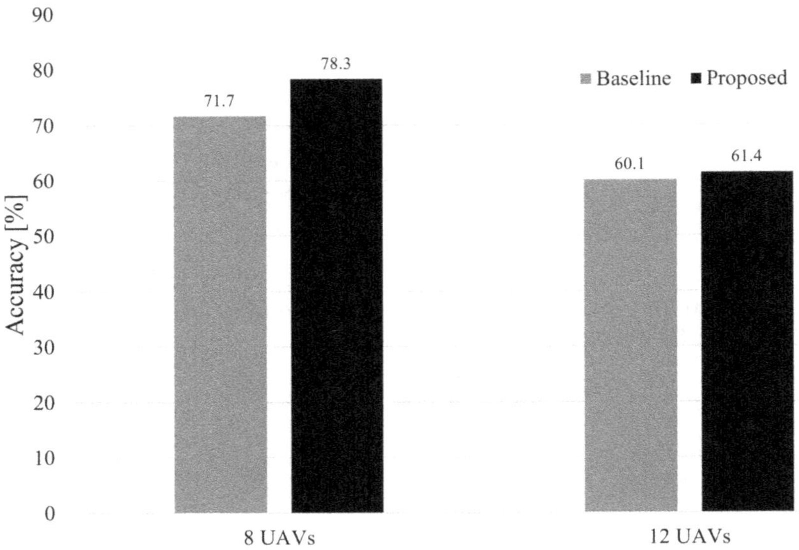

FIGURE 9.10 Comparison of the proposed method with the baseline [17].

To compare the performance of the proposed method, we consider the deep learning (DL) method proposed in [17] as the baseline method. Figure 9.10 shows that the proposed method outperforms the baseline DL method in both cases. However, the proposed method outperforms the DL method by a higher margin for 8 UAVs compared to the 12 UAV case. For 8 UAVs, while the DL method achieves 71.7%, the proposed method achieves 78.3%, thereby improving the accuracy by 7.3%. In the case of 12 UAVs, the improvement is 1.3%.

9.7 CONCLUSIONS

In this chapter, we proposed a machine learning model for allocating resource plans of SFR to various UAVs of the UAV-assisted networks. The purpose of the proposed method is to reduce interference by allocating resource plans among various UAVs in such a way that the distance between any two UAVs with the same resource plan is maximized. In the proposed method, we trained various machine learning models to allocate resource plans for the networks with 8 and 12 UAVs. We considered 9 classifiers: support vector machine, logistic regression, KNN, decision tree, random forest, gradient boosting, XGBoost, Adaboost, and Gaussian Naïve Bayes. Of these, the random forest is found to achieve the highest accuracy. It achieves 78.3% and 61.4% accuracy for the network with 8 and 12 UAVs, respectively. Furthermore, we also investigated whether the positions of all UAVs were required for effectively allocating resource plans among the UAVs. Through experiments, we found that the positions of all UAVs should be considered for the networks with 8 UAVs, while the positions of the 9 UAVs were enough to effectively allocate resource plans in the networks with 12 UAVs. With respect to the baseline deep learning-based solution, the proposed model achieved a significantly higher accuracy.

REFERENCES

[1] M. Mozaffari, W. Saad, M. Bennis, Y.H. Nam and M. Debbah, "A tutorial on UAVs for wireless networks: Applications, challenges, and open problems," *IEEE Communications Surveys & Tutorials*, vol. *21*, no. 3, pp. 2334–2360, 2019.

[2] G. Tuna, T.V. Mumcu, K. Gulez, V.C. Gungor, and H. Erturk, "Unmanned aerial vehicle aided wireless sensor network deployment system for post-disaster monitoring," in *Proceedings of the International Conference on Intelligent Computing*, 2012, pp. 298–305.

[3] X. Zhang and L. Duan, "Optimization of emergency UAV deployment for providing wireless coverage", in *Proceedings of the IEEE Global Communications Conference*, 2017, pp. 1–6.

[4] G. Liu, H. Shakhatreh, A. Khreishah, X. Guo, and N. Ansari, "Efficient deployment of UAVs for maximum wireless coverage using genetic algorithm," in *Proceedings of the IEEE 39th Sarnoff Symposium*, 2018, pp. 1–6.

[5] H. Hydher, D.N.K. Jayakody, K.T. Hemachandra, and T. Samarasinghe, "Intelligent UAV deployment for a disaster-resilient wireless network," *Sensors*, vol. *20*, no. 21, pp. 1–18, 2020.

[6] Q. Zhang, W. Saad, M. Bennis, X. Lu, M Debbah, and W. Zuo, "Predictive deployment of UAV base stations in wireless networks: Machine learning meets contract theory," *IEEE Transactions on Wireless Communications*, vol. 20, no. 1, pp. 637–652, 2020.

[7] R. Masroor, M. Naeem, and W. Ejaz, "Resource management in UAV-assisted wireless networks: An optimization perspective," *Ad Hoc Networks*, vol. 121, Art. no. 102596, 2021.

[8] W. Shi, H. Zhou, J. Li, W. Xu, N. Zhang, and X. Shen, "Drone assisted vehicular networks: Architecture, challenges and opportunities," *IEEE Network*, vol. 32, no. 3, pp. 130–137, 2018.

[9] X. Hu, X. Zhang, and D. Yang, "Deployment of UAV and Interference Coordination in UAV-assisted Cellular Networks," in *Proceedings of the IEEE/CIC International Conference on Communications in China (ICCC Workshops)*, 2018, pp. 153–157.

[10] M. Khabbaz, J. Antoun, and C. Assi, "Modeling and performance analysis of UAV-assisted vehicular networks," *IEEE Transactions on Vehicular Technology*, vol. 68, no. 9, pp. 8384–8396, 2019.

[11] T. Zhang, Z. Wang, Y. Liu, W. Xu, and A. Nallanathan, "Caching placement and resource allocation for cache-enabling UAV NOMA networks," *IEEE Transactions on Vehicular Technology*, vol. 69, no. 11, pp. 12897–12911, 2020.

[12] Y. Zeng, X. Xu, S. Jin, and R. Zhang, "Simultaneous navigation and radio mapping for cellular-connected UAV with deep reinforcement learning," *IEEE Transactions on Wireless Communications*, vol. 20, no. 7, pp. 4205–4220, 2021.

[13] M. Qian, W. Hardjawana, Y. Li, B. Vucetic, X. Yang, and J. Shi, "Adaptive soft frequency reuse scheme for wireless cellular networks," *IEEE Transactions on Vehicular Technology*, vol. 64, no. 1, pp. 118–131, 2014.

[14] G. Giambene, T. Bourgeau, and H. Chaouchi, "Soft frequency reuse schemes for heterogeneous LTE systems," in *Proceedings of the IEEE International Conference on Communications (ICC)*, 2015, pp. 3161–3166.

[15] M.S. Hossain, F. Tariq, G.A. Safdar, N.H. Mahmood, and M.R. Khandaker, "Multi-layer soft frequency reuse scheme for 5G heterogeneous cellular networks," in *Proceedings of the IEEE Globecom Workshops*, 2017, pp. 1–6.

[16] M. S. Hossain and Z. Becvar. "Flexible soft frequency reuse for interference management in the networks with flying base stations," in *Proceedings of the IEEE 91st Vehicular Technology Conference (VTC2020-Spring)*, 2020, pp. 1–7.
[17] M. S. Hossain and Z. Becvar, "Soft frequency reuse with allocation of resource plans based on machine learning in the networks with flying base stations," *IEEE Access*, vol. 9, pp. 104887–104903, 2021.
[18] GPP and Huawei, "Soft frequency reuse scheme for UTRAN LTE," in R1-050507, TSG RAN WG1 Meeting#41, Athens, Greece, 2005.
[19] Yu, Yiwei, et al. "Performance analysis of soft frequency reuse for inter-cell interference coordination in LTE networks." in *Proceedings of the IEEE 10th International Symposium on Communications and Information Technologies*, 2010.
[20] Hossain, Md S., Faisal Tariq, and Ghazanfar Ali Safdar. "Enhancing cell-edge performance using multi-layer soft frequency reuse scheme." *Electronics Letters*, vol. 51, no. 22, pp. 1826–1828, 2015.
[21] I. Guyon, S. Gunn, M. Nikravesh, and L. A. Zadeh, *Feature Extraction: Foundations and Applications*, Springer, pp. 145–146, 2008.

10 Computational Intelligence in Communication Networks

Classification, Clustering, Reinforcement Learning, Deep Learning

Kiran Muloor
LTIMindtree Limited and CHRIST (Deemed to be University), Bangalore, India

Somesh Kumar Sahu
LTIMindtree Limited, Bangalore, India

Tapan Kumar Behera
Forrester Research, Cambridge, MA, USA

Debabrata Samanta
Rochester Institute of Technology, Pristina, Kosovo

10.1 INTRODUCTION

Computational Intelligence (CI) is often referred to as the Intelligence of machines that perform intellectual tasks, which includes both Artificial Intelligence (AI) and Computational Intelligence (CI). The concept of Computational Intelligence is normally regarded as a subset of Artificial Intelligence, and, therefore often likened to one of its subsets in most cases. Machine intelligence systems can be classified into two main categories: Artificial Intelligence (AI) and Computational Intelligence (CI). As mentioned earlier, AI relies on hard computing techniques, while CI utilizes soft computing techniques, which allow it to adapt to different scenarios [1].

To illustrate the point more clearly, a good example is that all hard computing techniques rely on binary logic, which comprises solely two values, such as the Boolean

DOI: 10.1201/9781003303114-10

true or false or zero, which is the cornerstone of modern computers. It must be noted that in spite of a logical approach, there is one major problem with it, which is that there are many words we use every day in our personal and professional lives that cannot always easily be translated into absolute terms, such as zero and one, which we require to carry out our everyday activities [2]. It could be possible to design a soft computing approach based on the use of fuzzy logic to address this problem. One of the most important aspects of cognitive intelligence is the way in which the logic is based on the psychology of the human brain, which is one of the most important characteristics of this technology. Logical reasoning aims to aggregate data into partial truths, which can be either crisp or fuzzy, similar to the way the brain processes information [3].

Computational Intelligence (CI) is an area of research and application with three main branches: neural networks, fuzzy systems, and evolutionary computation, all of which fall under the umbrella of CI. It is essential to utilize Computational Intelligence when developing games and cognitive development systems to form successful, intelligent systems. Much research has been conducted on deep learning and deep convolutional neural networks over the past few years. As Artificial Intelligence has developed significantly in the past couple of decades, the concept of deep learning has become one of its most important aspects. As a result of the introduction of CI into Artificial Intelligence systems, it has been proven that its performance has been quite good [4].

Computational Intelligence (CI) is a term used to describe a computationally intelligent system characterized by several characteristics like its ability to withstand faults, its ability to compute fast, and its ability to adapt to changes in the environment. Adaptive mechanisms are defined as those mechanisms that enable or facilitate intelligent behavior in complex and changing environments, as well as the analysis of adaptive mechanisms in complex and changing environments. For instance, an example of computational adaptation would be the ability of a system to respond to changes in its inputs and outputs as they arise over the course of the computation. As part of the adaptive mechanisms, five AI paradigms have demonstrated the ability to adapt to new environments: swarm intelligence (SI), artificial neural networks (ANN), evolutionary computation (EC) [5], artificial immune systems (AIS), and fuzzy systems (FS) [6]. The goal of Computational Intelligence can only be achieved by understanding all these paradigms of Artificial Intelligence so that we can achieve CI. Computing has evolved in recent years in such a way that it has begun to parallel the incredible abilities of the mind, and it continues to do so in an ever-evolving way [7].

Computational Intelligence is an important component of Artificial Intelligence, which focuses on identifying adaptive mechanisms that facilitate or enable intelligent behavior in complex environments. Andries P. Engelbrecht introduced the concept of Computational Intelligence in his book, *Computational Intelligence: An Introduction*. Computing of this type is also known as soft computing, which refers to the ability of a computer to learn how to perform a specific task by using data or experimental observations. Concepts, paradigms, algorithms, and implementations that facilitate or enable intelligent behavior are employed to enable intelligent behavior in a complex and changing environment [8].

Computer-integrated methods investigate problems that lack practical algorithms because they can't be formulated or are complicated and thus can't be used in real-world applications. In cognitive science, research is focused on the study of low-level cognitive functions such as perception, object recognition, signal analysis, the discovery of patterns in data, simple associations, and control. Implementing these types of problems can be achieved in various ways, including supervised and unsupervised methods [9].

10.1.1 Neural Networks

Generally, the functions of a neural network can be summarized as follows: it is capable of analyzing and classifying data, of using an associative memory, of producing patterns, and of creating clusters. By adopting this method, medical data can be analyzed and categorized, and face recognition and fraud detection can be pursued. Nonlinearities can be addressed to control the system more efficiently. Another benefit of neural networks and fuzzy logic algorithms is that they can cluster data. Neural networks cannot cluster data, which is one of their most significant advantages [10].

10.1.2 Fuzzy Systems

A fuzzy system (FS) is a language model that uses human language as a source of inspiration and solves uncertain problems in a generalized way, which is a generalization of traditional logic. Based on our ability to use our human language as a source of inspiration, we are able to perform approximate reasoning. This area of research covers several topics, including fuzzy sets and systems, fuzzy clustering and classification, fuzzy controllers, linguistic summarization, fuzzy neural networks, and type 2 fuzzy sets and systems. Among the branches of knowledge within Computational Intelligence is a branch known as fuzzy systems that attempts to represent uncertainty in an uncertain environment. As its name suggests, fuzzy systems are inspired by the imprecision of human language, which fuels the idea behind them [11].

10.1.3 Evolutionary Computation

In Artificial Intelligence, evolutionary computation is considered one of the most widely used techniques for solving complex optimization problems and continuous optimization problems as a special branch of the field. These computational models utilize evolutionary algorithms to solve these complex problems. Their models are developed based on evolutionary principles, such as inheritance from previous generations of successful models, as well as natural selection, which is the process by which the characteristics of successful generations are passed on to the next generation of the model they are developing [12]. A theory influenced by the process of biological evolution can be utilized as a source of inspiration to solve optimization problems through an approach called evolving computation (EC). It is possible to generate, assess, and modify a population of possible solutions by using EC algorithms [13]. Currently, a wide variety of evolutionary computation methods are available, including genetic algorithms, evolutionary programming, evolution strategies, genetic programming, swarm intelligence, differential evolution, evolvable hardware, multi-objective optimization, and adaptive hardware closely related [14].

10.2 DIFFERENCES BETWEEN COMPUTATIONAL INTELLIGENCE AND ARTIFICIAL INTELLIGENCE

Artificial Intelligence (AI) is becoming one of the most exciting developments in history as it becomes mainstream in our day-to-day lives and augments human capabilities to assist us in solving some of the complex problems we have battled for many years. Artificial Intelligence, or AI, is a technology that enables machines to behave and think like humans by mimicking their intelligence. Despite the growing demand for machine learning solutions from industry, traditional methods of Artificial Intelligence are unable to meet all the demands of the growing market [15]. The limitations of AI have allowed non-conventional models to gain more exposure because of its limitations. Thus, Computational Intelligence (CI) has emerged as a result of these limitations. CI techniques are developed in a very different manner from AI techniques, and their evolution follows a completely different path. Machine intelligence can be categorized into two types, hard-computational and soft-computational, both of which can be applied to a variety of situations and adapted to many problems [16]. Figure 10.1 shows the relationship between Computational Intelligence and Artificial Intelligence.

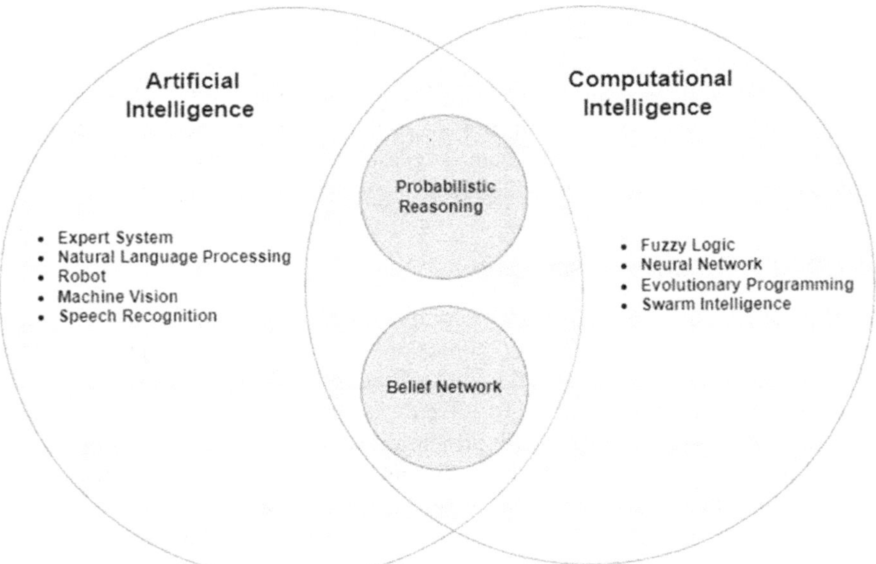

FIGURE 10.1 Relation between CI and AI.

Artificial Intelligence (AI) refers to computers or machines programmed to think and act like humans in a manner that mimics the actions performed by humans, allowing them to simulate the intelligence of humans. A machine can also be defined as having characteristics similar to those of a human mind based on this definition, which has several advantages. The use of virtual reality for learning and solving problems and gaining knowledge is an example. Artificial Intelligence (AI) is concerned with studying the intelligent behavior of machines and their ability to perform tasks better than humans. Artificial Intelligence is one of the major technological advancements in the digital age. The concept of Artificial Intelligence is believed to have emerged in part from the belief that computer programs could replicate the intelligence of humans in a computer environment. In general, intelligent machines, such as those that are as intelligent or more intelligent than humans, are not new concepts, but they have become an integral part of modern science with the advent of digital computers and the proliferation of the Internet. In Artificial Intelligence, computer programs can perform tasks on par with or better than those performed by humans. AI is a branch of science concerned with studying how machines exhibit intelligent behavior, rather than natural intelligence found in humans [17].

AI systems are based on the idea that human intelligence can be defined and implemented to simulate that intelligence and allow machines to complete tasks, ranging from the simplest to the most complex. A key goal of Artificial Intelligence is to mimic human cognitive activity and what humans do when they perform cognitive tasks. There have been surprisingly rapid advances in Artificial Intelligence in the area of replicating activities such as learning, reasoning, and perception. These activities are easily defined to maximize the effectiveness of these technologies. There have been predictions that innovators can develop systems that can learn or reason out any subject more effectively than humans [18]. Table 10.1 defines the comparison between Computational Intelligence and Artificial Intelligence.

TABLE 10.1
Comparison of Computational Intelligence and Artificial Intelligence

Computational Intelligence	Artificial Intelligence
In Computational Intelligence (CI), adaptive mechanisms are studied to enable intelligent behavior in complex and changing environments.	The study of Artificial Intelligence (AI) is the study of how machines demonstrate intelligent behavior, as opposed to how humans exhibit natural intelligence.
CI aims to understand how natural and artificial systems can act intelligently in complex and changing environments by understanding the computational paradigms that enable them.	To create an intelligent machine that exhibits intelligent behavior and can think and learn effectively like a human, an intelligent machine needs to exhibit intelligent behavior.
Applications of CI in the real world include intelligent household appliances, medical diagnosis, banking and consumer electronics, optimization applications, and industrial applications, among others.	The most common applications of Artificial Intelligence are speech recognition, handwriting recognition, optical character recognition, machine vision, natural language processing, and big data solutions.

10.3 DETAILS OF COMPUTATIONAL INTELLIGENCE

10.3.1 Branches of Computational Intelligence

Due to their robustness and flexibility, Computational Intelligence techniques are useful when dealing with complex nonlinear systems characterized by complexity and nonlinearity. There are many techniques in the area of Computational Intelligence, including artificial neural networks, genetic algorithms, fuzzy logic control, adaptive neuro-fuzzy inference systems, and particle swarm optimization [19]. Figure 10.2 shows the different types of techniques in Computational Intelligence.

10.3.2 Principles of CI and Application

As a branch of Artificial Intelligence, Computational Intelligence deals with soft computing, whereas Artificial Intelligence deals with hard computing. The foundation of hard computing theory lies in the utilization of existing mathematical algorithms to solve specific problems. With this method, you will be able to obtain an accurate and precise solution to your problem. It is possible to classify a variety of numerical problems as being hard computing problems because they are all related to numbers. The difference between a soft computing approach and a hard computing approach lies in the fact that the soft computing approach is based on a different perspective. With the aid of soft computing, it is possible to generate solutions to existing complex problems by computing solutions in the form of algorithms. A soft computing method also produces results that are less precise due to the way in which they are calculated and provided to the user [2]. Figure 10.3 explains the details of AI and CI based on the principles and applications.

In the past few decades, soft computing has been speculated as a type of computer model developed to solve non-linear problems that involve unsure, imprecise, and approximate solutions. In real-life situations, these are the types of issues that need to be resolved in a way similar to a person's thinking. A soft computing approach is based on the understanding that the human brain is a model of probability, fuzzy

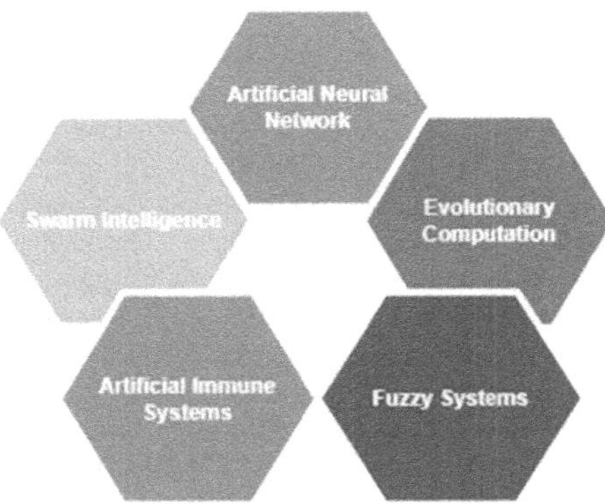

FIGURE 10.2 Branches of Computational Intelligence.

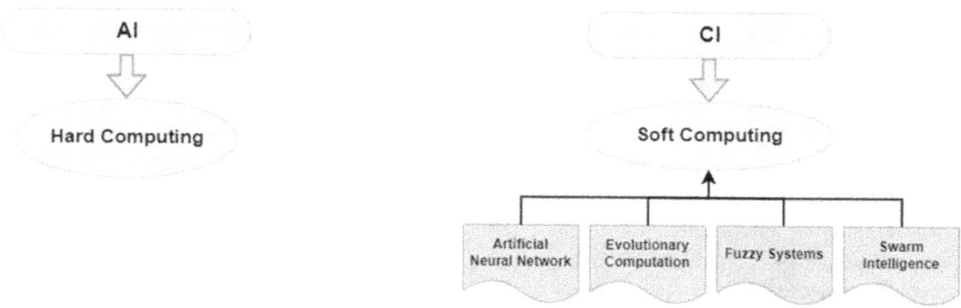

FIGURE 10.3 AI vs. CI for Hard and Soft Computing.

logic, and multivalued logic that receives information from the environment in real time [20].

E.g.:

- A computer-assisted diagnosis is one of the most commonly used methods of diagnosis in the medical field
- Handwriting recognition as a method of detecting fraud

A method of computing known as hard computing is an ancient approach to computing that involves using a mathematical model that is accurately defined to perform the computation. A hard computing approach might result in a warranted, finalized, accurate solution and a set of definite management actions based on a mathematical model or algorithm as the outcome of the hard computing approach. Specifically, the application is based on binary and crisp logic, which requires the input file to be exactly consecutive for it to run correctly. Using hard computing to solve real-world problems won't solve them in the long run if that is the approach taken. A hard computer uses predefined instructions, such as numerical analysis, rapid software, and binary logic. [21].

E.g.:

- Numerous examples of conventional algorithms include merge sort, quick sort, binary search, greedy algorithms, dynamic programming, etc.

Table 10.2 defines the difference between soft computing and hard computing under the principles of Artificial Intelligence.

Computational Intelligence is a concept based on five principles used in a wide range of fields, including computer science, engineering, data analysis, and biomedicine. Figure 10.4 shows the paradigms of Computational Intelligence.

10.3.2.1 Fuzzy Logic

The main principles of CI follow along with fuzzy logic, which is one of the primary principles of CI, so that measurements are taken and models are constructed that reflect the complexity of actual human processes, one of the main principles of CI.

TABLE 10.2
Difference between Soft and Hard Computing

Soft Computing	Hard Computing
An important aspect of soft computing is its liberal acceptance of inexactness, uncertainty, partial truths, and approximations.	Performing hard computations requires a model that represents the state of the world exactly.
Soft computing relies on formal logic and probabilistic reasoning techniques in many ways.	A hard computer relies on binary logic and crisp systems to operate.
Approximation and disposition are some of the characteristics of soft computing.	An important feature of hard computing is the accuracy (precision) and categorization it offers.
A soft computing system is a stochastic system in nature.	In contrast to soft computing, hard computing is deterministic.
Data that is ambiguous and noisy is dealt with by soft computing.	When it comes to hard computing, data must be exact.
Parallel computations can be performed using soft computing.	In hard computing, computations are performed sequentially.

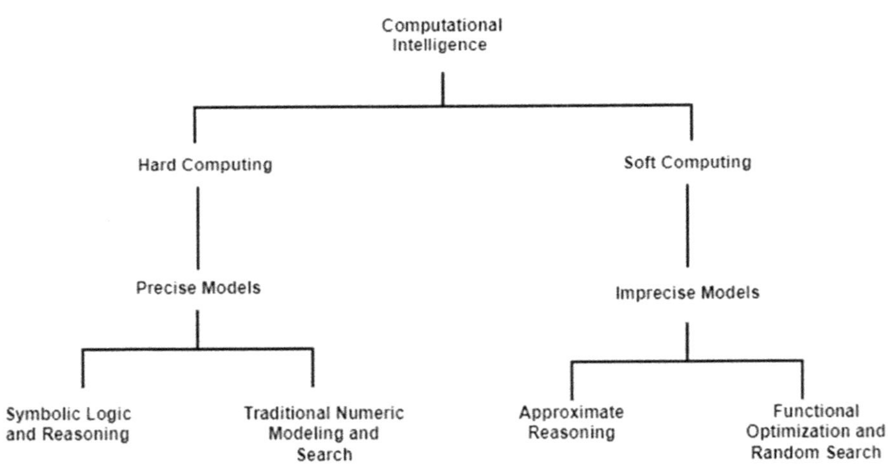

FIGURE 10.4 Computational Intelligence paradigms.

Process models can cope with incompleteness and ignorance of data, unlike Artificial Intelligence, which must be based on exact knowledge [22].

Over the years, the technique of hard computing has found numerous applications across various domains, including control systems, image processing, and decision-making. Additionally, this technology has been widely adopted in household appliances like washing machines and microwave ovens, which incorporate aspects of hard computing. Further, in scenarios where a video camera is used, there may arise a need for image stabilization to compensate for shaky camera movements during video recording [19]. Furthermore, this principle is also applicable to a wide range of fields, such as medical diagnostics, foreign exchange trading, and business strategy selection [23, 24].

Generally, fuzzy logic is used for approximate reasoning, as it lacks human abilities such as learning. By learning from past mistakes, they are able to improve themselves and not repeat the same mistakes in the future [25, 26].

10.3.2.2 Neural Networks

Artificial neural networks are being developed by CI experts to solve real-world problems. These are inspired by biological neurons, which are composed of three main components: the cell body, which processes information; the axon, which transmits signals; and the synapses, which act as controllers of those signals. Hence, artificial neural networks consist of decentralized information processing systems that facilitate the learning process by utilizing data acquired from past experiences. The main advantage of this principle is that it works just like a human being, and one of its biggest assets is its ability to tolerate faults [27].

The application of neural networks can be classified into six categories: data analysis, classification, associative memory, clustering, pattern generation, and control. The main objectives utilizing this method are to analyze and categorize medical data, perform facial recognition, detect fraud, and, importantly, address the nonlinearities within a system to achieve effective control. A further advantage of both neural networks and fuzzy logic techniques is the fact that they both allow clustering of data [10, 17].

10.3.2.3 Evolutionary Computation

The concept of evolutionary computation, which makes use of the natural selection process developed by Charles Robert Darwin, is used to produce new artificial evolutionary methods that incorporate the benefits of natural evolution. Furthermore, it also encompasses other areas of research such as evolution strategy and adaptive algorithms that play an important role in solving problems. According to the theory, its main applications include areas such as optimization and multi-objective optimization, to which traditional mathematical techniques are not sufficient anymore to apply to a wide range of problems including DNA analysis, scheduling problems, etc. [28, 29].

10.3.2.4 Natural Language Processing

In many cases, communicating with someone can be challenging in the event that you do not understand their language. Similarly, if you do not speak the same language as your computer system, you may be unable to communicate with it.

Since computers can only comprehend binary digits, they have difficulty understanding words. Natural Language Processing has been developed as part of computer science to address this challenge [30, 31].

Computer systems are modified through this process to enable them to understand the basic interactions between humans so they can work in harmony with them. A machine can receive the sounds emanating from human interaction as a result of this process. Upon receiving a sound, it converts that sound into text format to make it easier for a human to read and understand it. Through the conversion of these texts into components, the computer system can determine what the human was attempting to communicate [32].

10.3.2.5 Probabilistic Methods

In the 1970s, Paul Erdos and Joel Spencer introduced probabilistic methods to fuzzy logic, which are vital components of fuzzy logic. A major issue that needs to be considered in the context of a CI system that involves a large amount of randomness is how to assess the outcomes of the system [33]. Consequently, probabilistic methods can identify possible solutions to problems by using existing knowledge as a starting point to suggest possible solutions based on prior knowledge [34].

10.3.3 Implementation of CI

Throughout history, human beings have evolved, and one of our goals is to develop intelligent systems that will make our lives easier. There is a progressive nature to this goal, which means that as time goes on, our needs change, and something better will be required.

10.3.3.1 Using a Mobile Phone as an Example

Among the greatest inventions of the 20th century, the telephone was one of the most revolutionary. In the past few decades, we have developed more and more capabilities. Therefore, the size, the installation process, and the result are portable phones, which are essentially mobile phones today, due to their small size. Early on, there were only a limited number of functions that could be performed on portable phones, like making calls, sending text messages, adding contacts, and using utilities such as calendars, watches, alarms, and games. Hrlequin721-Rpnt smartphone provides us with an amazing array of features that are constantly growing.

These innovations require a strategy that can keep up with evolution, which is precisely where Computation Intelligence differs from other approaches, as it can adapt to change. Applied to real-life problems, Computational Intelligence, also known as soft computing, is beneficial for addressing them. CI is characterized by algorithms and approaches, such as fuzzy logic, neural networks, evolutionary theory, learning theory, and probabilistic theory, which make it well-suited to solving the kinds of complex problems encountered in real life. The possibilities for solving real-life problems with CI are endless.

10.4 COMPUTATIONAL INTELLIGENCE IN COMMUNICATION NETWORK

A structure and flow of information or communication between individuals or groups characterize a communication network. Communication networks serve the purpose of facilitating information exchange between various parts of an organization more efficiently. The general practice within organizations is to channel information through a system instead of flowing freely. Organizations have communication networks that facilitate the flow of information between different levels within the organization through regular interactions between individuals.

Current communication systems face performance requirements that pose significant challenges to existing technologies. Such requirements include high transmission rates, large data volumes, and fast response speeds, which the current theoretical framework has fundamental limitations in processing and data mining of massive data. Due to these limitations, deep learning technology has gained significant attention from researchers. Deep learning-based communication technologies have shown great potential in end-to-end communication systems, channel estimation, signal detection, and modulation identification. The process of communication involves reliably transmitting a message from a transmitter to a receiver over a channel. Two approaches have been identified for integrating deep learning with communication: the holistic approach, which considers communication as an end-to-end process; and the phase-oriented approach, which investigates the application of deep learning in certain phases of the communication process, such as modulation recognition, demodulation, compression, decoding, equalization, and channel modeling. Although these investigations provide a theoretical alternative to achieve performance bounds compared to traditional communication theory, they are yet to be implemented in practical applications.

Mobile applications with limited resources cannot be executed on mobile devices due to the ever-increasing volume of mobile traffic data. In the future, wireless communication is expected to become increasingly complex, primarily due to the presence of diverse radio access technologies, transmission links, and network slices. To navigate these intricate scenarios and effectively adapt to dynamic mobile environments, intelligent technologies are essential. Recently, there has been a surge of interest in the intersection of deep learning and wireless communication. By leveraging deep learning-driven algorithms and models, it becomes possible to enhance the analysis and management of wireless networks, enabling them to handle the expanding communication and computation demands associated with emerging mobile applications. However, the adaptation of deep learning techniques to function optimally within heterogeneous mobile environments is an ongoing area of exploration and development. Mobile wireless systems currently use several learning algorithms, but these algorithms are undeveloped and inefficient [35].

Computational intelligence techniques are increasingly being used in communication networks to optimize network performance, improve efficiency, and reduce costs. One example of the use of computational intelligence in communication networks is in the area of network management.

Network management involves monitoring, controlling, and optimizing the performance of communication networks. It involves tasks such as fault detection, network configuration, performance monitoring, and security management. Computational intelligence techniques such as artificial intelligence, machine learning, and data analytics can be applied to these tasks to automate and optimize network management [35].

For example, machine learning algorithms can be used for predictive maintenance in communication networks. These algorithms can analyze network performance data to detect patterns and anomalies that may indicate potential network failures. By detecting and addressing issues before they cause significant problems, machine learning algorithms can help reduce network downtime and improve overall network reliability [36].

Another example of the use of computational intelligence in communication networks is in the area of network security. Artificial intelligence techniques can be used to detect and prevent network attacks by analyzing network traffic and identifying potential threats. These techniques can help improve network security and reduce the risk of data breaches. Computational Intelligence techniques can also be used to optimize network resource allocation. For example, genetic algorithms can be used to optimize the placement of network nodes, reducing the distance that data needs to travel and improving network efficiency.

Scientists have developed a new way to detect faults in wireless sensor networks using redundant data from several sensors and meteorological elements. By using support vector regression, they have been able to detect faults among sensor observations. This will also help to solve the problems in supply chain, which uses RFID techniques. In addition, sensor data indexed in blockchain nodes bring transparency to the systems under consideration. The blockchain node information is used further to apply machine learning to enable the intelligence capability, such as prediction of the problems before they occur, speeding up the system, meeting customer demands in peak-period, and enabling the shortest route for transportation. Power-automated quality inspections reduce the chances of delivering defective or faulty goods to customers. We can use machine learning models to predict demand forecasting based on historical data [36]. In addition to reducing the communication between sensor nodes, this proposed algorithm offers high levels of detection accuracy and low false alarm rates, both of which are particularly useful for fault detection in meteorological sensor networks, regardless of the failure rate [21]. Figure 10.5 explains the demand forecasting for a supply chain using RFID technique.

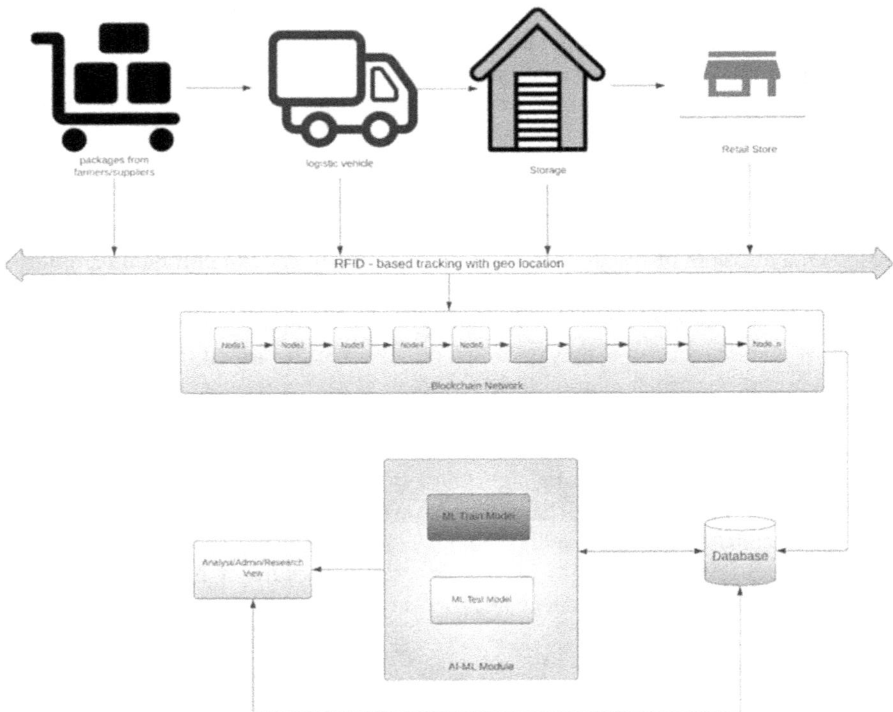

FIGURE 10.5 Supply Chain Model using RFID technique.

The optimal approach to address the challenge of increasing traffic congestion in hierarchical wireless networks seems to be the implementation of an edge caching strategy based on Double Deep Q-Network (double DQN). Considering the aforementioned factors, significant efforts are being made to maximize traffic offloading and minimize congestion by leveraging a device-to-device (D2D) communication system. The proposed cache replacement strategy, which utilizes deep reinforcement learning and Markov decision processes (MDP), is designed to effectively manage cache utilization by employing an MDP-based approach [37].

Integrated methods can also be used to increase the accuracy of indoor WiFi positioning through an integrated method. The accuracy of indoor WiFi positioning can be enhanced with a combination of Deep Neural Networks (DNNs) and K-Nearest Neighbors (KNNs). Based on these classifications, the final position is determined using the KNN algorithm after the DNN algorithm has been used to classify the WiFi RSSI Fingerprinting dataset [10, 38].

We present a framework for a personalized recommendation over mobile wireless networks that employ deep convolutional neural networks (CNN). The mobile app suggests potential visiting locations based on data from the users' sequences, user-shared images, and their contact information. In a CNN network, we are looking for ways to process the big data from the user's social and mobile trajectory using the sampled big data from the user so that we can process the big data [20].

A cognitive radio network is a type of communication network that utilizes computational intelligence to dynamically allocate spectrum resources in real time. This wireless communication system employs software-defined radio (SDR) technology to detect the available spectrum and identify unused frequencies. Computational Intelligence techniques are used in cognitive radio networks to optimize spectrum usage. For example, artificial neural networks (ANNs) can be used to learn and predict the availability of spectrum in different regions and adjust communication accordingly. Fuzzy logic can be used to decide which frequency band to use based on signal strength and noise levels. Genetic algorithms (GAs) can also be utilized to optimize channel allocation and minimize interference while maximizing the utilization of available spectrum resources. Machine learning algorithms can detect and mitigate interference and adapt to the changing conditions of the network.

In traditional wireless communication systems, the spectrum is allocated statically, leading to inefficient use of resources and spectrum wastage. In contrast, cognitive radio networks can dynamically and adaptively allocate spectrum resources in real time based on the communication requirements and spectrum availability. This is made possible using software-defined radio (SDR) technology, which enables the network to sense the available spectrum and identify unused frequencies. Spectrum sensing is accomplished using sensors to detect spectrum utilization in a particular frequency band. The sensors can identify which frequency bands are used by primary users (licensed users) and which are unoccupied, allowing the cognitive radio network to transmit data using the unoccupied frequency bands. However, since the sensing process is not foolproof, Computational Intelligence techniques such as ANNs can be employed to learn and predict the spectrum availability in different regions and reduce the likelihood of false detection and missed detection.

Fuzzy logic is a useful tool for determining the optimal frequency band for a cognitive radio network. It can consider imprecise and uncertain data such as signal strength and noise levels. Genetic algorithms are also useful for optimizing channel allocation, as they can find the optimal frequency allocation scheme that minimizes interference and maximizes utilization of available spectrum resources. Using a fitness function to evaluate the quality of the solution, the GA algorithm iteratively improves the solution until the optimal solution is found.

Machine learning algorithms are also valuable in a cognitive radio network for detecting and mitigating interference, and for learning and adapting to changing network conditions. These algorithms can learn from network data and adjust communication parameters to optimize network performance. In a cognitive radio network, Computational Intelligence techniques are critical for optimizing available spectrum resources and improving overall network performance. Examples of computational intelligence techniques that can be used in a cognitive radio network include ANNs, fuzzy logic, genetic algorithms, and machine learning algorithms.

10.5 CONCLUSION

We have discussed the architecture of Computational Intelligence (CI) in this paper, including neurocomputing, granular computing, fuzzy sets, evolutionary algorithms, and the design methodology underlying it. Artificial Intelligence (AI) is being used to improve communication efficiency. For successful communication network models, Computational Intelligence systems must be adaptable, fault-tolerant, fast, and error-resistant. An overview of neural networks, fuzzy systems, evolutionary computation, artificial immune systems, and swarm intelligence is presented in this review. Also discussed is how machine learning helps in supply chain to speed up the system and meet customer demands. The research fields are synthesized and compared to identify promising new directions. Computational Intelligence techniques are increasingly used in communication networks to optimize network performance, improve efficiency, and reduce costs. To automate and optimize network operations, these techniques can be applied to various tasks, including network management, security, and resource allocation. Additionally, this study offers insight into future research directions and better understanding of AI techniques for improving communication networks.

REFERENCES

[1] Kruse, R., Mostaghim, S., Borgelt, C., Braune, C., Steinbrecher, M. (2022). Introduction to Artificial Neural Networks. In: *Computational Intelligence. Texts in Computer Science*. Springer, Cham. https://doi.org/10.1007/978-3-030-42227-1_2

[2] Dhaliwal, P., Kaur, M., Thakur, H. K., Arya, R. K., & Lu, J. (2023). Computational Intelligence in Analytics and Information Systems: 2-volume set. In *International Conference on Computational Intelligence in Analytics and Information Systems*. CRC Press/Taylor Francis.

[3] E. Diamant, "Computational Intelligence: Are you crazy? Since when has intelligence become computational? 2016 IEEE Symposium Series on Computational Intelligence (SSCI)," *2016 IEEE Symposium Series on Computational Intelligence (SSCI)*, 2016.

[4] R. Kruse, S. Mostaghim, C. Borgelt, C. Braune, and M. Steinbrecher, *Computational Intelligence*, 2022, doi: 10.1007/978-3-030-42227-1
[5] S. Ding, H. Li, C. Su, J. Yu, and F. Jin, *Evolutionary artificial neural networks: a review*, 2011, doi: 10.1007/s10462-011-9270-6
[6] X. F. Lipo Wang, *Data Mining with Computational Intelligence – Lipo Wang, Xiuju Fu – Google Books*, Springer Science & Business Media, 1998. https://books.google.co.in/books?hl=en&lr=&id=eeDYyp098s8C&oi=fnd&pg=PA1&dq=computational+intelligence+and+artificial++intelligence+%2B+research+article&ots=TLA4h4vD9E&sig=DipEpka0MIxHX9eAk3MMO8DsSxQ#v=onepage&q&f=false (accessed February 15, 2023).
[7] R. C. Eberhart, "Overview of computational intelligence [and biomedical engineering applications]," pp. 1125–1129, 2002, doi: 10.1109/IEMBS.1998.747069
[8] Kruse, R., Mostaghim, S., Borgelt, C., Braune, C., & Steinbrecher, M. (2022). *Computational Intelligence - A Methodological Introduction, 3rd edition.* Springer Verlag. https://doi.org/10.1007/978-3-030-42227-1
[9] E. S. M. El-Alfy and W. S. Awad, "Computational Intelligence Paradigms: An Overview," https://services.igi-global.com/resolvedoi/resolve.aspx?doi=10.4018/978-1-4666-9426-2.ch001, pp. 1–27, Jan. 1AD, doi: 10.4018/978-1-4666-9426-2.CH001
[10] R. Kruse, S. Mostaghim, C. Borgelt, C. Braune, and M. Steinbrecher, "Introduction to Artificial Neural Networks," pp. 7–13, 2022, doi: 10.1007/978-3-030-42227-1_2
[11] R. Kruse, S. Mostaghim, C. Borgelt, C. Braune, and M. Steinbrecher, "Introduction to Fuzzy Sets and Fuzzy Logics," pp. 373–405, 2022, doi: 10.1007/978-3-030-42227-1_15
[12] P. A. Vikhar, "Evolutionary algorithms: A critical review and its future prospects," *Proc. - Int. Conf. Glob. Trends Signal Process. Inf. Comput. Commun. ICGTSPICC 2016*, pp. 261–265, 2017, doi: 10.1109/ICGTSPICC.2016.7955308
[13] J. Hu, X. Ou, P. Liang, and B. Li, "Applying particle swarm optimization-based decision tree classifier for wart treatment selection," *Complex Intell. Syst.*, vol. 8, no. 1, pp. 163–177, 2022, doi: 10.1007/S40747-021-00348-3/TABLES/9
[14] R. Kruse, S. Mostaghim, C. Borgelt, C. Braune, and M. Steinbrecher, "Introduction to Evolutionary Algorithms," pp. 225–254, 2022, doi: 10.1007/978-3-030-42227-1_11
[15] M. Cui and D. Y. Zhang, "Artificial intelligence and computational pathology," *Lab. Investig.*, vol. 101, no. 4, pp. 412–422, 2021, doi: 10.1038/S41374-020-00514-0
[16] A. Konar, *Computational Intelligence: Principles, Techniques and Applications*, ISBN 3-540-20898-4. Springer: Berlin Heidelberg New York, 2005. https://books.google.co.in/books?hl=en&lr=&id=NuvAERUGUAAC&oi=fnd&pg=PA1&dq=Computational+Intelligence+PDF&ots=YgQdFnfiT8&sig=igM9fbwhOFCqhD3ekGvGIWJtt-U#v=onepage&q&f=false (accessed Feb. 14, 2023).
[17] R. Kruse, S. Mostaghim, C. Borgelt, C. Braune, and M. Steinbrecher, "Neural Networks: Mathematical Remarks," pp. 213–221, 2022, doi: 10.1007/978-3-030-42227-1_10
[18] S. H. Alsamhi, O. Ma, and M. S. Ansari, "Survey on artificial intelligence based techniques for emerging robotic communication," *Telecommun. Syst.*, vol. 72, no. 3, pp. 483–503, 2019, doi: 10.1007/S11235-019-00561-Z/TABLES/3
[19] M. R. Abdi, "Fuzzy Sets and Analytical Hierarchical Process for Manufacturing Process Choice," 2008, Accessed: February 15, 2023. [Online]. Available: www.igi-global.com/chapter/fuzzy-sets-analytical-hierarchical-
[20] Y. Chen, Y. Li, H. Hu, J. Zhang, D. Gu, and P. Xu, "Computational intelligence approaches to robotics, automation, and control," *Math. Probl. Eng.*, vol. 2015, 2015, doi: 10.1155/2015/620275
[21] J. S. Raj, "A comprehensive survey on the computational intelligence techniques and its applications," *J. ISMAC*, vol. 01, no. 3, pp. 147–159, 2019, doi: 10.36548/jismac.2019.3.002

[22] A. Roshani, V. Derhami, and F. Aalam, "Applying hierarchical fuzzy systems to predict unplanned feeder outages in the Yazd," Jan. 2016. doi: 10.1109/CFIS.2015.7391651
[23] Y. P. Z. Mikhail and Z. Zgurovsky, "Fuzzy neural networks in classification problems," *Stud. Comput. Intell.*, vol. 652, pp. 179–219, 2017, doi: 10.1007/978-3-319-35162-9_5
[24] Y. P. Z. Mikhail and Z. Zgurovsky, "Application of fuzzy logic systems and fuzzy neural networks in forecasting problems in macroeconomy and finance," *Stud. Comput. Intell.*, vol. 652, pp. 133–178, 2017, doi: 10.1007/978-3-319-35162-9_4
[25] G. Zhang and H. X. Li, "An efficient configuration for probabilistic fuzzy logic system," *IEEE Trans. Fuzzy Syst.*, vol. 20, no. 5, pp. 898–909, 2012, doi: 10.1109/TFUZZ.2012.2188897
[26] R. J. Almeida, N. Verbeek, U. Kaymak, and J. M. Da Costa Sousa, "Probabilistic Fuzzy Systems as Additive Fuzzy Systems," *Commun. Comput. Inf. Sci.*, vol. 442 CCIS, no. PART 1, pp. 567–576, 2014, doi: 10.1007/978-3-319-08795-5_58/COVER
[27] "Neural networks with feedback and self-organization," *Stud. Comput. Intell.*, vol. 652, pp. 39–79, 2017, doi: 10.1007/978-3-319-35162-9_2
[28] R. Kruse, S. Mostaghim, C. Borgelt, C. Braune, and M. Steinbrecher, "Fundamental Evolutionary Algorithms," pp. 287–341, 2022, doi: 10.1007/978-3-030-42227-1_13
[29] R. Kruse, S. Mostaghim, C. Borgelt, C. Braune, and M. Steinbrecher, "Elements of Evolutionary Algorithms," pp. 255–285, 2022, doi: 10.1007/978-3-030-42227-1_12
[30] D. Khurana, A. Koli, K. Khatter, and S. Singh, "Natural language processing: state of the art, current trends and challenges," *Multimed. Tools Appl.*, vol. 82, no. 3, pp. 3713–3744, 2023, doi: 10.1007/S11042-022-13428-4
[31] D. Meurers, "Natural Language Processing and Language Learning," *Encycl. Appl. Linguist.*, 2012, doi: 10.1002/9781405198431.WBEAL0858
[32] M. Ahmadlou and H. Adeli, "Enhanced probabilistic neural network with local decision circles: A robust classifier," *Integr. Comput. Aided. Eng.*, vol. 17, no. 3, pp. 197–210, 2010, doi: 10.3233/ICA-2010-0345
[33] "Probabilistic Reasoning in Artificial Intelligence – Javatpoint." https://www.javatpoint.com/probabilistic-reasoning-in-artifical-intelligence (accessed February 14, 2023).
[34] W. Pedrycz and A. V. Vasilakos, "Computational intelligence: A development environment for telecommunications networks," *Comput. Intell. Telecommun. Networks*, pp. 1–28, Jan. 2000, doi: 10.1201/9781315220178-1/COMPUTATIONAL-INTELLIGENCE-DEVELOPMENT-ENVIRONMENT-TELECOMMUNICATIONS-NETWORKS-WITOLD-PEDRYCZ-ATHANASIOS-VASILAKOS
[35] Q.-V. Pham et al., "Swarm intelligence for next-generation networks: Recent advances and applications," *J. Netw. Comput. Appl.*, vol. 191, p. 103141, 2021, doi: 10.1016/J.JNCA.2021.103141
[36] B. I. Seraphim, S. Palit, K. Srivastava, and E. Poovammal, "A survey on machine learning techniques in network intrusion detection system," *2018 4th Int. Conf. Comput. Commun. Autom. ICCCA 2018*, December 2018, doi: 10.1109/CCAA.2018.8777596
[37] M. Moghaddam and B. Zohuri, "Business Resilience System Integrated Artificial Intelligence System," *International Journal of Theoretical & Computational Physics*, 2022. https://www.researchgate.net/publication/358279313_Business_Resilience_System_Integrated_Artificial_Intelligence_System (accessed Feb. 15, 2023).
[38] S. X. Wu and W. Banzhaf, "The use of computational intelligence in intrusion detection systems: A review," *Appl. Soft Comput.*, vol. 10, no. 1, pp. 1–35, 2010, doi: 10.1016/J.ASOC.2009.06.019

Index

Pages in *italics* refer to figures and pages in **bold** refer to tables.

For Product Safety Concerns and Information please contact our EU representative GPSR@taylorandfrancis.com
Taylor & Francis Verlag GmbH, Kaufingerstraße 24, 80331 München, Germany

www.ingramcontent.com/pod-product-compliance
Lightning Source LLC
LaVergne TN
LVHW010553110826
845149LV00003B/642

* 9 7 8 1 0 3 2 3 0 0 2 6 9 *